CONTINUITY OR REPLACEMENT

SELECTED PAPERS FROM THE SYMPOSIUM ON CONTROVERSIES IN *HOMO SAPIENS* EVOLUTION / ZAGREB / JULY 1988

CONTINUITY OR REPLACEMENT

Controversies in Homo sapiens evolution

Edited by

GÜNTER BRÄUER

Institute for Human Biology, University of Hamburg

FRED H. SMITH

Department of Anthropology, Northern Illinois University

A.A. BALKEMA / ROTTERDAM / BROOKFIELD / 1992

Published by
A.A. Balkema, P.O. Box 1675, 3000 BR Rotterdam, Netherlands
A.A. Balkema Publishers, Old Post Road, Brookfield, VT 05036, USA

ISBN 90 6191 149 4

Printed in the Netherlands

To our families
Ellen, Maria, Burton, and Mary Kate
with our gratitude and our love

Table of contents

Preface

In recent years, the question of modern human origins has become a topic of tremendous interest outside the community of specialists who have studied this phenomenon for almost 150 years. Readers of such popular scientific magazines as *Discover* and *Natural History* have had the opportunity to follow this issue very closely. Even major news magazines like *Newsweek, US News* and *World Report* and *Aera* have run feature articles dealing with the beginnings of modern people. The reasons for this new interest in an old question certainly stem from new discoveries pertinent to modern human origins, but probably more significant are the rather dynamic controversies these new discoveries have generated among the experts concerning exactly how modern humans originated in the various regions of the Old World.

Some experts have long favored a classic monocentric origin for modern humans. Over the last several years, a combination of new fossil discoveries, new dating and analyses of previously-known specimens, and interpretations of human genetic variability has, for many of these experts, pointed to Africa as the most likely source for modern people. Supporters of an 'Out of Africa' model suggest that the transition to modern humans occurred only in Africa. Some argue that the dispersal of these modern people throughout the Old World resulted in the total, or virtually total, replacement of the previous inhabitants of these regions, most notably the Eurasian Neandertals. Others allow for hybridization during replacement, but clearly view the African influence as being the catalyst and major element in the emergence of modern Eurasians.

Other experts support what has come to be called the 'Multiregional evolution' model, which suggests that local populations of archaic humans throughout Eurasia played significant, if not the major, roles in the origins of modern people in their respective regions. Although agreeing on the importance of local continuity, some purveyors of multiregional views believe that the fundamental modern human anatomical pattern may well have developed initially in one region and strongly influenced the emergence of this morphological pattern elsewhere. Others, however, assert that this type of gene flow played no particular 'catalytic' role in modern human origins.

The idea behind this volume and the symposium on which it is based was to examine the competing models for modern human origins within the framework of present-day knowledge in the areas of human paleontology, paleolithic archaeology, and molecular genetics. Our goal in organizing the symposium on '*Homo sapiens* evolution' for the 12th International Congress of Anthropological and Ethnological Sciences in Zagreb was to assemble as wide a range of perspectives on issues pertinent to the controversy as possible. To that end we invited some 40 scholars from around the world, covering the full gamut of pertinent topics, to participate in the symposium. Unfortunately, several colleagues were, for various reasons, not able to participate in the symposium. This led to unequal coverage of certain regions and topics. Nevertheless, we believe that the participants in the symposium will long remember the sparkling atmosphere that characterized the presentation of the papers and the discussions that took place long into the nights in many Zagreb restaurants and on the excursion to the famous Neandertal sites of Vindija and Krapina.

The presentations made in Zagreb contained so much up-to-date information and new perspectives that we had little doubt that the proceedings of the symposium would mark another major contribution to our understanding of the complexities of *Homo sapiens* evolution. Consequently, we invited the presentors to submit their presentations for publication. To fill in some gaps, we also invited a few scholars who were unable to be present in Zagreb to contribute to this volume. A total of 27 authors submitted revised manuscripts during 1989, and further necessary revisions continued until mid-1990.

We asked all contributors to deal with the central controversy concerning the mode of modern human origins to the fullest extent possible in their papers. The result is a volume which demonstrates, in our opinion, that it is increasingly difficult to propose models that fit easily all of the available evidence from all portions of the Old World. It is becoming increasingly evident that while replacement can perhaps be justified in some areas, other areas exhibit more defensible cases for varying degrees of continuity. This leads to the exceedingly difficult task of trying to determine the relative importance of indigenous and extraneous contributions to modern human gene pools in specific regions, and such attempts are certain to lead to continued debate. The evidence further suggests that reality, as it pertains to modern human origins, is unlikely to be found in 'extreme' approaches to this issue.

Although the majority of the 20 chapters in this volume are written by physical anthropologists, we have tried to cover both the behavioral and genetic aspects of modern human origins. While only two contributions deal specifically with genetics, both present some interesting new information pertinent to the interpretation of patterns in these data. The majority of archaeological and paleontological papers focus on either Europe or Africa. Indeed, only three focus on Asia and Australia. However, these regions and the genetic data are also discussed by other authors in some of the more general papers. Within unavoidable limitations, we believe that present perspectives and evidence pertinent to *Homo sapiens* evolution are well-covered in this volume.

We would like to express our gratitude to all of the contributors to this volume, all of whom have patiently put up with the editors' general confusion and still managed to provide us with an excellent set of manuscripts. We also thank Professor Pavao Rudan for his consistent help with the organization of the symposium. To Dr Jakov Radovčić of the Croatian Natural History Museum in Zagreb, we offer our thanks for organizing the excursion to Krapina and Vindija and for acting as 'host' for many of our discussions in the Museum and around town. The Alexander von Humboldt-Stiftung, the Deutsche Forschungsgemeinschaft (DFG), Northern Illinois University, and the University of Hamburg provided various forms of financial support. Moreover, G.B. was granted a Heisenberg Fellowship by the DFG, which covered his work on this book; and F.H.S.'s participation in Zagreb was made possible by the US National Academy of Sciences and the Jugoslavenska Akademija Znanosti i Umjetnosti. We are grateful to all of these agencies for their support.

We are especially grateful to Mr A.T. Balkema for immediately agreeing to publish this symposium, even before it was held. Many thanks also go to Ms K. Vogels and Ms S. Peters for their help with editorial aspects and preparation of the index. To Jutta Nehring and Dr Tal Simmons for proofreading and particularly to our wives, Ellen and Maria, for constant encouragement and many hours of proofreading, we are very grateful.

Finally, it is with great sadness that we note the death of one of the participants in our symposium. Professor Mirko Malez of Zagreb died in August of 1990. Malez directed the excavation of several important sites around Zagreb, including Vindija and Velika Pećina. He was a positive force in Yugoslav science and a friend for many years to one of us. He will certainly be missed.

Günter Bräuer
Fred H. Smith
April 1991

Cladistics and later Pleistocene human evolution

Erik Trinkaus
Department of Anthropology, University of New Mexico, Albuquerque, USA and CNRS, Laboratoire d'Anthropologie, Université de Bordeaux I, Talence, France

It was partly his own fault for wanting a clear-cut answer rather than the frayed ends of reality.
Brett (1976:168)

1 INTRODUCTION

The past decade has seen considerable interest in deciphering the possible phylogenetic roles of the later Pleistocene late archaic humans,[1] such as the Neanderthals of Europe and western Asia, especially with respect to their relationships to the emergence and spread of modern humans during the Upper Pleistocene. This discussion has been joined by the increasing application of the principles of cladistics to the human fossil record, presumably with the hope that the clear identification of the phylogenetic polarities and affinities of various late archaic human morphological attributes will enable us to reach a consensus on their places in the human family tree.

The following discussion has been inspired primarily by phylogenetic issues concerning the Neanderthals. Mainly, to what extent do the Neanderthals possess uniquely derived ('specialized') traits, as opposed to shared ancestral ('generalized') traits with other later Pleistocene hominids and shared derived traits with modern humans, and what might be the implications of such an assessment of Neanderthal morphology to our understanding of their phylogenetic role in later Pleistocene human evolution?

The use of cladistics with respect to the Neanderthals in particular and later Pleistocene humans in general (e.g. Condemi 1988, Hublin 1988, Stringer 1984, 1987, this volume, Stringer & Andrews 1988, Tillier 1988, Trinkaus 1988b) has, without a doubt, encouraged human paleontologists to be more explicit about how they are delimiting the Neanderthal morphological pattern and those of their temporal and spatial neighbors. It has allowed considerable clarification of Neanderthal morphology beyond the general Neanderthal trait lists previously provided (e.g. Boule 1921, LeGros Clark 1955). And, perhaps more importantly, it has emphasized the need to distinguish between ancestral and derived, unique and shared morphological traits.

However, the use of cladistics and its application to the evolution of late archaic humans (such as the Neanderthals) and early modern humans has its limitations; limitations which are seldom considered in the use of the methodology. If we are to hope to clarify eventually the phylogenetic relationships of these later Pleistocene hominids and reach something resembling a consensus on those relationships, it is incumbent upon us to consider the strengths and, especially, the weaknesses of our methodologies. The following discussion is a consideration of those strengths and weaknesses of cladistics, particularly with respect to the application of the methodology to the intrageneric, and possibly intraspecific, human groups of the later Pleistocene.[2]

2 AN ASSUMPTION

It is assumed here that the ultimate goal of later Pleistocene paleoanthropology is to understand the actual past events and processes which led to the evolutionary emergence of modern humans from more archaic members of the genus *Homo*. Models, however elegant in their simplicity or presumed theoretical sophistication, are taken to be only of heuristic utility unless they can take into consideration, at some level, all of the currently known information from the paleoanthropological record.

In this, it is fully recognized that our current paleoanthropological record is incomplete and that our interpretations of later Pleistocene human evolution will change as further discoveries are made. Furthermore, the information available at one time to any one researcher or group of researchers is an incomplete set of the potentially available data set. Therefore, all interpretations can be realistically seen as no more than state-of-the-art conclusions, inevitably to fall or be modified as the field progresses.

3 THE APPLICATION AND ADVANTAGES OF CLADISTICS

The introduction and spread of cladistic methodology in paleoanthropology during the past decade has served greatly to make phylogenetic arguments more explicit. It has become increasingly routine for paleontologists to reconstruct branching patterns (cladograms) and resultant phylogenies (with the addition of geological time and paleogeography) starting with the similarities and differences between groups in their neontologically and paleontologically documented morphologies. Scenarios, while remaining important as syntheses of our knowledge, interpretive frameworks, and sources of hypotheses for further analysis of the paleoanthropological record, have played less of a role in actual phylogeny building than previously.

In addition, cladistics has emphasized the need to determine the polarities of morphological patterns, to distinguish between ancestral (plesiomorphous) and derived (apomorphous) traits and between shared (symplesiomorphous [shared ancestral] and synapomorphous [shared derived]) and unique (autapomorphous [uniquely derived]) traits. In this way, the relative phylogenetic utilities of various identified traits are clearly indicated, with synapomorphous traits having greater relevance for indicating shared ancestry than plesiomorphous ones. As a result, once the polarities of traits at any given branching point have been determined, the traits can be differentially weighted in the resultant analysis.

Cladistics has assumed that parallelisms and convergences, if correctly identified, are unusual in evolution. Therefore, the cladogram with the smallest number of required parallelisms and convergences is the preferred one. In addition, evolutionary reversals are considered to have been rare, and groups with significant numbers of autapomorphous traits are considered to have been more likely to have become extinct than to have experienced evolutionary reversals of previous morphological trends. In this, the principle of parsimony (frequently taken as a given) is usually invoked to arrive at the desired cladogram.

It should be mentioned that little of the core of cladistics is really new. Issues of polarity, parsimony, etc. have long been recognized in paleontology. It is primarily the emphasis on these methodologies which has been the major contribution of cladistics in the past decade.

4 LIMITATIONS OF CLADISTICS

Despite these advantages, the universal and uncritical application of cladistics has its limitations and pitfalls. Much as the uncritical application of multivariate distance studies during the 1970s (facilitated by package software programs and the availability of mainframe computers) generated considerable work of dubious value, the widespread and casual application of cladistics to the paleontological and/or neontological records has considerable potential to generate work of minimal utility (particularly since cladistic analysis computer programs [e.g. PAUP: Phylogenetic Analysis Using Parsimony] are now generally available [Swofford 1985]). This is especially true since cladistics, like multivariate analysis, is only a methodology.

4.1 *A question of parsimony*

Virtually all current cladistic analyses employ the principle of parsimony, usually to limit the number of homoplasies necessary to produce the desired cladogram and the resultant phylogenetic tree. Parsimony is a very useful tool in such analyses, since it has clearly defined properties and explicitly generates testable interpretations of given data sets. Yet, all too often parsimony is taken as a given of the real world, rather than as a heuristic device for the ordering of data.

Parsimony is no more than an optimization model, one which assumes that the simplest solution has the greatest probability of accurately representing the real world; it is a set of algorithms and not an aspect of biology (Felsenstein 1983). Like all optimization models, it is a mathematically-derived theoretical ideal. It is of utility for the preliminary ordering of data, for the formulation of hypotheses, and for the identification of deviations in the real world from the ideal world of the model. It can never be more than an approximation of reality.

This issue is additionally important in cladistics, since the most parsimonious solution is, by definition, the one with the fewest homoplasies. Yet, in closely related biological forms, such as the congeneric and possibly conspecific later Pleistocene hominids, all of the populations share a substantial biological baseline upon which evolutionary processes operate. Therefore, similar evolutionary pressures (likely in such behaviorally similar populations) are very likely to produce independently similar morphologies. Can we hope to sort out consistently such homoplasies from symplesiomorphies or synapomorphies?

If we do not consistently sort them out (likely given that the methodology operates on the adult phenotype [the paleontologically observed morphology] to reconstruct phylogenetic history and that the ancestral baseline is frequently unknown or poorly known), to what extent are we biasing both our phylogenetic trees and our interpretations of the evolutionary processes involved? Furthermore, at the intrageneric or especially intraspecific level, most of the identified apomorphous (derived) traits are ones of unknown or apparently minor functional significance. Given the high probability of evolutionary reversals and parallelisms at the genetic (nucleotide) level and the probably low adaptive valence of most of the morphological traits considered, is an apparent trait reversal or parallelism likely to be all that unusual?

4.2 *Units of analysis*

Cladistic analysis requires the explicit definition of units of analysis, groups of organisms to be compared and arranged into cladograms and eventually into phylogenies. These groups, samples or OTUs (operational taxonomic units) are relatively easy to delimit in the modern world, since they are usually represented by biologically-defined populations, species or higher taxonomic units. Paleontologically, the exercise becomes more complex.

Traditionally and currently, some combination of morphology and temporal (stratigraphic) and spatial (regional) criteria are employed to define paleontological taxonomic groups. Yet, time and space are not biological criteria for distinguishing groups, and using morphology to distinguish groups can become tautological, prejudging the outcome of the subsequent cladistic analysis.

This is particularly relevant to the caldistic analysis of later Pleistocene hominids. Current consensus is that they all belong in one species, *Homo sapiens*, even though it may be reasonable, on morphological grounds, to restrict *H. sapiens* to modern humans and use another species designation (*H. neanderthalensis*) for late archaic humans. However, regardless of the level of taxonomic distinction preferred for subsets of later Pleistocene humans, it is apparent that they are all closely related, possibly conspecific. Therefore, temporal and geographic clines should have been common, as the available fossil record indicates. Consequently, the delineation of the units of analysis becomes largely arbitrary, determined by temporal and geographic gaps in the fossil record and by *a priori* morphological decisions as to where the breaks should occur.

4.3 *The role of variation*

All current modern syntheses of biological evolution are in agreement that the existence of normal ranges of genotypic and phenotypic variation within populations is the baseline of biological evolution and that the conversion of within-population variation to between-population variation is the core of the evolutionary process. Whatever directional or random processes are primary in producing a given evolutionary change or apparent stasis, that observed pattern is the product of the presence/absence of shifts in allelic (and by extension phenotypic) frequencies.

These elementary principles bear repeating since virtually all cladistic analyses assume a single character state per variable per analytical unit. Yet, any casual observation of reasonable samples from recent populations (or otherwise delimited paleontological groups [see above]) illustrates that there is considerable variation in most observed features. This is evident even in small paleontological samples, in which the limited number of specimens would tend to reduce the number of character states represented.

The central tendencies of observations from closely related groups may be significantly different in a probablistic sense, yet their ranges of variation may overlap considerably. Merely using some nominal characterization of their central tendencies denies the variations and range overlaps, whereas full considerations of the ranges of variation might disguise significant frequency shifts between the samples. Clearly both the central tendencies and the variances of the samples are relevant to any consideration of morphological affinities between samples.

4.4 *The identification of traits*

The use of cladistics, while it has made paleontologists be more explicit about the morphological characters used, has raised several questions about trait definition and utility.

First, to what extent do the traits employed need to have any biological relevance, as opposed to being merely frequently preserved and occurring in easily distinguishable character states? Since the presumed goal here is the decipherment and understanding of human evolutionary events and processes, it is expected that some biological relevance should exist for the traits employed. Our ability to identify the biological significance of various traits may be limited, but that does not mean that we should not attempt, to the best of our ability, to employ morphological traits which have biological relevance.

Second, to what extent might different traits be intercorrelated? It is methodologically possible to assess the degree of statistical correlation between variables and their character states (given adequate sample sizes), but it is seldom known whether such correlations are spurious or real, sample specific or of broader biological relevance. If they are biologically meaningful, are they due to pleiotropy and/or functional interrelationships? Are we identifying the primary biological features or merely some convenient secondary reflections of them? If the intercorrelations are determined from the samples in question, how does one correct for them without becoming tautological?

Third, to what extent do the traits employed exist in discrete character states, as opposed to being continuous variables? Since most cladistic analyses require that there be a single character state per variable per group, the common practice is to define a series of states based on one's previous knowledge of the paleontological record. Yet, most variables are designed to describe regional anatomical shape, and virtually all aspects of morphology (especially skeletal and dental morphology) exist as continuous ranges of variation.

Some of these ranges of variation (e.g. dental occlusal morphology, occipito-mastoid configurations, carpal and tarsal articular morphology) seem to fall into discrete forms and therefore are more easily quantified as a series of character states rather than as continuous metric variables. Yet, analyses of large samples of recent skeletal remains invariably show continuous ranges of variation from absent, to trace, to present, to pronounced for the feature in question. The presence/absence of teeth (dental agenesis) is one of the few skeletal features which truly exists as a discrete character, and even it is the product of a threshold effect acting on a continuous range of size variation.

Other aspects of morphology are frequently quantified only as continuous metric variables (e.g. endocranial capacity, cranio-facial shape, postcranial robusticity). In order to utilize them in cladistic analyses, researchers establish arbitrary metric boundaries, with character states being defined as less than or greater than a specific value. Yet, how are those boundary values determined? They are the products of our current knowledge of the fossil record and our *a priori* definitions of paleontological samples. Given boundary values just happen to separate (completely or largely) samples which we already 'know' are distinct groups. How much within group variation is obscured and how many unknown intermediate forms are denied by this method are uncertain.

4.5 *The issue of polarity*

One of the major contributions of cladistics has been to refocus attention on the phylogenetic polarities of traits and the differential roles of derived versus ancestral morphologies (assuming one can identify traits appropriately) in phylogenetic reconstruction. Yet, there are problems as well with determining the polarities of traits.

First, there are two non-mutually-exclusive standard means of determining polarity, however explicitly they are stated. In one method, the character states for neontological and/or paleontological groups are analyzed, using principles of parsimony, and the hypothetical ancestral forms are reconstructed. This method contains the limitations of extreme assumptions of parsimony, and it only identifies 'hypothetical' ancestors, not necessarily morphological forms which ever existed. The second method uses an outgroup to provide information on the ancestral form. However, if that outgroup is a neontological group (such as the African apes for the Hominidae), the degree of evolution within the outgroup's lineage since the common ancestor must be assessed and the common ancestor hypothetically reconstructed. If the outgroup is a paleontological sample, then assumptions about the ancestor-descendant relationships of the paleontological groups must be made. Since the goal is phylogenetic reconstruction, this can easily become tautological.

Second, the way in which traits are defined can greatly influence their perceived polarity. All morphologically identified groups, whether neontological or paleontological, are to some extent unique. Otherwise we would be unlikely to identify them as discrete groups. Their uniqueness can be due to their possession of truly unique morphological patterns. However, among closely related forms, such as later Pleistocene members of the genus *Homo*, the number of truly unique morphological patterns is likely to be small. More probably, their unique 'total morphological patterns' will be produced by unique combinations of traits, each of which is shared with one or more closely related groups. Therefore, whether a group is identified as possessing a large number of uniquely derived traits (autapomorphies) or some combination of shared (symplesiomorphous and synapomorphous) traits will depend upon how the traits are defined and to what extent they are subdivided into biologically relevant units. All of the issues of trait definition discussed above apply here.

5 CONCLUSION

It is therefore apparent that the application of the cladistic methodology at the intrageneric or even intraspecific level, such as to issues concerning later Pleistocene human evolution, has some advantages and a series of limitations. Primarily, it helps us to be more explicit about our morphological criteria and methodological assumptions. However, it requires us to make certain assumptions that cannot be substantiated and forces us to place data into categories which can considerably obscure and distort information of evolutionary relevance.

Yet, it is evident that cladistics is here to stay. It is too seductive in its simplicity and apparent methodological rigor to fade away. For those more interested in the evolutionary insights contained in 'the frayed ends of reality', it will remain little more than a heuristic device for preliminary sortings of the known paleontological record.

ACKNOWLEDGEMENTS

I would like to thank F.H. Smith and G. Bräuer for inviting me to contribute to this volume, C.B. Stringer for attempting to convince me of the advantages of a cladistic approach, S. Condemi, J.-J. Hublin and A.-M. Tillier for forcing me to re-examine the biological bases of several supposed Neanderthal autapomorphies, and J.C. Long and K.M. Trinkaus for helpful comments during the preparation of this paper.

NOTES

1 The 'later Pleistocene' is here taken to include all of the Upper Pleistocene and the more recent phases of the Middle Pleistocene; it therefore includes the last 200 000 to 300 000 years. 'Late archaic humans' includes those Old World hominids from the later Pleistocene which are morphologically more archaic than modern humans; the term is roughly equivalent to 'archaic *Homo sapiens*' but lacks the formal taxonomic implications of this latter designation. The Neanderthals of the early last glacial of western Asia and Europe are one temporal and geographical subset of late archaic humans.

2 Many of the points made here have been made previously in discussions of cladistics and phylogenetic reconstruction, even if they have frequently been conveniently forgotten by paleontological practitioners of cladistics. This paper is not intended to be a review of the scientific and philosophical issues relevant to cladistics and its application to the paleontological record; it is intended only to emphasize some of the assumptions and pitfalls associated with the application of the methodology to the later Pleistocene human and similar fossil records. Nor is this discussion intended to resolve issues concerning later Pleistocene human evolution; such issues have been discussed extensively recently (e.g. Mellars & Stringer 1989, Smith & Spencer 1984, Stringer & Andrews 1988, Trinkaus 1986, 1988a, 1989, Vandermeersch 1988) with little apparent resolution, due more to differing approaches and assumptions than to the state of the paleoanthropological record.

REFERENCES

Boule, M. 1921. *Les Hommes Fossiles*. Paris: Masson et Cie.

Brett, S. 1976. *So Much Blood*. New York: Scribner's Sons.

Condemi, S. 1988. Caractères plésiomorphes et apomorphes de l'os temporal des néandertaliens européens würmiens. In E. Trinkaus (ed.), L'Homme de Néandertal 3: L'Anatomie. *Etudes Rech. Archéol. Univ. Liège* 30: 49-52.

Felsenstein, J. 1983. Parsimony in systematics: Biological and statistical issues. *Ann. Rev. Ecol. Syst.* 14: 313-333.

Hublin, J.-J. 1988. Caractères dérivés de la région occipito-mastoïdienne chez les Néandertaliens. In E. Trinkaus (ed.), L'Homme de Néandertal 3: L'Anatomie. *Etudes Rech. Archéol. Univ. Liège* 30: 67-73.

LeGros Clark, W.E. 1955. *The Fossil Evidence for Human Evolution*. Chicago: Univ. of Chicago Press.

Mellars, P., & C.B. Stringer (eds) 1989. *The Human Revolution: Behavioural and Biological Perspectives in the Origins of Modern Humans*. Edinburgh: Univ. of Edinburgh Press.

Smith, F.H. & F. Spencer (eds) 1984. *The Origins of Modern Humans*. New York: Alan R. Liss.

Stringer, C.B. 1984. The definition of *Homo erectus* and the existence of the species in Africa and Europe. In P. Andrews & J.L. Franzen (eds), The Early Evolution of Man. *Cour. Forsch-Inst. Senckenberg* 69: 131-143.

Stringer, C.B. 1987. A numerical cladistic analysis for the genus *Homo*. *J. Hum. Evol. 16: 135-146.*

Stringer, C.B. & P. Andrews 1988. The genetic and fossil evidence for the origin of modern humans. *Science* 239: 1263-1268.

Swofford, D.L. 1985. *PAUP: version 2.4*. Champaign: Illinois Natural History Survey.

Tillier, A.-M. 1988. A propos de séquences phylogénique et ontogénique chez les Néandertaliens. In E. Trinkaus (ed.), L'Homme de Néandertal 3: L'Anatomie. *Etudes Rech. Archéol. Univ. Liège* 30: 125-135.

Trinkaus, E. 1986. The Neandertals and modern human origins. *Ann. Rev. Anthropol.* 15: 193-218.

Trinkaus, E. (ed.) 1988a. L'Homme de Néandertal 3: L'Anatomie. *Etudes Rech. Archéol. Univ. Liège* 30.

Trinkaus, E. 1988b. The evolutionary origins of the Neandertals or, why were there Neandertals? In E. Trinkaus (ed.), L'Homme de Néandertal 3: L'Anatomie. *Etudes Rech. Archéol. Univ. Liège* 30: 11-29.

Trinkaus, E. (ed.) 1989. *The Emergence of Modern Humans: Biocultural Adaptations in the Later Pleistocene*. Cambridge: Univ. of Cambridge Press.

Vandermeersch, B. (ed.) 1988. L'Homme de Néandertal 7: L'Extinction. *Etudes Rech. Archéol. Univ. Liège* 34.

Replacement, continuity and the origin of *Homo sapiens*

Christopher B. Stringer
Department of Palaeontology, Natural History Museum, London, UK

1 INTRODUCTION

The last ten years have seen remarkable advances in our understanding of the origin of *Homo sapiens* (used here in the restricted sense of Stringer & Andrews 1988a, to represent living and fossil humans displaying specific morphological synapomorphies). Five of these advances are summarised below:

1. Neanderthals were present in western Europe at a surprisingly young date ('Wurm III' of the old usage, probably less than 35 000 radiocarbon years BP). These Neanderthals are apparently associated with the Upper Palaeolithic Châtelperronian industry, and other Neanderthals may well have been responsible for comparable industries elsewhere in Europe, such as the Szeletian and Uluzzian. The chronological implications are that there was, broadly speaking, contemporaneity between these late Neanderthals and early *H. sapiens* (assumed to be responsible for the Aurignacian) in various parts of Europe after 40 000 radiocarbon years ago (see various papers in Mellars & Stringer 1989). Documentation of the full extent of the inferred period of coexistence is hampered at present by practical limits on radiocarbon dating.

2. The hominids from Skhūl and Qafzeh (Israel) represent a somewhat archaic *H. sapiens* (*sensu stricto*), rather than a transitional or hybrid form between Neanderthals and *H. sapiens*. Just as significantly, there are growing indications of an early (oxygen isotope stage 5 equivalent) date for these specimens (Valladas et al. 1988, Schwarcz et al. 1989, Stringer et al. 1989). They may thus lie chronologically between the earlier, poorly-known, but apparently archaic hominids from sites such as Gesher Benot Ya'acov and Zuttiyeh, and Neanderthals from sites such as Tabun C-D and Kebara. However, uncertainty will remain about the sequence of populations in the area until a better absolute dating framework can be established for all the important sites. If the Skhūl and Qafzeh hominids are now correctly dated, they seem too ancient to serve as the direct and exclusive ancestors for the earliest *H. sapiens* of Europe implied by their colloquial name 'Proto-Cro-Magnons'. It is unclear whether their presence in south west Asia represents a brief incursion into the area from elsewhere (north Africa?) which was part of an episodic first appearance of *H. sapiens* in Eurasia, without wider consequences. Alternatively, their appearance might presage a major dispersal of *H. sapiens* across Asia and into the Far East. In either case, these specimens provide us with our best samples of early *H. sapiens*, at perhaps three times the age of European examples, and an age comparable to that inferred for less complete African specimens from sites such as Klasies River Mouth Caves, Omo-Kibish and, less certainly, Border Cave and 'Guomde' (Bräuer, this volume). They also

seem to confirm that Neanderthals *sensu stricto* were not ancestral to the earliest *H. sapiens* (Stringer 1988).

3. *H. sapiens* was present in Africa during the time span of oxygen isotope stage 5 (prior to 70 kyr BP). Each part of the morphological, archaeological, stratigraphic or chronological evidence from the African sites mentioned in 2 is subject to some degree of uncertainty, and it is right that the evidence should be scrutinised carefully. However, it would be wrong to be especially sceptical of the African evidence simply because it does not fit with particular models developed elsewhere (Wolpoff 1989). Otherwise there will seem to be double standards operating when comparing evidence from these African sites with some in Europe and elsewhere (see Mellars & Stringer 1989, Bräuer, this volume, Bräuer et al., this volume, Morris, this volume).

4. The geographic spread of humans across Australia had occurred by about 30 kyr BP, and the first arrival of humans in Sahul must have been at least 10 kyr before this. The early Australians already displayed considerable morphological variation, and the significance of this remains in dispute (Brown 1987, Wolpoff 1989).

5. The significance of new genetic evidence on present human variation to interpretations of the origin of *H. sapiens* has been much debated (see, for example, Saitou & Omoto 1987, Moritz et al. 1987, Darlu & Tassy 1987, Cann et al. 1987, Excoffier et al. 1987, Bowcock et al. 1987, Sanchez-Mazas & Langaney 1988, Pamilo & Nei 1988, Cann 1988, Spuhler 1988, Stoneking & Cann 1989). Although the peculiar features of mitochondrial DNA (mtDNA) make it especially useful for intraspecific phylogeographic studies (Avise et al. 1987), it also requires evidence from the many different genetic polymorphism and nuclear DNA data to build up a realistic overall picture of the genetic history of *H. sapiens*. The different bodies of evidence are now complementing each other to show the fundamental patterns of relationship between major geographical populations of humans, and the probability of an African origin for *H. sapiens* (Cann et al. 1987, Cann 1988, Cavalli-Sforza et al. 1988, various papers in Mellars & Stringer 1989). The pattern which is emerging is not simple (cf. the relationships of Ethiopian and San populations discussed in Cavalli-Sforza et al. 1988), but it cannot easily be reconciled with the global fossil record from the middle Pleistocene, suggesting that recent human population relationships do not have a long Pleistocene history (see below). This is so even if direct calibration of past events remains problematic from the genetic evidence alone (Stringer & Andrews 1988a, Wolpoff 1989).

2 INTERPRETING THE FOSSIL RECORD: MODELS AND GEOGRAPHIC PATTERNING

My co-authorship of a controversial paper in *Science* supporting the 'Out of Africa' model for *H. sapiens* origins (Stringer & Andrews 1988a) was not intended to imply that there was no evidence against the model or no problems with the supporting palaeontological and genetic evidence cited. Indeed I do not pretend that the evidence in support of the 'Out of Africa' model is unequivocal. We certainly need a lot more fossil material and genetic data from many parts of the world before we can fill in the details of the process of *H. sapiens* origin. Therefore we cannot yet say whether the single process of origin and dispersal implied by the model will be adequate when applied universally. The predictions from the multiregional and 'Out of Africa' models laid out in our paper remain to be fully explored against the growing body of palaeontological, genetic and behavioural evidence. However, I do consider that 'Out of Africa' is the most appropriate explanatory model yet available to us. Moreover, even if the complete 'Out of Africa' scenario were to be falsified, this would not necessarily represent proof for a general multiregional model.

Figures 1 and 2 present a comparison of the multiregional and 'Out of Africa' models, which clearly provide quite different expectations for past and present human population patterns (see also Table 1 in Stringer & Andrews 1988a). Comparing these figures, only the multiregional model posits consistent population relationships, maintained by gene flow, through time. The 'Out of Africa' model instead posits a major change in population relationships at the appearance of modern humans, with a break in patterns of regional continuity outside of Africa. Next, Figure 3 compares representations of past and present population relationships in order to see which of the two models is most compatible with the fossil and genetic evidence. Genetic data from many different systems now appear to support a closer relationship between European and Asian populations than between either of these and sub-Saharan African populations. Furthermore, various genetic distance data support the relationship of Australasian populations to those of Africa and Eurasia shown in Figure 3d (Nei 1982, Cavalli-Sforza et al. 1988), although it is not implied that there was total genetic separation between these regional populations. Such total separation, for example in the case of the aborigines of Tasmania (Jones 1979), has probably been only a rare occurrence during the last 20 000 years.

The multiregional evolution hypothesis proposes that human geographic variation dates to the initial emergence of *H. erectus* from Africa (Wolpoff et al. 1984, Wolpoff 1989). Thus if the genetic pattern is accurate, there should be a consistency of pattern through much of the Pleistocene, and this may have been initiated through a common founding Eurasian/Australasian population over 750 000 years ago (Fig. 3a). This original pattern could have been modified to the present one (Fig. 3d) by greater gene flow occurring between the European and Asian populations compared with the other populations. Thus under the multiregional model, modern patterns of population relationships might already have been developing at the beginning of the middle Pleistocene, but should certainly have become evident later in the middle Pleistocene.

Although population relationships inferred for the later middle Pleistocene show increasing evidence of regional differentiation, there is neither the pattern expected from maintenance of the supposed original early Pleistocene pattern forwards in time, nor the one expected from extrapolation of the modern pattern backwards in time (Fig. 3b). Instead it is evident that during the middle part of the middle Pleistocene, East Asian and Australasian hominids assigned to late *H. erectus* were rather similar morphologically, while those of Europe and Africa were also similar morphologically, but distinct from those in the Far East (Santa Luca 1980, Stringer 1984). This is usually recognised by their classification as 'archaic *H. sapiens*' rather than *H. erectus*. In other words, the main geographical differentiation of hominids, whether regarded as specific or subspecific, appears to have been an east-west one, and this is certainly not consistent with the present population relationships depicted in Figure 3d. Of course, poorly known or uncertainly dated material such as that from Narmada and Jinnuishan remains to be integrated into the overall picture, but it will surely remain true that specimens such as Petralona and Broken Hill are more similar to each other phenetically and cladistically than either are similar to the Zhoukoudian or Hexian fossils, which in turn seem to more closely resemble the Sambungmachan or Ngandong crania from Indonesia (Santa Luca 1980, Stringer 1984, Wolpoff et al. 1984).

The situation was apparently even more distinctive during the later middle and early late Pleistocene with the evolution of the Neanderthals in western Eurasia, seemingly from western

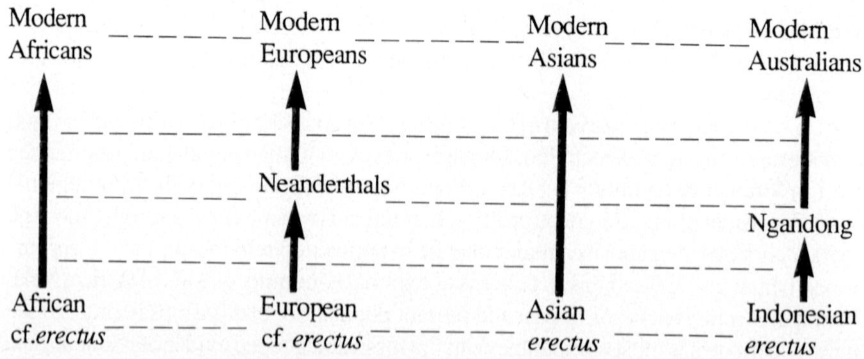

Figure 1. A diagrammatic representation of the multiregional model (with gene flow). There is regional continuity from archaic to modern humans.

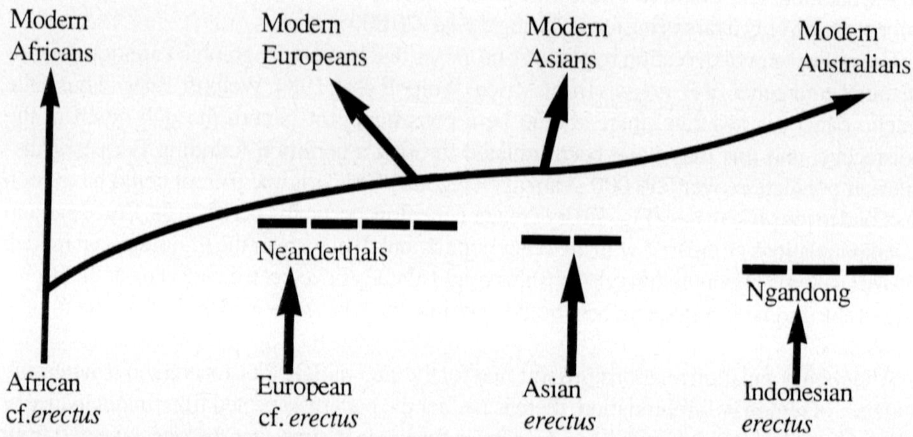

Figure 2. A diagrammatic representation of the recent African origin model. Local archaic lineages outside Africa become extinct with little or no gene flow with modern humans.

'archaic *sapiens*' ancestors (Stringer et al. 1984, Vandermeersch 1985). This should have maintained the closer African-European relationship (Fig. 3c). There was evidently extensive gene flow or population continuity between Europe and western Asia during that time period, but Asian Neanderthals show little affinity with known hominids further East, and there is thus little evidence of gene flow and the closer relationship to east Asian populations which is indicated in recent genetic analyses. However we regard an enigmatic fossil such as Dali, it does not appear to be closely related to the Neanderthals of western Eurasia.

This pattern of close relationship across Eurasia is in fact something which only emerges in the fossil hominid record when the Neanderthals are leaving the scene, and such an observation is most consistent with models which posit that modern population relationships were established quite recently. An example is provided by the Zhoukoudian Upper Cave fossils, recently

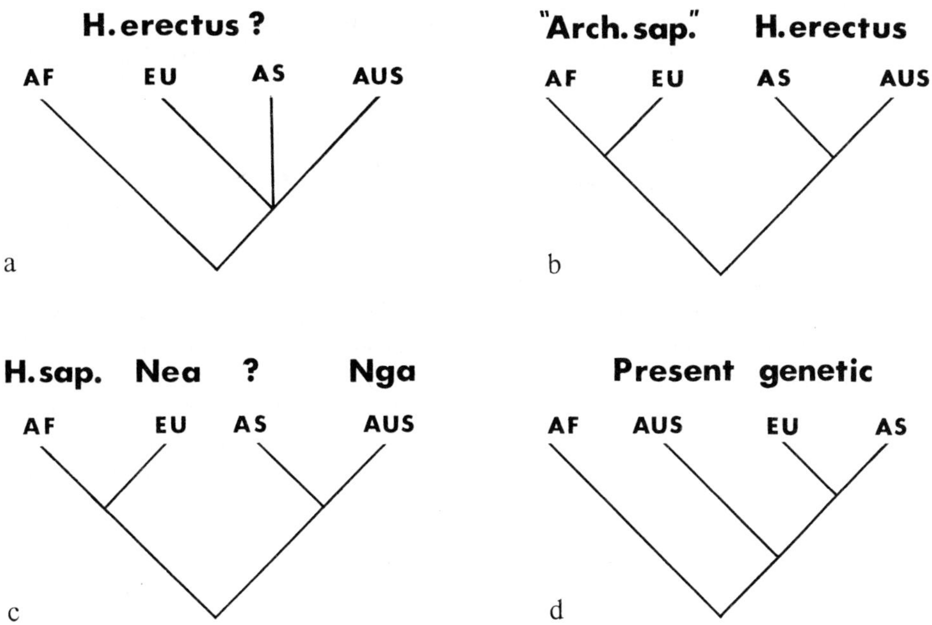

Figure 3. Geographic patterning of human populations through time. AF = Africa; EU = Europe; AS = Asia; AUS = Australasia. NGA = Ngandong (dated as in Bartstra et al. 1988). a) Inferred pattern for the early-early middle Pleistocene radiation of *H. erectus* or related hominids. b) Inferred pattern for the later middle Pleistocene. c) Inferred pattern for the terminal middle Pleistocene-early late Pleistocene. d) Present population relationships suggested by genetic distance analyses.

dated to about 20 kyr BP by radiocarbon (Hedges et al. 1988). These specimens show numerous synapomorphies with early *H. sapiens* fossils from Europe (see, for example, Figure 1 in Stringer & Andrews 1988a) and are surely more similar cladistically or phenetically to their penecontemporaneous western counterparts than we would find comparing, say, Petralona with Hexian in the middle Pleistocene, or Saccopastore with Dali in the early late Pleistocene.

Furthermore, for a multiregional scheme which argues that many 'Mongoloid' features were established in the middle Pleistocene, it is problematic that few regional continuity features are shared between the Upper Cave crania and their supposed local *H. erectus* ancestors, as was recognised by Weidenreich (1939). Just as surprising, given their relatively young age, is the lack of resemblance to their modern 'Mongoloid' counterparts, as was recognised by Wolpoff (1980). In recent multivariate assessments, the 101 cranium shows no special affinity to Asian 'Mongoloids' (unless one includes the atypical Ainu) (Kamminga & Wright 1989, Howells 1989). In fact from Wright's latest multivariate work, the cranium seems to bear most resemblance to recent Australasian and African samples (Wright, pers. comm.). However, there is dental evidence of generalised Mongoloid affinities in the Upper Cave material (C. Turner 1986, but compare Kamminga & Wright 1989), and analyses I performed using facial shape indices do indicate a general placement with Asian/Amerindian populations (Stringer 1989a). What seems to be emerging is that many of the most distinctive present day 'racial' features were still evolving at the end of the Pleistocene, and this is certainly true for the European Upper

Palaeolithic samples, which differ in facial and postcranial form from both earlier (Neanderthal) and later (recent European) samples (Trinkaus 1981, Stringer 1989a). Thus it seems unlikely that such features had a middle Pleistocene antiquity in the region.

3 REPLACEMENT AND CONTINUITY IN EUROPE AND AUSTRALASIA

The fossil record of Europe is critical to our understanding of the emergence of *H. sapiens*, even if the area was not central to the original evolution of our species. In Europe we have a unique record of apparent regional continuity for the establishment of the Neanderthals over 200 000 years or more, and the critical evidence of their disappearance/replacement over a much shorter period of time during the middle of the last glaciation. Why, though, is it so important to the global pattern of modern human origins if the Neanderthals were not significant contributors to the ancestry of modern humans in Europe (even allowing for the possibility of some limited gene flow at the time of their replacement – see below). There are several reasons for this, but most importantly it is because the western Eurasian fossil evidence of archaic-modern relationships is so much better than that of any other area. If it can be shown that the construction of detailed gradualistic evolutionary trends from Neanderthals to moderns is misleading where the fossil record is most complete, it should give a clear signal about the reliability of similar constructions based on less complete material from south west Asia and the Far East (and equally, of course, even in Africa). Additionally, the degree of chronological and geographical overlap between Neanderthal and early modern populations is important evidence about their degree of taxonomic distinctiveness (see below).

Moreover, if there was replacement of 'archaic' hominids in Europe and western Asia, this should give us a yardstick by which to judge genetic data concerning the origin of modern humans. If it were accepted that Neanderthals played little or no role in the evolution of modern humans, then we do have a clear calibration point (c. 30-50 kyr) by which to assess the probability of similar replacement events elsewhere. Evidence of a close genetic relationship between Europeans and Asians (Fig. 3d) suggests that these populations did not have radically

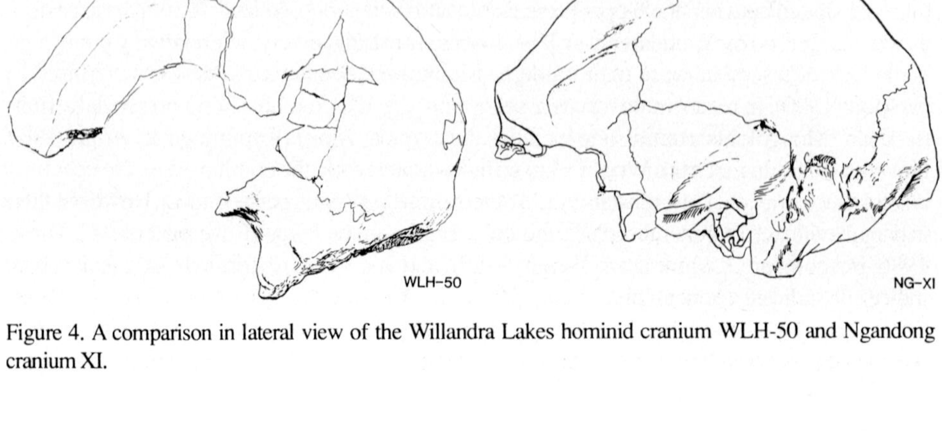

Figure 4. A comparison in lateral view of the Willandra Lakes hominid cranium WLH-50 and Ngandong cranium XI.

different evolutionary histories, and work from mtDNA and nuclear DNA suggests that the antiquity of their regional differentiation is comparable.

One of the distinguishing features of the multiregional and 'Out of Africa' models concerns their radically different predictions about the ubiquitous or restricted presence of 'transitional' fossils (see Stringer & Andrews 1988a, Table 1, Stringer & Andrews 1988b). Such discussion begs the questions which should be asked, but are rarely answered, of how we recognise transitional fossils, and how we distinguish regional features which uniquely characterise a clade from those evolved in parallel in different clades, or those independently retained from a more distant common ancestor.

Regarding transitional forms, we should first propose an appropriate ancestor-descendant relationship, then characterise the relevant morphologies, and only then investigate what would be an acceptable chronological and morphological intermediate between the proposed ancestor and the assumed descendant. Once we employ these criteria, both the Saint-Césaire and Hahnöfersand specimens become poor transitional forms because both are probably contemporaneous with or even post-date the appearance of non-transitional anatomically modern hominids in Europe. Furthermore, one is clearly within the general Neanderthal range in every important respect (apart, perhaps, from nasal breadth) while the other bears a great resemblance to the more complete anatomically modern Podbaba specimen and also falls within the morphological range of even recent European crania. Elsewhere in Eurasia and Australasia, there are several other possible examples of transitional fossils such as those from Vindija, Tabun 2 and Willandra Lakes 50 (WLH-50), whose status remains unclear. Here it becomes important to differentiate *in situ* evolutionary change from the effects of possible hybridisation and the retention of symplesiomorphies. The last mentioned fossil, WLH-50, is an especially interesting specimen (Fig. 4) and I will now briefly discuss the Australasian evidence which still seems to me to present the greatest challenge to the 'Out of Africa' model.

From the late Pleistocene and early Holocene of Australia we are presented with a variation in cranial morphology which seems far greater than would be expected under either a simple multiregional or 'Out of Africa' model. Proponents of the multiregional model have therefore proposed that there were at least two distinct regions of origin for these early Australians – the more robust samples were descended from Indonesian *H. erectus* while the more gracile ones descended from the main Asian *H. erectus* stock. As an alternative, proponents of the 'Out of Africa' model have, up to now, only been able to speculate about the possibility of one original anatomically modern founding population, and the subsequent development of greater cranial variation within Australia.

However, things are becoming a little clearer. First, an electron spin resonance date of 25-30 kyr is now published for WLH-50 (Caddie et al. 1987), which contradicts previous guesses that the specimen could be as much as 50 kyr old. If this first published date is approximately correct, the WLH-50 individual was not living earlier than the nearby Mungo hominids of 25-30 kyr, but at about the same time. Furthermore, although a full morphological description of the specimen has not yet been published, it is evident that some of its supposed 'archaic' characters may be individual features rather than specific population links with Indonesian hominids. For example, the indisputably thick cranial vault consists at maximum thickness of about 90% diploic bone, rather than the thickened inner and outer tables characteristic of *H. erectus*, leading both Peter Brown and Stephen Webb to suggest a possible pathological cause (Brown, pers. comm., Webb, pers. comm.).

Furthermore, Brown (1987:61) has recently concluded from the most wide-ranging survey of Pleistocene and early Holocene Australian skeletal material yet published that:

In contrast to Thorne (1977), combination of the craniometric, tooth size and vault thickness results suggest a single, homogenous, Pleistocene population...Variation within the Pleistocene group is present at an individual level...and this is combined with the effects of artificial cranial deformation on cranial morphology and size relationships at Coobol Creek and Kow Swamp.

Thus it remains possible that there was essentially only one anatomically modern founder population for all late Pleistocene Australians, followed by the rapid development of local variation, and this seems to be in line with recent genetic analyses. New late Pleistocene uranium-series dates for the Ngandong material make an *in situ* transition to *H. sapiens* seem even less likely, but leave open the possibility of gene flow from these Indonesian late *H. erectus* (Bartstra et al. 1988). However, there are still marked differences between the Ngandong and WLH-50 crania, with the latter specimen showing more modern parietal and occipital proportions, as well as showing the non-homologous cranial thickening already mentioned (Fig. 4).

4 THE USE OF 'REGIONAL' CHARACTERS AND OUTGROUP COMPARISONS

There remains the problem of the supposed 'regional' characters said to link many of the Australian hominids with their Indonesian predecessors, rather than with an 'Out of Africa' founding population. Additionally, such characters have been recognised by proponents of the multiregional model for the European and Asian evolution of modern humans, where they are also said to characterise morphological clades (Wolpoff et al. 1984, Wolpoff 1989). There is a great need for comprehensive critical reviews of these characters, their precise definition, variation and regional significance, and the first such reviews have recently been conducted (D. Hodgson, pers. comm., Groves 1989, Habgood 1989). It is evident that, for regional analyses to be effective, the characters chosen must be clearly defined, homologous, and additionally it must be demonstrated by outgroup comparisons that the characters are indeed especially characteristic of the particular 'lineage' chosen.

One of the first problems to be considered is the character composition taken as typifying a particular recent population. It is evident that the percentage occurrence of some of the supposed 'regional' characters is in fact highly variable within recent 'Mongoloid' and 'Australoid' cranial samples, and this gives considerable flexibility in deciding which characters will be emphasised in regional comparisons. Moreover, where there is evidence of late Pleistocene differentiation within the 'clades', this should be examined in order to reconstruct the appropriate ancestral morphotype.

For example, it is generally agreed that Amerindians and east Asian populations have diverged in the last 30 kyr. Comparing these groups it is evident that, since both groups show high frequencies of shovelled upper incisors (C. Turner 1986), this was probably present in the common ancestral population. However, the same groups show different proportions of frontal flattening, M3 agenesis, nasal prominence and subnasal projection (with respectively lower, higher, lower and lower frequencies in many Asian samples). Thus is should not be concluded that the original east Asian 'clade' characters were identical to those of present Asian samples. Careful outgroup comparison should be used in order to reconstruct the appropriate ancestral Amerindian/Asian morphotype for use in regional comparisons further back in time.

Similarly, mainland Australians and Tasmanians have probably been isolated from each other since the late Pleistocene (Jones 1979), and are certainly distinct in their frequencies of supposed regional characters, for example in the lower frequencies of flattened frontals, massive supraorbital tori and robust malars in the Tasmanian samples. It should not automatically be assumed that the mainland Australian pattern was the ancestral one. Construction of an ancestral morphotype using New Guinean, Australian and Tasmanian data, together with outgroup comparisons where there is character conflict, are thus essential prerequisites for earlier Australasian regional comparisons. Additionally, such comparisons show that the frequency or degree of expression of these 'regional' characters can change over quite short periods of time, as also noted by Groves (1989) from other regional data.

Where outgroup comparisons have already been made (D. Hodgson, pers. comm., Groves 1989, Habgood 1989), they are instructive about the reliability of supposed 'clade' indicators. All the regional characters used in modern cranial comparisons occur outside the appropriate groups, and many can be recognised in earlier hominid samples from other regions. Moreover, they can occur at high frequencies outside the 'clade' (e.g. in recently described late Pleistocene material from Gough's Cave, England, there is apparent M3 agenesis in 2 out of 3 mandibles, giving an 'ultra-Mongoloid' frequency of 67% for this character: Currant et al. 1989. Furthermore, M3 agenesis was evidently present in middle-late Pleistocene African hominids, such as those from the Cave of Hearths and Klasies 41815: Tobias 1971, Singer & Wymer 1982).

When Pleistocene fossil material is being examined there is a particular problem caused by assessing frequencies of polymorphic characters in small or damaged samples. For this reason I wish to present a preliminary analysis of the occurrence of many of these regional characters outside the regions of Europe, Asia or Australasia. The left side of Table 1 lists some suggested 'clade' characters said to typify archaic-modern lineages in Europe, Asia and Australasia (Weidenreich 1943, Wolpoff et al. 1984, Wolpoff 1989). For Australasia there is a particular emphasis on the face and frontal bone, where characters are claimed to have been retained over about 700 000 years between a specimen like Sangiran 17 and robust Australian hominids such as the Kow Swamp sample and WLH-50 (Wolpoff et al. 1984, Wolpoff 1989).

For outgroup comparison, I have chosen to document the frequency of these so-called regional characters in a large cranial sample which is not considered to be part of any of these clades, but which covers about the same time-span (c. 10-30 kyr) as the early modern fossils of Europe, eastern Asia and Australia. The chosen material comprises the samples from the late Pleistocene of north Africa (Afalou, Taforalt, Wadi Halfa and Jebel Sahaba). I have studied some of the material myself, but there are also excellent publications which provide documentation of many of these characters in the samples (Arambourg et al. 1934, Ferembach 1962, Anderson 1968, Greene et al. 1967, Greene & Armelagos 1972). The right side of Table 1 provides frequency assessments (only approximate in some cases and subject to confirmation by further study) of the supposed regional characters in the samples, and it is apparent that most of these characters are present to some degree in these African samples. Not only that, some of the Eurasian/Australasian characters are actually at a higher frequency in the north African sample than in their supposed original clades, and there is a particular resemblance to the Australian 'clade'. Since it is highly unlikely that these north African specimens were ever part of an east Asian or Australasian clade (although they may well be closely related to the early moderns of western Eurasia), how can we explain the presence of so many of the 'regional' characters?

I think the answer lies partly in the retention of primitive characters which were present in the earliest modern human populations (possible examples are indicated by a 'P' in Table 1), and partly through parallel evolution. There is no better comment on this than in the conclusions of Greene & Armelagos's (1972:78) study of the Wadi Halfa sample which provides us with an excellent summation of the situation:

The preceding analyses indicate that the crania from 6B36 (Wadi Halfa) and the SMU (Jebel Sahaba) sites are very similar. Both populations show high frequencies of heavy supra-orbital ridges, marked temporal lines, gonial eversion, zygo-maxillary tubercles, massive jaws and teeth, and other features most likely related to heavy masticatory stress. The existence of this stress is documented in the heavy degree of dental attrition found in both populations. Other similar features are wide nasal apertures, dolichocranic skulls, high incidence of wormian bones, and a flattening in the lambdoid region producing what we have called an occipital bun effect.

While some of these characters were certainly widespread in earlier African or south west Asian hominids (and therefore probably represent primitive retentions), others were uncommon, and it would appear that their high frequency may have evolved locally during the late Pleistocene. If such characters could evolve locally and relatively recently in north Africa, why could they not have evolved locally and relatively recently in other regions as well? Given this situation, what is needed are further reviews of these so-called regional features to ascertain their phylo-

Table 1. Selected 'regional' characters, and their occurrence in late Pleistocene north African cranial samples. Frequencies derived from [1]Arambourg et al. 1934 (Afalou); [2]Ferembach 1962 (Taforalt); [3]Anderson 1968 (Jebel Sahaba); [4]Greene et al. 1967; Greene & Armelagos 1972 (Wadi Halfa); [5]Stringer, general compilation or personal observation. Some frequency assessments are provisional and require futher study. Proper determination of incisor shovelling and prognathism frequencies in the Afalou and Taforalt samples is not feasible because of maxillary tooth evulsion. Characters which may be primitive for *Homo sapiens* are indicated by P.

'European'		
Prominent nasal bones		>50%[1-5]
Occipital 'hemibun'		c.30%[1-5]
'Mongoloid'		
Small frontal sinus		Common[4]
Low prognathism		c.20%[3-5]
Transversely flat face	P	Common[1-5]
Shovelled upper incisors	P	>70%[3-4]
Wormian bones (Inca)		Occurs but rare[1-5]
(Other)		>50%[1-5]
Metopic suture		Occurs but rare[1-5]
Ear exostoses		c.10%[3-4]
Mandibular/maxillary tori		>10%[1-5]
M3 agenesis		Not present?[1-5]
'Australoid'		
High prognathism	P	c.50%[3-5]
Flat frontal	P?	>15%[1-5]
Strong supraorbital region	P	>50%[1-5]
Horizontal supraorbital border	P?	c.50%[1-5]
Rounded orbital margin		Common[5]
Occipital torus	P	10-45%[1-5]
Everted malars		>50%[1-5]
Zygomaxillary ridge/tuberosity		>50%[1-5]
Large teeth	P	Common[1-5]

genetic value in properly delimiting local clades. Until that is done, I would prefer to regard many of these characters as primitive retentions or late Pleistocene local modifications to a fundamental modern morphology which was probably African in origin.

5 THE RECOGNITION OF SPECIES AND HYBRIDISATION

There has been much, very welcome, discussion of species recognition in palaeoanthropology in the last few years (see e.g. Tattersall 1986, Wolpoff 1986, Rightmire 1986, A. Turner 1986). However, there is still widespread confusion between different concepts of species, in particular the biological and evolutionary (phylogenetic) concepts (compare Wolpoff et al. 1988, Stringer & Andrews 1988b). My use of *H. sapiens* is evolutionary, (although other workers seem to have assumed it is biological, Wolpoff et al. 1988), and is based on the recognition of specific skeletal synapomorphies between recent and fossil human material. It has been suggested that cladistics is misapplied when used within species (Wolpoff et al. 1989), but this is not necessarily so (Mishler 1990), and it should be remembered where it was first proposed that regional 'clades' existed in *H. sapiens* (Thorne & Wolpoff 1981, Wolpoff et al. 1984). Additionally, if this restriction on cladistic analysis was rigorously applied, it would be impossible under conventional classifications to even study the relationship between Neanderthals and modern humans cladistically! This is not to deny that great care must be exercised when working cladistically at low taxonomic levels. There is always a danger of confusing intraspecies polymorphisms with characters which differentiate between species, and there are also problems with the use of continous data in such analyses, including my own (Pimentel & Riggins 1987).

The unsatisfactory nature of the biological concept of species when applied to the past should now be well understood (Wiley 1981, Ax 1987) and we cannot use a hybridisation test to investigate the relationship between Neanderthals and modern *H. sapiens*. Moreover, even if there was some hybridisation between Neanderthals and *H. sapiens* (which is still not established), this would not necessarily indicate that they were conspecific, as pointed out by Simpson, writing before the rise of cladism (1961:153):

It is also clear that two species may interbreed to some extent without losing their distinction in evolutionary roles and that this is the really important point for evolutionary taxonomy. The amount of interbreeding allowable by definition is then precisely as much as does not cause their roles to merge.

Wiley (1981:62-64) discussed such problems from a biological and phylogenetic perspective, and looked at species recognition in cases of 'partial sympatry'. He distinguished 3 types of occurrence:

1. Each geographic population is a distinct epiphenotype, no intermediate epiphenotypes are found in the area of sympatry: Indicates separate species.
2. Each geographic population is a distinct epiphenotype in allopatry, sporadic hybridisation or local (limited) introgression in part(s) of the area of sympatry: Indicates separate species.
3. Each population is a distinct epiphenotype in allopatry, but there is complete introgression throughout the zone of sympatry.

Recognition of distinct species depends on the perceived phylogenetic relationships of the populations and the extent of the zone of introgression or intergradation. If the zone of interaction is narrow and parental epiphenotypes predominate outside the zone, these are probably distinct species.

Given the close genetic relationship between the African apes and *H. sapiens*, even following an evolutionary divergence of several million years, it is evident that Neanderthals and early *H. sapiens* must have been very closely related genetically. If they did represent distinct species, they were probably siblings. Thus hybridisation probably could have occurred. But it remains to be determined (if it ever can be) what effect such hypothesised hybridisation would have had on the discreteness of the populations concerned. Suggestions of a long-term coexistence of Neanderthals and early modern humans in southwest Asia (Bower 1988) would argue strongly for a greater rather than lesser degree of taxonomic separation between the groups (Stringer 1988), as indicated from Wiley's (1981) categories 1 and 2 given above. The case for gene flow and hybridisation between Neanderthals and early modern humans is stronger in central Europe (Frayer, this volume, Smith, this volume, Smith & Trinkaus, in press), but even under Wiley's category 3, it could be argued that beyond the claimed zone of introgression the populations maintained their separate identities (e.g. as shown by the occurrence of an unequivocal Neanderthal at Saint-Césaire at a late date in western Europe). So even if the existence of hybridisation between Neanderthals and early modern humans were demonstrated, this would by no means prove that they were conspecific, nor would it automatically mean that the Neanderthals contributed significantly to the modern human gene pool beyond (possibly mobile) hybridisation zones.

6 FUTURE PROSPECTS AND CONCLUDING REMARKS

Thus to conclude, I would like to reaffirm my preference for the 'Out of Africa' model, but emphasise that we still lack much evidence for the real origin of the first modern humans. Specimens such as those from Florisbad, Ngaloba, Omo Kibish 2 and Jebel Irhoud are suggestive of a late middle-early late Pleistocene 'transition' which took place in Africa (Bräuer, this volume), but these specimens are too few and scattered to give us any real idea of the evolutionary processes involved (compare Rightmire 1989). Although population and chronological relationships during the Mousterian of south west Asia are becoming clearer, we cannot really link the Skhul-Qafzeh samples to specific later human groups, and given their apparent antiquity, it remains possible that no special links exist. Evidence from eastern Europe, together with archaeological indications of possible 'acculturation' of late Neanderthals (Mellars 1988, Mellars & Stringer 1989) leaves open the possibility of some gene flow between Neanderthals and early *H. sapiens*, if on a very limited scale, and with undetermined significance for the ancestry of later Europeans. The lack of good data from eastern Asia and Australasia between 30-100 kyr means that scenarios of *H. sapiens* evolution cannot yet be tested satisfactorily there, and the genetic data assume particular importance in reconstructing late Pleistocene events in these regions.

Many recent data still need to be integrated into models of *H. sapiens* origins, and these include multivariate cranial analyses which, although indicative of a close relationship between all recent samples and the late evolution of many 'regional' features, do not reflect the pattern of relationships expected from recent genetic analyses (Howells 1973, 1989). Similarly, a series of dental analyses by C. Turner (1986) are also indicative of a late origin for modern *H. sapiens*, but his 'Mongoloid' samples are most distinctive compared with a surprisingly close African-Australasian relationship. While it is tempting to see the latter as due to the retention of symplesiomorphies from original African ancestors, the accompanying divergences of other groups, especially the 'Sinodonts', would need to be explained.

Perhaps the continuing growth in our knowledge of the genetic relationships between different human populations will provide the stimulus for further development of these valuable, but primarily phenetic, studies of cranial and dental variation. One essential task is the construction of morphotypes of hypothetical ancestors for recent geographical variants, as discussed earlier. Assuming that the population relationships shown in Figure 3d are accurate, it should be possible to construct a cranial and dental hypothetical common ancestor for the European + Asian/Amerindian groups, then to construct a common ancestor with Australasian groups, and finally, to construct an original *H. sapiens* morphotype by adding the recent African data, utilising a hominid outgroup at this last stage. This should also enable us to escape from the idea that any of these common ancestors should somehow have especially resembled modern Africans (see e.g. the cover of *Newsweek* 111(2), January 11th, 1988, Excoffier et al. 1987, Wolpoff 1989). It will then be possible to interpolate our fossil samples more usefully into such a framework.

C. Turner's (1986) dental data suggested to him a possible Asian origin for modern *H. sapiens*, a view also supported from early mtDNA analyses (Johnson et al. 1983) and some more recent genetic polymorphism studies (Excoffier et al. 1987, Sanchez-Mazas & Langaney 1988). Habgood (1988) has proposed a dual origin for modern humans, which bypasses the Neanderthals, but incorporates archaic-modern regional continuity in both Africa and eastern Asia. This latter model is more realistic than a widely-based multiregional model, given the presence of rather generalised 'archaic *sapiens*' populations in both Africa and eastern Asia in the middle-early late Pleistocene, but each of these alternative proposals must at least come to terms with the results of the other palaeontological and genetic research summarised in points (2), (3) and (5) of the Introduction.

A primary African origin for *H. sapiens* in the late middle-early late Pleistocene still seems to fit best with the available palaeontological and genetic evidence, whether we adopt a complete 'Out of Africa' scenario or an 'African hybridisation and replacement model' (Bräuer, this volume). Whether it is also compatible with the behavioural record remains to be seen (Mellars 1988), but it is evident that important behavioural changes may lie behind the origin and subsequent dispersal of *H. sapiens* (Lewin 1987, Mellars & Stringer 1989). Study of these should allow us to go beyond the use of unnecessary phrases such as 'holocaust' (Wolpoff et al. 1988) or 'killing off' (Spuhler 1988) to examine the real mechanisms behind what was anything but an overnight establishment of early *H. sapiens* across the Old World.

I would now like to add a few more general comments. In future I think it would be helpful to give greater consideration to the wider field of palaeontology when discussing the origin of *Homo sapiens*. Occurrences of apparent punctuational change, cladogenesis and parallel evolution are common evolutionary messages from the fossil record, and yet many palaeoanthropologists still focus on unilinear gradualism as the mode of hominid evolution. It should also be remembered that most specimens in the fossil record have no living descendants (and this statement does not just apply to Palaeozoic or Mesozoic material, since it is as valid for the Pleistocene). Thus extinctions are a normal part of the process of evolution, and exact ancestors are difficult or impossible to recognise. Yet 'ancestor-searches' still seems to dominate palaeoanthropological discussion.

I think the view has developed that in some way the 'Out of Africa' model is non-evolutionary, that belief in it means that the Pleistocene hominid record of the rest of the world is somehow devalued, and that evolution stopped with the first appearance of modern humans. These views are mistaken, most notably because we should not value fossils only by their status as possible ancestors for us. They will teach us much more about human evolution if we can take a broader palaeontological view of them, liberated from their particular relevance to the origin of modern humans. The Neanderthals certainly changed morphologically through time and space. Are they any less interesting and informative about evolutionary processes if they did not become modern humans?

What proponents of multiregional/gene flow models perceive as non-evolutionary about the 'Out of Africa' model is, in fact, its requirement that we abandon previous models for a widespread origin of the shared characters of modern humans. If the new model is correct, the selective or other forces behind our species origin must be sought in the midde-early late Pleistocene record of Africa, and nowhere else, while it is the present-day regional differences which were widespread in their origins. Moreover, if behavioural superiority did lie behind the dispersal of *H. sapiens* and the replacement of archaic humans, it is certainly possible that many 'modern' skeletal features became established globally not because of their inherent superiority under local selection, but because they 'hitch-hiked' along with the spread of new behavioural patterns (Stringer 1989b).

Finally, I believe that as the genetic data increase in scale and refinement, we will be better able to use our fossils to reconstruct the actual recent history of the human species, including an increasingly fine resolution of the morphological and genetic changes which have occurred during the last 100 000 years of human history.

ACKNOWLEDGEMENTS

First I would like to thank Günter Bräuer and Fred Smith for inviting me to participate in the session on the origins of modern *H. sapiens* at Zagreb in 1988, and for the opportunity to prepare this paper. I would also like to thank my colleagues Robert Kruszynski and Rosemary Powers for their help in the preparation of the figures, and Peter Andrews, Robert Kruszynski and Günter Bräuer for their useful comments on an earlier version of this manuscript.

REFERENCES

Anderson, J.E. 1968. Late Paleolithic skeletal remains from Nubia. In F. Wendorf (ed.), *The Prehistory of Nubia*. Vol. II. Dallas: Southern Methodist University Press: pp. 996-1040.

Arambourg, C., M. Boule, H.V. Vallois & R. Verneau 1934. Les grottes Paléolithiques des Beni-Segoual (Algérie). *Archives Inst. Paléont. Hum.* 13 (whole volume).

Avise, J.C., J. Arnold, R.M. Ball, E. Bermingham, T. Lamb, J.E. Neigel, C.A. Reeb & N.C. Saunders 1987. Intraspecific phylogeography: The mitochondrial DNA bridge between population genetics and systematics. *Ann. Rev. Ecol. Syst.* 18: 489-522.

Ax, P. 1987. *The phylogenetic system*. Chichester: Wiley.

Bartstra, G-J., S. Soegondho & A. van der Wijk 1988. Ngandong man: Age and artifacts. *J. Hum. Evol.* 17: 325-337.

Bowcock, A.M., C. Bucci, J.M. Hebert, J.R. Kidd, K.K. Kidd, J.A. Friedlaender & L.L. Cavalli-Sforza 1987. Study of 47 DNA markers in five populations from four continents. *Gene Geog.* 1: 47-64.

Bower, B. 1988. An earlier dawn for modern humans? *Science News* 133: 138.

Brown, P. 1987. Pleistocene homogeneity and Holocene size reduction: The Australian human skeletal evidence. *Archaeol. Oceania* 22: 41-67.

Caddie, D.A., D.S. Hunter, P.J. Pomeroy & H.J. Hall 1987. The ageing chemist – can electron spin resonance (ESR) help? In W.R. Ambrose & J.M.J. Mummery (eds), *Archaeometry: Further Australasian studies*. Canberra: Australian National University: pp. 167-176.

Cann, R.L. 1988. DNA and human origins. *Ann. Rev. Anthrop.* 17: 127-143.

Cann, R.L., M. Stoneking & A.C. Wilson 1987. Disputed African origin of human populations. *Nature* 327: 111-112.

Cavalli-Sforza, L.L., A. Piazza, P. Menozzi & J. Mountain 1988. Reconstruction of human evolution: Bringing together genetic, archaeological and linguistic data. *Proc. Nat. Acad. Sci. USA* 85: 6002-6006.

Currant, A.P., R.M. Jacobi & C.B. Stringer 1989. Excavations at Gough's Cave, Somerset 1986-7. *Antiquity* 63: 131-136.

Darlu, P. & P. Tassy 1987. Roots (a comment on the evolution of human mitochondrial DNA and the origins of modern humans). *Hum. Evol.* 2: 407-412.

Excoffier, L., B. Pellegrini, A. Sanchez-Mazas, C. Simon & A. Langaney 1987. Genetics and history of sub-Saharan Africa. *Yrbk. Phys. Anthrop.* 30: 151-194.

Ferembach, D. 1962. *La Nécropole Epipaléolithique de Taforalt* (Maroc Oriental). CNRS: Rabat.

Greene, D.L. & G.J. Armelagos 1972. The Wadi Halfa Mesolithic population. *University of Massachusetts Research Report* 11: Amherst.

Greene, D.L., G.H. Ewing & G.J. Armelagos 1967. Dentition of a Mesolithic population from Wadi Halfa, Sudan. *Am. J. Phys. Anthrop.* 27: 41-55.

Groves, C.P. 1989. A regional approach to the problem of the origin of modern humans in Australasia. In P.A. Mellars & C.B. Stringer (eds), *The Human Revolution*. Edinburgh: Edinburgh University Press: pp. 274-285.

Habgood, P.J. 1988. A morphometric investigation into the origin(s) of anatomically modern humans. Ph.D. thesis, University of Sydney.

Habgood, P.J. 1989. The origin of anatomically modern humans in Australia. In P.A. Mellars & C.B. Stringer (eds), *The Human Revolution*. Edinburgh: Edinburgh University Press: pp. 245-273.

Hedges, R.E.M., R.A. Housley, I.A. Law, C. Perry & E. Hendy 1988. Radiocarbon dates from the Oxford AMS system: Archaeometry date list 8. *Archaeometry* 30: 291-305.

Howells, W.W. 1973. Cranial variation in Man. *Pap. Peabody Mus.* 67: 1-259 (whole volume).

Howells, W.W. 1989. Skull shapes and the map. Cambridge/Mass.: Harvard University Press.

Johnson, M.J., D.C. Wallace, S.D. Ferris, M.C. Rattazi & L.L. Cavalli-Sforza 1983. Radiation of human mitochondrial DNA types analysed by restriction endonuclease cleavage patterns. *J. Mol. Evol.* 19:255-271.

Jones, R. 1979. The fifth continent: Problems concerning the human colonization of Australia. *Ann. Rev. Anthrop.* 8: 445-466.

Kamminga, J. & R.V.S. Wright 1989. The Upper Cave at Zhoukoudian and the origins of the Mongoloids. *J. Hum. Evol.* 17: 739-767.

Lewin, R. 1987. The origin of the modern human mind. *Science* 236: 668-670.

Mellars, P.A. 1988. The origin and dispersal of modern humans. *Curr. Anthrop.* 29: 186-188.

Mellars, P.A. & C.B. Stringer (eds) 1989. *The Human Revolution*. Edinburgh: Edinburgh University Press.

Mishler, B.D. 1990. Species, speciation, and phylogenetic systematics. *Cladistics* 6: 205-209.

Moritz, C., T.E. Dowling & W.M. Brown 1987. Evolution of animal mitochondrial DNA: Relevance for population biology and systematics. *Ann. Rev. Ecol. Syst.* 18: 269-292.

Nei, M. 1982. Evolution of human races at the gene level. In *Human genetics, part A: The unfolding genome*. New York: Liss: pp. 167-181.

Pamilo, P. & M. Nei 1988. Relationships between gene trees and species trees. *Mol. Biol. Evol.* 5: 568-583.

Pimentel, R.A. & R. Riggins 1987. The nature of cladistic data. *Cladistics* 3: 201-209.

Rightmire, G.P. 1986. Species recognition and *Homo erectus*. *J. Hum. Evol.* 15: 823-826.

Rightmire, G.P. 1989. Middle Stone Age humans from eastern and southern Africa. In P.A. Mellars & C.B. Stringer (eds), *The Human Revolution*. Edinburgh: Edinburgh University Press: pp. 109-122.

Saitou, N. & K. Omoto 1987. Time and place of human origins from mtDNA data. *Nature* 327: 288.

Sanchez-Mazas, A. & A. Langaney 1988. Common genetic pools between human populations. *Hum. Genetics* 78: 161-166.

Santa Luca, A.P. 1980. The Ngandong fossil hominids: A comparative study of a far eastern *Homo erectus* group. *Yale Univ. Publ. Anthrop.* 78: 1-175 (whole volume).

Schwarcz, H.P., R. Grün, B. Vandermeersch, O. Bar-Yosef, H. Valladas & E. Tchernov 1989. ESR dates for the hominid burial site of Qafzeh. *J. Human Evol.* 17: 733-737.

Simpson, G.C. 1961. *Principles of Animal Taxonomy*. New York: Columbia University Press.

Singer, R. & J.J. Wymer 1982. *The Middle Stone Age at Klasies River Mouth in South Africa*. Chicago: University of Chicago Press.

Smith, F.H. & E. Trinkaus (in press). Modern human origins in central Europe: A case of continuity. In J.-J. Hublin & A.M. Tillier (eds), *Aux origines de la diversité humaine*. Paris: Presse Universitaires de France.

Spuhler, J.N. 1988. Evolution of mitochondrial DNA in monkeys, apes and humans. *Yrbk phys. Anthrop.* 31: 15-48.

Stoneking, M. & R.L. Cann 1989. African origin of human mitochondrial DNA. In P.A. Mellars & C.B. Stringer (eds), *The Human Revolution*. Edinburgh: Edinburgh University Press: pp. 17-30.

Stringer, C.B. 1984. The definition of *Homo erectus* and the existence of the species in Africa and Europe. *Cour. Forsch-Inst. Senckenberg* 69: 131-144.

Stringer, C.B. 1988. The dates of Eden. *Nature* 331: 565-566.

Stringer, C.B. 1989a. The origin of Early Modern Humans: A comparison of the European and non-European evidence. In P.A. Mellars & C.B. Stringer (eds), *The Human Revolution*. Edinburgh: Edinburgh University Press: pp. 232-244.

Stringer, C.B. 1989b. Documenting the origin of modern humans. In E. Trinkaus (ed.), *The Emergence of Modern Humans: Biocultural Adaptation in the Later Pleistocene*. Cambridge: Cambridge University Press.

Stringer, C.B. & P. Andrews 1988a. Genetic and fossil evidence for the origin of modern humans. *Science* 239: 1263-1268.

Stringer, C.B. & P. Andrews 1988b. Modern human origins. *Science* 241: 773-774.

Stringer, C.B., J.-J. Hublin & B. Vandermeersch 1984. The origin of anatomically modern humans in western Europe. In F.H. Smith & F. Spencer (eds), *The Origins of Modern Humans*. New York: Liss: pp. 51-135.

Stringer, C.B., R. Grün, H.P. Schwarcz & P. Goldberg 1989. ESR dates for the hominid burial site of Es Skhul in Israel. *Nature* 338: 756-758.

Tattersall, I. 1986. Species recognition in human paleontology. *J. Hum. Evol.* 15: 165-175.

Thorne, A.G. 1977. Separation or reconciliation? Biological clues to the development of Australian society. In J. Allen, J. Golson & R. Jones (eds), *Sunda and Sahul*. London: Academic Press: pp. 187-204.

Thorne, A.G. & M.H. Wolpoff 1981. Regional continuity in Australasian Pleistocene hominid evolution. *Am. J. Phys. Anthrop.* 55: 337-349.

Tobias, P.V. 1971. Human skeletal remains from the Cave of Hearths, Makapansgat, Northern Transvaal. *Am. J. Phys. Anthrop.* 34: 335-368.

Trinkaus, E. 1981. Neanderthal limb proportions and cold adaptation. In C.B. Stringer, *Aspects of Human Evolution*. London: Taylor and Francis: pp. 187-224.

Turner, A. 1986. Species, speciation and human evolution. *Hum. Evol.* 1: 419-430.

Turner, C.G. II 1986. The first Americans: The dental evidence. *Nat. Geog. Res.* 2: 37-46.

Valladas, H., J.L. Reyss, J.L. Joron, G. Valladas, O. Bar-Yosef & B. Vandermeersch 1988. Thermoluminescence dating of Mousterian 'Proto-Cro-Magnon' remains from Israel and the origin of modern man. *Nature* 331: 614-616.

Vandermeersch, B. 1985. The origin of the Neandertals. In E. Delson (ed.), *Ancestors: The hard evidence*. New York: Liss: pp. 306-309.

Weidenreich, F. 1939. On the earliest representatives of modern mankind recovered on the soil of East Asia. *Peking Nat. Hist. Bull.* 13: 161-174.

Weidenreich, F. 1943. The skull of *Sinanthropus pekinensis*: A comparative study of a primitive hominid skull. *Palaeont. Sinica* 10 (New series D, whole volume).

Wiley, E.O. 1981. *Phylogenetics*. New York: Wiley.

Wolpoff, M.H. 1980. *Paleoanthropology*. New York: Knopf.

Wolpoff, M.H. 1986. Describing anatomically modern *Homo sapiens*: A distinction without a definable difference. *Anthropos (Brno)* 23: 41-53.

Wolpoff, M.H. 1989. Multiregional evolution: The fossil alternative to Eden. In P.A. Mellars & C.B. Stringer (eds), *The Human Revolution*. Edinburgh: Edinburgh University Press: pp. 62-108.

Wolfpoff, M.H., J.N. Spuhler, F.H. Smith, J. Radovčić, G. Pope, D.W. Frayer, R. Eckhardt & G. Clark 1988. Modern human origins. *Science* 241: 772-773.

Wolpoff, M.H., X. Wu & A.G. Thorne 1984. Modern *Homo sapiens* origins: A general theory of hominid evolution involving the fossil evidence from East Asia. In F.H. Smith & F. Spencer, *The Origins of Modern Humans*. New York: Liss: pp. 411-484.

Theories of modern human origins

Milford H. Wolpoff
Paleoanthropology Laboratory, University of Michigan, Ann Arbor, USA

1 INTRODUCTION

Two markedly contrasting theories of modern human origins have come into focus because of a new meld of morphological and genetic information and a renewed appreciation of race as a valid basis for partitioning human variation. These widely differing attempts to provide a historic element for modern human variation are not new, but each involves a novel combination of comparative morphological and paleontological observations, and each relies on a different interpretation of newly available nuclear and cytoplasmic (i.e. mitochondrial) DNA data. These theories are referred to by a number of names but are perhaps best described as the 'Garden of Eden' and 'Multiregional Evolution' models, titles that may well be more accurate than many of the statements that have been published about them. Perhaps the most visual expression of the contrast is accurately given by Leakey & Lewin (1977:137):

...one can think of taking a handful of pebbles and flinging them into a pool of water. Each pebble generates outward-spreading ripples that sooner or later meet the oncoming ripples set in motion by other pebbles. The pool represents the Old World with its basic sapiens population; the place where each pebble lands is a point of transition to Homo sapiens sapiens; and the outward-spreading ripples are the migrations of truly modern humans. This image is in contrast to the misleading idea of one pebble landing in the pond, and of the ripples that eventually cover the whole surface arising from that single source.

The 'Garden of Eden' theory mixes two older ideas. The first of these was developed by Protsch von Zieten (a.k.a. Protsch) in his 'African Origins' hypothesis, in which he detailed the claim that modern humans evolved first in Africa and then spread throughout the world (Protsch 1975). This was later named the 'Afro-European *sapiens*' hypothesis (Bräuer 1984a) and more dates and other specifics were added. The second idea developed around the precept of a complete replacement of archaic humans by modern humans. According to this sudden replacement theory there was a single recent origin for all living populations, in Africa. The 'modern humans' who 'already displayed most of the anatomical characters shared by living people' spread out of Africa and completely replaced indigenous populations. As summarized by the two leading advocates of complete replacement (Stringer & Andrews 1988:1263):

Proponents of this model have proposed... an origin for the species during the early part of the late Pleistocene, followed by an initiation of African regional differentiation, subsequent radiation from Africa, and final establishment of modern regional characteristics outside Africa.

The main elements of this 'Garden of Eden' theory are (1) an African origin for modern humans (from the Protsch/Bräuer model and certain inferences made from mtDNA distribution, mostly in Black Americans)], (2) the presence of regional continuity only in Africa and the appearance of African racial features first, and (3) the complete replacement of indigenous peoples by the Africans (from two paleoanthropologists [Stringer 1986, Stringer & Andrews 1988] and from one interpretation of the mtDNA [Stoneking & Cann 1989]). If the visual analogue of modern human origins according to the multiregional evolution is to throw many pebbles into a pond, the equivalent analogue for the 'Garden of Eden' theory is to throw in one big rock!

Multiregional Evolution also has a historic basis, in the 'Polycentric Evolution' hypothesis of Franz Weidenreich. The Multiregional model proposes that the conditions associated with the initial migrations of humans from Africa, approximately one million years ago, created the central and peripheral contrasts that affected the early establishment of regional features at the peripheries of the human range (Thorne 1981). The persistence of these peripheral distinctions, referred to as 'regional continuities', through much of the remainder of the Pleistocene was a consequence of long lasting clines maintained by a balance of gene flow (Wolpoff et al. 1984, Van Valen 1986), often from the center toward the edges, and local selection and in some cases drift. It is the idea of an active balance of the evolutionary forces, rather than static parallel evolution across the polytypic species, that distinguished this model from earlier attempts to provide an evolutionary explanation for Weidenreich's insights. The clinal model for variation of polytypic species is discussed in detail by a number of authors (Birdsell 1972, Charlesworth et al. 1982, Endler 1977, May et al. 1975, Mayr 1963, Lande 1982, Slatkin 1987, Thorpe 1985, Wright 1980, Wolpoff et al. 1984).

Multiregional Evolution theory shares several underlying assumptions with the 'Garden of Eden' model. Three critical shared assumptions are the positing of an ultimately African ancestry for all populations, the expectation that when demonstrably modern humans appeared in Africa they had at least some uniquely African regional (i.e. racial) features, and the assumption of widespread movements of genes throughout the range of Pleistocene human populations. Interestingly, both models also agree in predicting a fairly late appearance for uniquely regional African features (i.e. with modern populations), although for very different reasons and according to very different patterns (a speciation according to the 'Garden of Eden' model and gradual evolution according to Multiregional Evolution. However, the gene flow and population movement assumptions of these two models are not synonyms for each other (Weiss 1988). Moreover, while both models assume (actually depend on) significant population movements in the past, there is an important difference in whether or not these are expected to have resulted in the complete replacement of indigenous natives. In fact, apart from the above areas of agreement these models differ in numerous significant ways.

Paralleling the different evolutionary models, there are also different explanations for the observed mitochondrial DNA (mtDNA) variation. It was first assumed that mutation accumulation in a mostly neutral genome was the main driving force for the observed magnitude and pattern of the mtDNA differentiation. However, the likelihood that local human populations were small for most of the Pleistocene (Weiss 1984), and the dramatic population size fluctuations that probably characterized the end of the last glaciation and certainly are associated with the consequences of agriculture combines to provide the basis for the contention that differential mtDNA lineage survivorships, and the dramatic population increases and population crashes of historic times, must also be considered as part of the explanation for the mtDNA variation seen in living populations (see especially Avise et al. 1988). Mutation, selection, and random loss of mitochondrial lineages combine to produce dynamic equilibria for the number of mtDNA lines surviving at any time.

The 'Garden of Eden' hypothesis for modern human origins has come to be associated with the mutation accumulation (i.e. historical) explanation for mtDNA variation, while the Multiregional Evolution theory associates with the more complex equilibrium explanation of mtDNA variation. Both the two models accounting for modern human origins and the two explanations accounting for limited mtDNA variation are contradictory over a vast range of predictions. This provides the opportunity to test the conflicting predictions when data are available, which is often the case. This paper focuses on the most significant of the testable contrasting predictions, comparing them to evidence from the human fossil record. There are two aspects of these comparisons. First, the predictions that can be derived from the alternative explanations of limited mtDNA variation in living populations are described and summarized in Table 1, and will be discussed in the context of their differing implications for understanding modern populational origins. Second, the contrasting predictions of the 'Garden of Eden' and Multiregional Evolution models provide a specific series of mutually contradictory expectations about the distribution and morphology of human fossil remains from different regions of the world, before and after the origins of modern populations. These contrasting predictions are reviewed in Table 2. In the course of comparing them, the presentation of paleontological data will follow the organization of this table.

2 COMPETING EVOLUTIONARY MODELS OF MITOCHONDRIAL DNA VARIATION

The discovery of limited mtDNA variability in human populations was unexpected (Brown 1980), and its recognition led almost immediately to the simplest explanation; based on the assumption that random mutations were the main source of variation, the small number of variants was taken to indicate a short time span of mutation accumulation and therefore a recent divergence from a common ancestor (Cann et al. 1982, Wilson et al. 1985). However soon thereafter an alternative explanation for limited variability was proposed by Avise and his colleagues (Avise 1986, Avise et al. 1984, Latorre et al. 1986). Their hypothesis is based on the consequences of maternal inheritance for cytoplasms and their mitochondrial organelles. When generations without daughters are encountered (one quarter of all cases in a stable population), mitochondrial lineages are not passed on. In fact, the extinctions of mitochondrial lineages are normal and all existing mitochondrial variants will eventually become extinct for this reason. Therefore, lineage extinctions must be considered along with the rate and duration of accumulating mutations to account for limited variation in the existing number of mitochondrial lineages.

These two different explanations cannot be collapsed into a single one. For instance, when Cann (1987:37) stated 'There is no evidence to suggest that people today have retained any of Eve's physical features. We know only that we have inherited her mtDNA; she might have contributed very little to the ancestry of human nuclear DNA', she was proposing a contradiction. This is because the evolutionary model that Cann and her colleagues use to explain why observed mitochondrial variants are limited creates the need to believe that there actually was a mitochondrial 'Eve'. Both the existence of a mitochondrial 'Eve' and the specific details of how she came to found the mtDNA lines of all living people follow from their explanation of why no older mtDNA variants are ever found in non-African populations. It is reasoned that the mito-

Table 1. Comparison of two explanations for observed difference in mtDNA variation in living humans. Each of the 'origins' models compared in Table 2 assumes a different explanation for these cytoplasmic data.

	Garden of Eden Historical explanation	Multiregional evolution Equilibrium explanation
The explanation for limited mitochondrial DNA variation	There is a single unique ancestry for the one surviving mtDNA lineage and the genetic variation in its living members is limited by the mutation rate and the length of time since it originated	A combination of stochastic losses of some polymorphisms and mutational origins of others creates the conditions for equilibria in mtDNA polymorphisms, whose numbers are thereby controlled by the balance of evolutionary forces creating the equilibria: mutations,selection, and stochastic loss (drift)
The specific causes of mtDNA variation in living populations	The rate of mutation and the length of time since the origin of the existing lines	The past and present magnitudes of the equilibrium-creating forces
The explanation of a single ancestry for cytoplasmic DNA	A single unique ancestral population for all living populations which replaced earlier indigenous natives without mixture, and therefore prevented the continuation of all of their cytoplasmic DNA variants	A single unique ancestral cytoplasm, marked by a mitochondrial polymorphism, spread by gene flow and population movements, and unique because other DNA variants were lost stochastically
The mtDNA 'clock'	Closely times speciation events through the regular ticking of randomly accumulating neutral mutations, but may be less accurate for timing: population splits	There is no clock that works accurately for humans of the last million years , with their small interconnected populations and dramatic population size fluctuations that came with the changing adaptations at the end of the Pleistocene. This is because three processes effect the size and form of reconstructed lineage divergence trees. While mtDNA variation is created by mutations, it is limited by selection and unpredictably depleted by random losses when daughterless generations are encountered
Nuclear and mitochondrial DNA variations	Are necessarily linked because they were spread around the world by the same mechanism – complete populational replacement	Are independent of each other

chondrial variants of today descend from a single ancestral form that spread throughout the world by expanding populations of Africans who did not mix with the indigenous natives they replaced. After discussing several less probable explanations of why the more ancient mtDNA forms cannot be found, Stoneking & Cann (1989:28) put it this way:

the rather staggering implication is that the dispersing African population replaced the non-African resident populations without any interbreeding.

The lack of population mixing explains why older, non-African mtDNA lineages are not found by positing that they never got into the mtDNA gene pool to begin with. As Table 1 summarizes, this explanation links mitochondrial and nuclear DNA variants through the replacement process.

These biochemical geneticists hope to rely on paleoanthropological interpretations to support their model, for instance citing work by Bräuer, Stringer & Andrews. But only the latter authors are unequivocal in their support. Bräuer, for instance, says 'hard evidence which could be interpreted as the result of mixing is relatively rare' (1989:134) and 'I cannot at present recognize any unequivocal facts which would disprove even a global 'Out of Africa' model as has been suggested from recent genetic analyses' (p. 148), but at the same time admits 'the fossil record does not favour a radical replacement process' (pp. 148-149). It might be better to consider the genetic interpretations on their own merits.

So it happens that just as there are competing paleoanthropological hypotheses to explain modern human origins, there are equally distinct competing hypotheses to explain the limited variability in modern human mitochondrial lineages. It has been very difficult to gather biochemical data, cytoplasmic or nuclear, to distinguish between these hypotheses, but their predictions for patterning in the fossil record of human evolution are quite clear. Yet, the fact that there are conflicting hypotheses about the causes of variation in mitochondria (Table 1) does not always seem to be appreciated.

2.1 *Explanation of limited variation in mtDNA*

The mutational explanation for limited mtDNA variation derived from the then widely accepted understanding that accumulating neutral mutations played the pre-eminent role in causing this variation. If mutations alone are the major cause of this variation, then a low magnitude of variation must imply a short mutational history. Yet, the contention of a recent ancestry for these genetic variants is not totally in agreement with the genetic evidence from other laboratories, or with observations made on living populations (Honeycutt & Wheeler 1987, Spuhler 1988).

In fact there are at least two additional factors determining the amount of observed mtDNA variation. One of these is natural selection. Since the first publications, there has been a more realistic assessment of a significant role for selection; perhaps as much as 30% of the mitochondrial genome may be subject to selection according to Cann and her colleagues (Cann et al. 1987b). There may well be as much selection acting on the mitochondrial genome as on nuclear DNA (Hale & Singh 1986, Whittam et al. 1986, Holt et al. 1988). One study suggests that selection acting on each of these genomes can influence the other, even further complicating the issue (MacRae & Anderson 1988), and it seems clear that the neutrality assumption is incorrect.

The second factor is critical because of its role in reducing mtDNA variation. This is the drift or random loss process, enormously multiplied in its effect on mtDNA variation because of its mode of inheritance, since an mtDNA line is lost whenever a daughterless generation is encountered. This random loss process is so significant that it can be easily predicted that every existing mtDNA variant will eventually be lost. The process, unlike mutation, is irregular since populations are continually changing in size, thereby altering the frequency of daughterless generations.

This process is very similar to the loss of family names that comes when sonless generations are encountered. Consider what would happen if a future anthropologist were to encounter an urban immigrant neighborhood in which all the families had the same family name. The anthropologist could assume that all the families descended from a single immigrant family that replaced their neighbors and multiplied to fill the neighborhood. This would be equivalent to the 'Garden of Eden' hypothesis. However, a better although less dramatic explanation would be that many families immigrated to the city, but over time family names came to be lost as generations without sons were encountered. This 'Multi-immigrant' model is equivalent to the Multiregional Evolution hypothesis.

Random loss could be an extremely powerful evolutionary cause of the loss of mtDNA variation. For instance, in a study of three different North American species (eels, hardhead catfish, and redwinged blackbirds), the effects of random loss were found to slow down the effective rate of evolution by 200-900 fold (Avise et al. 1988), and it was concluded that (p. 341):

the intraspecific genetic architecture of species appear to be shaped to a large degree by historical demographic factors.

Probability models with stochastic survivorship assumptions show that virtually all existing mitochondrial lineages will sooner or later become extinct, even in a stable population. With selection or population size fluctuations, the process would be considerably more rapid. According to one set of calculations, a stable population founded by 15 000 unrelated females would have a 50% chance of appearing to descend from a single female within 18K generations, perhaps as little as 300 kyr years for humans given the current estimates of a short lifespan for premodern humans (Trinkaus & Thompson 1987), under the assumption of stochastic lineage extinctions. The significant role for selection in this process and the unpredictable influence of demographic histories of populations combine to destroy the assumption of regularity that makes the mtDNA 'clock' potentially valid.

2.2 *Explanation of a single ancestry for existing variants*

Random loss (i.e. differential lineage survivorship), rather than an ancestry in a single unique source population, is almost certainly the cause of the mitochondrial pattern of limited variation, whether in fruitflies (Latorre et al. 1986), eels, catfish, blackbirds (Avise et al. 1988) or humans (Wolpoff 1989, Excoffier & Langaney 1989), especially if populations in the past were locally small, fluctuating, and linked by interconnected lines of gene flow throughout their evolutionary histories as Weidenreich (1946, 1947) suggested for humans in the development of his Polycentric hypothesis, and Avise and his colleagues suggested for the North American species in the study cited above. In fact, this precept is part of the analytical model that a number of workers have brought to the interpretation of today's mitochondrial distributions. For instance, Excoffier & Langaney (1989:82) reflect the same understanding when proposing that

generally, human populations have not been isolated since their constitution and have exchanged more or fewer genes with surrounding populations, increasing their own diversity.

The pattern of distributions for mitochondrial variants suggests that the source population concept is invalid:

It seems difficult to determine that a particular set of populations may be the source of the differentiation of all others, since ancient types are observed in each continental group and because the time necessary to produce the present molecular diversity largely exceeds 100 000 years. Excoffier & Langaney (1989:81-82)

There are two more problems in the single ancestry supposition. First, it assumes that the ancestral population was monomorphic for mtDNA variation. Yet, if the population was a founding source for all living people, replacing earlier peoples, one must consider the effect of mtDNA polymorphisms in it. The problems arise from the fact that splits in mtDNA lineages may not correspond to population splits. But not all polymorphisms are populational. A second problem results from the fact that even a single individual may have a number of different mtDNA variants.

2.3 *Is there a clock?*

There are numerous reasons to suppose that a mitochondrial 'clock' for timing recent events in human evolution is not valid (Saitou & Omoto 1987, Honeycutt & Wheeler 1987, Lewin 1990, Wolpoff 1989). The importance of viewing mtDNA variation as a balance influenced by a variation creating mutations, variation limiting selection, and variation destroying random losses as daughterless generations are encountered, is in the expectation that the rate of observed mtDNA lineage divergences is not constant, and not even predictable, unless the details of all three of these processes can be known. The number of divergences that seem to characterize the surviving lineages in such a balance is independent of time over periods when mtDNA variation can not be related directly and virtually uniquely to a single constant process as in the neutral mutation model. The existing variation is a consequence of the magnitudes of these forces and not necessarily the length of time they have been in effect.

Thus, when lines are lost as daughterless generations are encountered, the reconstructed tree of mtDNA divergences will be shorter (fewer mutations will be counted along its reconstructed branches because the record of the missing divergence will have been lost), and according to which parsimony assumptions are used in determining the characteristics of the tree, it may also be quite different in branching pattern. When divergent lines are not represented among the living, record of the mutations that caused them is lost, which makes observed divergence rates much slower than what could be calculated from estimated mutation rates, in some cases by as much as 200-900 fold as cited in Avise et al. (1988) who discuss this in detail. Disturbances to a predictable, or clock-like, regularity in divergences are most pronounced in species such as our own, combining gene flow between small populations and significant population size fluctuations. It is little wonder that a recent review article (Lewin 1990) was entitled '*Molecular clocks run out of time: the theory that we can date the birth of new species by charting the steady accumulation of mutations over evolutionary time is in serious trouble*'. It's not much of a clock that cannot distinguish between 5 minutes after the hour and 50 minutes after the hour (Stoneking & Cann [1989] give a date range for 'Eve' as 50 000 to 500 000 years ago). The only possible conclusion, in my view, is that as far as late Pleistocene human evolution is concerned there is no molecular clock.

2.4 *Relation of nuclear and mtDNA variations*

The 'Eden' view requires an 'Eve' who is ancestral to both the cytoplasms and the nuclear DNA of living people because of the link between these forged by population movements in the complete replacement process. In the stochastic loss model, however, there is no such link. This lack of a linkage between the cytoplasmic and the nuclear DNA requires neither a series of replacements without admixture, nor any speciation events, to account for a common origin for cytoplasms because there was no common recent origin for nuclear DNA.

Female exchanges between indigenous populations of pre-modern humans, combined with a constant rate of stochastic lineage extinctions, may well have resulted in a homogeneous distribution of a few mtDNA lineages across widely spread human populations in the past (Saitou & Omoto 1987, Excoffier & Langaney 1989). This ancient distribution of genetic polymorphisms, perhaps itself the consequence of a previous long span of differential stochastic lineage extinctions, could have provided by chance a single surviving mtDNA line for descendant populations as the result of continued lineage extinctions. Therefore, the pattern of today may simply show random survivorship of what earlier was a single very common widespread mitochondrial lineage. The late Pleistocene provides more than sufficient time to allow for the possibility that selection and stochastic lineage extinctions could (and in my opinion do) account for the limited mtDNA variation reported by Cann et al. and others.

There is some independent evidence to support the contention that pre-divergence mitochondrial polymorphisms (Horai et al. 1986, Harihara et al. 1988) and subsequent populational mixing has significantly altered the interpretation of mtDNA variation. Mixing is ubiquitous in the historic record of human population movements, regardless of how different the physical characteristics or cultural variations of the populations might be. The effects of mixing are to invalidate branching analysis, and confuse both the clusters of relationships and their magnitudes based on mtDNA variation. Thus, for instance, in an analysis of the mtDNA based genetic distances presented by Cann and her colleagues (Cann et al. 1987a), the New Guineans seem to be most closely related to Africans and only distantly related to Australians on the resulting phylogenetic trees (Saitou & Omoto 1987). Yet, by virtually every other measure (including nuclear DNA analysis) New Guineans and Australians are the same people, only separated by the rises in sea level since the end of the Pleistocene. Clearly, mixing takes place in non-human populations as well. In fact, it is the analysis of mtDNA itself that provided evidence of significant hybridization between two deer species, white-tailed and mule deer (Carr et al. 1986). A study of two closely related mice species with ranges that overlap in southern Denmark shows that while the nuclear DNA of each species penetrates only a few kilometers into the range of the other, the mtDNA variants of one are widespread within the other throughout Scandinavia (Ferris et al. 1983). A tree analysis based on mtDNA for these two species, similar to the tree analysis said to show an African origin for human mtDNA variants, would be totally misleading (Jones 1986).

In sum, there is a viable genetic model to explain the limited mtDNA variation that does not depend on population replacements linking mitochondrial and nuclear DNA. If the human fossil record refutes the 'Garden of Eden' model of 'modern *Homo sapiens*' origins, it refutes the single recent origin model for mtDNA variation as well. Contrary to a number of authors, ranging from biochemical geneticists such as Cann and her colleagues (Cann 1987, 1988, Cann et al. 1987a) to paleoanthropologists such as Bräuer (1989) and Stringer & Andrews (1988), genetic data do not support any paleoanthropological model because it is not the genetic data that help choose between the conflicting paleoanthropological models, but rather the fossil data that help choose between the conflicting genetic models!

3 'ANATOMICALLY MODERN' *HOMO SAPIENS*

3.1 *Why a new species?*

The basis for the 'Garden of Eden' hypothesis with its associated recent 'Out of Africa' migration date, is the interpretation that the existing mtDNA lineages reflect the consequences of a history of tree-like divergences from a single common mitochondrial form in the cytoplasm of a woman who lived in Africa at some time in the past: 100 000-200 000 years ago by one interpretation (Cann et al. 1987a); 50 000-500 000 by another (Stoneking & Cann 1989). This interpretation of mitochondrial variation requires that 'Eve' also be the last common nuclear DNA mother – a point often misunderstood by authors who support the model (for instance Gould 1987) but believe that the ancestral mitochondrial variant can have a populational rather than an individual origin. Descent from a single female is not the only interpretation of the mitochondrial data, as detailed above. However, it is the only valid interpretation of mitochondrial data when they are used in support of the 'Garden of Eden' model, given the other constraints that this model brings.

Those other constraints on the 'Garden of Eden' model that are related to the single Eve interpretation derive from the logical requirements following from the association of nuclear and cytoplasmic genes in descent from the single Eve. The reasons that the association of mtDNA ancestry and nuclear DNA ancestry is thought to be required by biochemical geneticists supporting the 'Garden of Eden' model are twofold: (1) only mtDNA forms that can be traced to 'Eve' are now found throughout the world; and, (2) there are no mtDNA variants to be found that could have been passed down from the indigenous natives of other regions of the world who were replaced by 'Eve's' descendants.

> *If there was hybridization between the resident archaic forms in Asia and anatomically modern forms emerging from Africa, we should expect to find extremely divergent types of mtDNA in present-day Asians, more divergent than any mtDNA found in Africa. There is no evidence for these types of mtDNA among the Asians studied. Although such archaic types of mtDNA could have been lost from the hybridizing population, the probability of mtDNA lineages becoming extinct in an expanding population is low.* Cann et al. (1987a:35)

What then could have become of these older lines if there wasn't complete replacement without mixture? One possible answer to this question has been proposed by Cann:

> *perhaps 'African female lineages have a selective advantage in the genetic background of Asians';* (1988:139)
>
> *perhaps 'random extinction of female lineages (was) higher in Asia';* (1988:139)
>
> *perhaps gene flow between the invaders and the natives was only one way, between the males of the indigenous natives and the invading females (consider Cann's explanation of why the higher mitochondrial variability in American Blacks is not due to population mixing – the 'many non-African nuclear genes' in Black Americans were 'probably contributed mainly by Caucasian males'.* Cann et al. (1987a:32)

In my opinion these 'explanations' are improbable, and posit behavior unknown in any mammalian species. Especially unlikely is the idea that there was significant one-way-only hybridization between invading females and indigenous males of the natives who were being replaced. It is not possible to validly combine the suppositions of a mutational (branching) origin for mitochondrial variants in a single source population, the worldwide distribution of these and only these mitochondrial variants, and the outward migrations of populations with only these variants from the source area to all other regions of the world where the migrants mixed and interbred with indigenous natives.

What does link invasion and replacement is invasion without mixture, as Cann and her

Table 2. Specific contrasting predictions of the two models for modern human origins that can be compared using data from the fossil record.

	Garden of Eden	Multiregional evolution
'Anatomically modern' *Homo sapiens*	Necessarily is a new species	Is not a new species, nor are modern populations necessarily temporal subspecies in all geographic regions.
	The species can be clearly defined and unambiguously distinguished from earlier human populations by a single set of criteria that validly apply in all regions of the world	No single morphological definition of 'modern' *Homo sapiens* will validly apply in all regions of the world
Transitional specimens	Are limited to Africa and should not be found elsewhere	Are expected in many if not most regions of the world, not necessarily with the same date at different places
The earliest modern human populations	Must be found in Africa	May be found in any region of the world, and may date earlier or later than the African occurrences
	Will have distinctly African features	When found outside of Africa or the immediately contiguous areas, will lack unique African features, instead sharing some unique regional features with their local antecedents
Regional or racial characteristics – (alternative character states that in combination distinguish geographic variants)	Must appear late in human evolution, after the spread of distinct modern populations out of Africa	Will appear early in human evolution, with at least some features preceding the first appearance of the earliest modern populations
	Modern racial features should appear first in Africa, an appearance linked to the new species ancestral to all living populations (Eve, or her immediate ancestors, according to this theory)	Modern racial features will be first identifiable in the most peripheral regions of the human range, before the appearance of modern humans, and the racial features of modern Africans will be distinguishable latest in the fossil record
	The links between these early appearing uniquely African features with earlier populations should be possible to establish	Links between the earliest groups with uniquely African features and their forebearers may be most difficult to establish
Outside of Africa the evolutionary sequence involving 'anatomically modern' humans	Will appear more African but not necessarily more archaic earlier in time	Will appear more archaic but not more African earlier in time

colleagues claimed in numerous other publications (especially Cann et al. 1987a, Stoneking & Cann 1989). These biochemical geneticists have detailed the reasons why support of the 'Garden of Eden' hypothesis necessarily requires the assumption that the mitochondrial and nuclear DNA's were spread by the same process – the movements of populations.

Therefore, the unavoidable implication of the 'Garden of Eden' explanation for modern human origins that the earliest modern population first appeared in Africa as a new species (Gould 1987, Stringer & Andrews 1988: 1263). The indigenous populations who were replaced at this time left no female descendants and their unique mitochondrial and nuclear DNA variants have been lost to modern humanity.

3.2 *Can 'modern' Homo sapiens be defined?*

If modern *Homo sapiens* is a new species that had a single origin and spread throughout the world by migration, completely replacing indigenous populations without mixing with them, this new species should be unambiguously definable with a set of morphological criteria that apply everywhere. The common recent ancestry for all of the populations that are supposed to fit the definition should not result in marked divergences in the earliest extra-African *Homo sapiens* populations.

Such a definition has been proposed several times, most recently by Stringer & Andrews (1988:1263) but in its most workable form by Day & Stringer (1982) who provide both unambiguous criteria for applying their definition to human skeletal remains and a clear rule for interpreting the results of their application. This provides an opportunity to determine whether a morphological species definition can be applied with consistency and accuracy to the subfossil and recent skeletal remains from this region. Day and Stringer define a taxonomic group 'anatomically modern *Homo sapiens*' that corresponds to a temporal subspecies, and then equate their subspecies definition with the species, *sensu stricto*, stating (1982:832):

... we will equate our definition of Homo sapiens (sensu stricto) with the term a.m. H. sapiens, and leave the question of whether various archaic forms rightly belong within the species H. sapiens to future research.

As is quoted above, subsequent to this, Stringer has taken the position that none of the 'archaic forms' belong in *Homo sapiens*.

'Anatomically modern *Homo sapiens*' crania are defined as short but high, with a high frontal bone that lacks a continuous supraorbital torus, a long and high parietal arch that is inferiorly narrow and superiorly broad, and an occipital bone that is well curved (rather than angulated) in the sagittal plane.

The five metric criteria and a sixth morphological observation developed for this discrimination use measurements and angles defined by Howells (1973). 'Anatomically modern *Homo sapiens*' crania are said to be characterized by:

1. A vault that is high but short, with a ratio of basion-bregma height to the maximum cranial length that exceeds 0.70;

2. A high frontal bone, as indicated by a frontal angle of less than 133°;

3. Supraorbitals either absent or represented by a degenerated torus with a division into lateral and medial (i.e. superciliary arch) sections;

4. A parietal with a long, high arch that has a parietal angle of less than 138°;

5. A parietal shape that is superiorly broad and inferiorly narrow, as reflected by a ratio of the bregma-asterion chord to the biasterionic breadth that exceeds 1.19;

6. A well curved occipital bone, as indicated by an occipital angle of greater than 114°.

The conditions and criteria for including particular specimens in 'anatomically modern *Homo sapiens*' are also explicitly stated (Day & Stringer 1982:832):

(If crania) can be reasonably classified as a.m. (anatomically modern) H. sapiens, we should expect over 50% of the 'modern' characteristics to be present of those which can be tested.

The criterion that over 50% of the preserved cranial characteristics in a specimen must be 'modern' for it to be considered 'anatomically modern *Homo sapiens*', can be restated operationally as follows. If three or more of the six criteria given above are not met by a specimen, it cannot be considered to represent 'anatomically modern *Homo sapiens*' according to this definition.

Using the definitions of characteristics and the criteria for their interpretation described above, samples of recent and living human populations can be examined to see if the definition applies accurately. One might expect that the 'anatomically modern *Homo sapiens*' definition works for European samples since it was developed at the British Museum. The question is whether it also works in other regions of the world, far from Europe, as one would expect if all living populations have a common recent origin as a new species. In fact, when applied to the indigenous populations of Australia, whether recent or living, the definition is problematic.

An immediate problem can be found within Howells' (1973) massive compilation of cranial dimensions and angles (one source of the data the definition was based upon). Howells provides data for the angles of the three cranial bones in two Australian groups: Lake Alexandrina tribes from south Australia and a collection of specimens believed to come from Tasmania. In both samples the means for the three cranial angles were quite close to the limits for 'anatomically modern *Homo sapiens*' set by the Day and Stringer definition.

Under the assumption that the data are distributed normally, the information that Howells provides can be used to estimate how many individuals in each sample probably fall beyond the 'anatomically modern *Homo sapiens*' range for these three angles (Table 3). For the females only a few individuals (7% of the south Australians and 9% of the Tasmanians) have frontal angles out of the modern range. The percentages with parietal angles outside of this range are higher (19% and 13% respectively), while high occipital angles outside the range are even more common (43% and 22%). For the males, the percentages of individuals from south Australia with parietal and occipital angles beyond the limits for 'anatomically modern *Homo sapiens*' given by Day and Stringer are about the same as the female percentages – 21% and 40%. These

Table 3. Percentage of Australian Aborigines determined to have cranial bone angles outside of the range of 'anatomically modern *Homo sapiens*' presented by Day & Stringer (1982). Data are from Howells (1973), based on calculations assuming that the distributions of cranial angles in the two Aboriginal groups are each normal.

Sex	Cranial bone	Lake Alexandrina	Tasmanian
Female	Frontal	7	9
	Parietal	19	13
	Occipital	43	22
Male	Frontal	30	30
	Parietal	21	9
	Occipital	40	34

percentages differ somewhat for the Tasmanian sample (9% and 34%). Moreover, 30% of the frontal bones of both male samples show flattening that is well beyond the limits given for 'anatomically modern *Homo sapiens*'.

In sum, the existence of so many modern individuals with particular features that extend beyond the 'anatomically modern *Homo sapiens*' ranges provided by Day and Stringer makes one wonder whether their definition can be validly applied in this geographic region. Of course, Howells' data are reported for samples by trait, and not for individuals, so they cannot be used to ascertain whether the same individuals are excluded from 'anatomically modern *Homo sapiens*' on the basis of all of the features. In fact, no previous publication provides sufficient data to examine all six of the cranial features detailed in the definition. Thus, it is necessary to examine another population. I collected data for a recent prehistoric sample of Aborigines from the Murray River Valley of southeast Australia.

I examined Australian specimens from the closely related sites of Coobool Crossing and Kow Swamp, a large sample of cranial remains from a limited span of time. Coobool Crossing and Kow Swamp are two fossil hominid sites within the valley of the Murray river. The sites are situated respectively just north and just south of the river and lying some 90 km apart (Thorne 1972, Brown 1981). Both sites have yielded large skeletal samples dated by either radiocarbon (Kow Swamp) or by more indirect methods to the terminal Pleistocene/earliest Holocene. Studies of these collections to date have indicated that they are very similar to each other, the Kow Swamp specimens (published first), falling within the range of variation of the Coobool sample. Fossil Australians not included within this combined sample are in some cases as gracile as any individual within the Murray valley fossil sample (i.e. Mungo 1 and 3). But also not included are crania from other sites with morphologies at the robust end of the Kow Swamp/Coobool Crossing range (i.e. Cossack, Mossigiel, and WLH 50) which like many in the Murray valley sample are distinguishable from 'anatomically modern *Homo sapiens*' by the criteria of the Day and Stringer definition. Thus, the Murray Valley sample can be considered indicative of many if not most Australian fossil hominids, encompassing most of their range, and generally what is true of this sample is likely to also be true of the fossil specimens not included within it.

The combined Kow Swamp/Coobool Crossing fossil sample was examined for the cranial criteria described above. A total of 31 adult crania were complete enough for accurate assessment of character states for all of the criteria to be made; 28 from Coobool and 3 from Kow Swamp (including Cohuna, see Thorne 1976). Of these, 9 (or 29%) were not anatomically modern *Homo sapiens* according to these criteria. Of these 9, one specimen (Coobool 50.37) had 1 of the 6 characteristics of 'anatomically modern *Homo sapiens*', a second specimen had 2, and the remaining 7 specimens had 3. Of these 7 only two had exactly the same 3 characters within the modern range, suggesting that the morphological bases for excluding the various specimens from *Homo sapiens* were not the same in all cases. Since all these specimens are 'anatomically modern *Homo sapiens*' by any conceivable interpretation of what this category might mean, the question is why the morphological criteria fail to apply accurately.

If robustness is the basis for this failure, it might be expected that being excluded from 'anatomically modern *Homo sapiens*' would be a uniquely male feature. However, two of the 9 specimens so excluded are females (22%). Because 9 of the total of 31 crania examined were assessed to be female (29%), the proportion of females in the sample excluded from 'anatomically modern *Homo sapiens*' is not greatly different from the proportion of females in the total sample. Since all the sample sizes are small, a Fisher exact probability test was used to compare the female/male ratios. For this test, comparison of sex ratios is between the subsamples defined as 'anatomically modern' (32%) and not 'anatomically modern' (22%) *Homo sapiens*. The probability calculated for rejecting the null hypothesis is .30 even in this case; therefore, it is reasonable to conclude that there is no significant difference between the sex ratios and that sex is not causal in the misclassifications in the Coobool Crossing/Kow Swamp sample. It seems likely that the failure of the Day and Stringer criteria to validly identify members of *Homo sapiens* in this sample is not a simple consequence of robustness or some other manifestation of 'maleness'.

Because it has been suggested that some of both the Kow Swamp and the Coobool crania show evidence of artificial deformation (Brown 1981), it is also important to determine whether such deformation, or more generally the morphological evidence that supports the interpretation of deformation, could have a significant effect on the results of this analysis. After all, it has been asserted (although never demonstrated) that those aspects of the morphology of the Kow Swamp/Coobool group, which have been interpreted as archaic or Ngandong-like might actually be a consequence of the alleged deformation (Tattersall & Delson 1984, Delson 1988). Brown considers six of the crania in the total combined sample to be 'definitely deformed' according to his criteria. The full data set could be collected for five of these six. An additional six are regarded as showing at least some of the deformation related characteristics. Thus, in the sample of crania allowing the full set of observations to be taken, there are eleven crania that are at the very least, according to Brown, 'possibly deformed'.

The crania that are 'definitely deformed' according to Brown. Of the total sample, 29% of the crania have three or more characteristics that lie outside the 'anatomically modern *Homo sapiens*' range, according to the Day and Stringer criteria. However, of the sample of five 'certainly deformed' crania, only one (20% of the sample) is not characterized as 'anatomically

Table 4. Comparison of the 'definitely deformed' sample with the 'definitely undeformed' sample, with regard to the question of how many of each are not regarded as 'anatomically modern *Homo sapiens*' according to the Day & Stringer criteria. Of the total Coobool Crossing/Kow Swamp cranial sample, some are regarded as definitely being deformed according to Brown, while others are not.

	Total sample	Definitely deformed	Definitely undeformed
Sample size	31	5	26
Number not characterized as *Homo sapiens*	9	1	8
Percent not characterized as *Homo sapiens*	29%	20%	31%

Table 5. Comparison of the 'possibly deformed' Coobool/Kow Swamp sample with the crania that definitely are undeformed, with regard to the question of how many of each are not regarded as 'anatomically modern *Homo sapiens*' according to the Day & Stringer criteria.

	Total sample	Possibly deformed	Definitely undeformed
Sample size	31	11	20
Number not characterized as *Homo sapiens*	9	3	6
Percent not characterized as *Homo sapiens*	29%	27%	30%

modern *Homo sapiens*' (Table 4). Of the remaining sample of 26 probably undeformed specimens, 31% cannot be characterized as 'anatomically modern *Homo sapiens*'. To ascertain the significance of this difference, a chi-square was calculated comparing the number of non-anatomically modern *Homo sapiens* characters for each of the six criteria in the 'definitely deformed' sample and the remaining probably undeformed crania. The difference is significant at the 0.001 level.

The crania that are 'possibly deformed' according to Brown. Of these 11 specimens, 27% are excluded from 'anatomically modern *Homo sapiens*', compared with 30% excluded in the remaining cranial sample of definitely undeformed crania (Table 5). Evidently, the same relation holds and a smaller percentage of 'deformed' crania are excluded from 'anatomically modern *Homo sapiens*'. However, in this case the percentages are much more similar, and a chi-square corresponding that described above is significant only at the 0.06 level.

Therefore, the specimens that Brown considers 'certainly deformed' are much less likely to be excluded from 'anatomically modern *Homo sapiens*' than are the other crania. It seems reasonable to conclude that characteristics that result in the 'deformation' interpretation are, at the very least, independent of those that exclude specimens from the Day and Stringer definition of 'anatomically modern *Homo sapiens*'. In fact, deformed crania are more likely not to be excluded from the '*Homo sapiens*' sample. The features that contribute to the interpretation of 'deformation' in this sample seem to make it more likely that the specimens will be included in the 'anatomically modern *Homo sapiens*' group. This directly contradicts claims made by both Delson (1988:206) and Stringer & Andrews (1988:1267). These authors are demonstrably incorrect in their assertions that the regional morphology of Australasia is a consequence of artificial deformation. To the contrary, whether or not some of the Pleistocene specimens are deformed, deformation does not create or contribute in any way to the problem in the definition.

It is not the Australians who fail to be *Homo sapiens*, but rather it is the definition of *Homo sapiens* that fails. The reasons for the failure of this definition, and others, are discussed in a number of publications (Wolpoff et al. 1984, Wolpoff et al. 1988, Wolpoff 1986). Summarized, they can be derived from what appears to be the confusion of grade and clade characteristics, which according to the Multiregional evolution model is probably inevitable in the Pleistocene hominids. The consequent 'identification' of so-called 'primitive' features (meaning almost invariably features not like modern Europeans) in modern non-European samples is common, even ubiquitous, and perhaps helps underlie the 'scientific racism' of several decades ago, which modern paleoanthropological theory totally rejects. If the implication stands that hominid evolution in each region can only be accurately understood in the context of clade as well as grade, features that could be regarded as archaic in the European sequence, should they be found in recent (or living) people from other areas, cannot simply be regarded out of context as indicative of 'primitive', or a pre modern human status. Yet, the converse to this understanding is that the early appearance of so called 'modern' features outside of Europe is not necessarily a marker of populations 'more advanced' than the Europeans of the time.

4 TRANSITIONAL SPECIMENS

Specimens transitional between modern *Homo sapiens* and populations ancestral to it could reflect the geographic source for this species, especially if they are found to be restricted to a single region. The 'Garden of Eden' model clearly predicts that they will be, and that this region is sub-Saharan Africa. But are there transitional specimens in Africa, linking the earliest 'anatomically modern' populations with their predecessors? Both models discussed here predict transitional specimens in sub-Saharan Africa, and it seems widely agreed that such transitional forms do exist although Rightmire has said (1984:316):

Information presently available does not satisfactorily settle the question of whether Homo sapiens sapiens evolved gradually, in several parts of Africa, or whether anatomically modern people arose from a single source in the sub-Saharan region or outside of Africa entirely.

The issue is whether or not there are transitional forms between 'modern *Homo sapiens*' and their immediate but more archaic predecessors, transitional forms which can be found outside of Africa and therefore could be potential non-African ancestors.

4.1 *Europe*

Beyond Africa there is the question of transitional specimens at the peripheries of the human range. According to the 'Garden of Eden' model, the complete populational replacement required by the mtDNA interpretation unambiguously predicts that transitional specimens must not be found out of Africa, if the model is to retain its validity. This is clearly understood by those who support the model. For instance, Stringer & Andrews assert (1988:239):

Although Europe and southwest Asia have the most complete fossil record for this period, there is an absence of Neanderthal-modern Homo sapiens transitional fossils in either area.

In fact, however, beginning as early as the writings of the Croatian paleoanthropologist Gorjanović-Kramberger (Radovčić 1988), many authors have found evidence that there was such a transition in Europe (Brose & Wolpoff 1971, Brace 1964, Frayer 1986, Hrdlička 1927, 1930, Jelínek 1969, 1983, 1985, Smith 1982, 1984, 1985, Trinkaus & Smith 1985, Weidenreich 1943a, Wolpoff et al. 1981) and have published details supporting the contention that *in situ* evolution, with or without hybridization with new populations, accounts for the morphological intermediates, including Bräuer (1981). Moreover, whether hybridization or another form of transition accounts for the appearance of the post-Neanderthal populations of Europe, there is yet another problem for the 'Garden of Eden' model as an explanation for the transition: the date of modern human origins is inexplicably late in this region, especially since the earliest transitions might be 'right next door' in the Levant (Valladas et al. 1988).

More recently, workers specifically familiar with the European fossil record are preparing or have published detailed evidence of transitional sample characteristics in hominids of the latest Middle Paleolithic, and in hominids of the earliest Upper Paleolithic, in samples regarded as 'late Neanderthals' as well as 'early moderns' (for instance Wolpoff 1980, 1989, Frayer 1986, Frayer et al. 1991, Smith and Frayer papers in this volume). The very complexity of the relation between the Middle Paleolithic and Neanderthals, and the Upper Paleolithic and 'modern *Homo sapiens*' suggests that transition rather than replacement characterized the region since the lack of any direct relation between hominids and industries shows, if nothing else, that there were not new people invading with a behavioral superiority resulting from novel and distinguishable cultural and technological differences.

The breakdown of what was once thought to be a causal relationship between 'new people' and 'new industries' results from the realization that some of the earliest Upper Paleolithic associated specimens are Neanderthals while some of the latest Middle Paleolithic associated specimens are not Neanderthals but rather similar to the post-Neanderthal Europeans. Stringer

(1982) seems to believe that the lack of linkage between hominids and industries argues against a hypothesis of changing selection, and instead supports a replacement model. However, in my view it is the demonstration of synchrony in the archaeological and morphological changes showing that there was a linkage of behavior and morphology that would best argue against local evolution of the post-Neanderthal Europeans, by presenting evidence for invasion by new people with new behaviors as the basis of the association. The 'link', of course, would be the fact that the new people brought the new behaviors and the new behaviors resulted in the successes of the new people, at the expense of the natives – a sort of a 'Prehistoric Imperialism' model. In fact, the hominids and industries are juxtaposed in all possible combinations, and there is no such link.

Leroi-Gourhan (1958) first proposed that there was a retention of archaic morphology in the hominids of the Châtelperronian at Arcy-sur-Cure, describing the teeth as having 'preserved in part paleoanthropine characteristics' (see also Brose & Wolpoff 1971). Subsequent discoveries have shown these comments to have been very insightful, given the characteristics of the human remains associated with the Châtelperronian at Saint Césaire (Vandermeersch 1984, Stringer et al. 1984). The cranium, mandible, and elements of postcranial skeleton indicate that the specimen is a Neanderthal. Nevertheless, several cranial features are transitional. One of these is the flattening of the posterior surface of the mastoid, which reflects an expansion of the nuchal musculature onto this surface that is unknown in other Neanderthals but common throughout the early Aurignacian sample (Frayer et al. 1991) because of the combination of expanded nuchal plane breadth (muscular expansion) and stability of occipital (biasterionic) breadth. Other important transitional characters include the narrow nasal aperture, and the combination of greater medial and lesser midorbit supraorbital torus height, compared with the earlier Neanderthals with more even supraorbital height development.

At Vindija (Wolpoff et al. 1981, Smith et al. 1985, Smith & Ranyard 1980) the latest Neanderthals are characterized by a number of features which in combination are indicative of a more modern European morphology, including relatively vertical mandibular symphyses with incipient to moderately pronounced mental eminence development, reduced facial size and midfacial prognathism possibly coupled with some expression of the canine fossa, somewhat smaller (especially narrower) anterior teeth, narrower nasal apertures, thinner and less projecting supraorbital tori (on the average) with a few specimens showing only very weak torus expression, higher foreheads, vault thinning, and the absence of occipital bunning. The Kůlna specimen (Jelínek 1967) also shows some expression of the canine fossa, while the anterior teeth of the Šipka mandible (Vlček 1969) and all of the Hortus anterior teeth (de Lumley 1973) are markedly reduced, especially in breadth.

Some of the latest Mousterian folk do not appear to be Neanderthals at all. At Krapina, in level 8 which is the uppermost portion of the rockshelter's Mousterian deposits, at the shelter's roof (precluding the possibility of burial from above), cranium A is a juvenile calvaria. Once very fragmentary, continued reconstruction reveals the specimen to be dissimilar to Neanderthals of similar age (see Radovčić et al. 1988, Wolpoff 1980). Also from this level, occipital 11 has an occipital plane that is somewhat more curved in lateral view than the other Krapina occiputs, circumlambdoidal flattening is clearly absent, and although not preserved to the midline the specimen probably lacked a suprainiac fossa.

The post-Neanderthal populations of the region are also characterized by transitional features. For instance, as reviewed by Frayer et al. (1991), the Mladeč 5 and 6 males retain a number of seemingly unique Neanderthal features, from a very Neanderthal-like vault form (dimensionally as well as in the shape of the lateral cranial contour (Frayer 1986), supraorbital projection, lambdoidal flattening, etc.) to features that obviously cannot be related to the greater robustness that is expected in any ancient population such as small mastoid processes and significant paramastoid development. The Mladeč site reveals specimens with Neanderthal-sized canines as well, and 'unique' Neanderthal features on the postcranial remains include a number of specimens with very large relative and absolute joint sizes. A number of non-metric features of the Neanderthals extend into the post-Neanderthal populations, with frequencies decreasing over time, including the H-O mandibular foramen form (Smith 1978), bisulcate axillary borders of the scapulae (Matiegka 1934, Trinkaus 1977, Smith 1985), and certain endocranial characters reviewed by Saban (1980). Frayer (this volume) reviews these and details additional so-called autapomorphic Neanderthal features extending into the post-Neanderthal European populations, decreasing in frequency over time. It is difficult to envision an explanation for this situation that conforms to the 'Garden of Eden' model.

In sum, when taken as a whole the European specimens from at or near the Middle to Upper Paleolithic shift are transitional between the earlier Neanderthal sample and the Aurignacian associated populations that followed them in time. Of course, this does not, and indeed cannot, mean that every specimen was transitional, or that each of the specimens was transitional in the same way. A transition is a concept that can only apply to a population.

4.2 *Australasia*

Transitional samples are not at all limited to Europe. For instance, in Australasia there are firm links between the Ngandong hominids and the fossil and living Australians. Weidenreich (1943b:248-250) was able to relate the Ngandong specimens to the Australian Aborigines of today, with features that included a well-developed supraorbital torus which uniquely combined a discontinuity at glabella with the lack of a supratoral sulcus, and a long, flat, receding forehead. Additional evidence was found in prelambdoidal depressions, sharp angulations between the occipital and nuchal planes, and short or even nonexistent sphenoparietal articulations in the region of pterion. Weidenreich's contention was quantified through the implications of a study conducted by Larnach & Macintosh (1974). These authors compared a number of Australian and New Guinea crania with Europeans and Africans, scoring them for the eighteen characters that Weidenreich (1951) claimed were unique for the Ngandong hominids. Six of these were absent in all modern samples, while nine of the twelve other features were found to attain their highest frequencies in the Australian and New Guinea natives and thereby fit the definition of 'regional' as indicated above. These are as follows: the large rounded zygomatic trigone; absence of a supraorbital sulcus; suprameatal tegmen; transverse squamo-tympanic fissure; angling of the petrous to tympanic in the petro-tympanic axis; lambdoidal protuberance; marked ridge-shaped occipital torus; external occipital crest emerging from the occipital torus; marked supratoral sulcus on the occiput. The authors did not consider frontal flattening. If they had, this would emerge as a tenth Solo character found at its highest frequency in modern Australians.

Therefore, from the Australian perspective, an Indonesian link (in the words of Macintosh, 'the mark of Java') has long been recognized for the living aboriginal populations. The comparisons of the Ngandong fossils with the living Australian Aborigines are further borne out by Australian fossil remains uncovered since Weidenreich and Larnach & Macintosh published their observations. Beginning with the temporally earliest of these, the WLH 50 hominid from the Willandra Lakes of Australia is a very convincing morphological and temporal intermediary between the Ngandong specimens and the recent and modern aboriginals of the continent. Even

Stringer & Andrews (1988) agree to this specimen's importance as a potential regional link, citing its pattern of robust and specifically Australasian features; the dramatic metric similarity to the Ngandong males, the position of the maximum cranial breadth, the morphology of the frontal (its shape and the details just superior to the supraorbitals), the details of the superior nuchal line and inion prominence and the associated nuchal torus morphology, the thickness of the vault (although in WLH 50 the thickening is caused by diploic expansion as in the European Neanderthals [Hublin 1978] and not cortical thickening as in Sangiran *Homo erectus* [Weidenreich 1945]), and a number of additional distinct morphological features (Thorne 1984, Flood 1983). The specimen is not especially like any of the transitional or early 'modern' Africans such as the Omo crania, Jebel Irhoud 2 or Ngaloba (contra Day & Stringer 1982, the discussion in Delson 1985, and see below), and the lack of any important resemblance to African specimens, archaic or modern, is decisive evidence against the 'Garden of Eden' hypothesis. Of all the Australian fossils discussed in the numerous publications on Australasian evolution, Stringer & Andrews find WLH 50 alone to be 'the only credible morphological intermediary between middle Pleistocene Indonesian hominids and late Pleistocene Australians'. Bräuer (1989) also focuses on the importance of this specimen in establishing continuity. He is more willing to conclude that hybridization was the case in this region, although like Stringer & Andrews (1988) he uses hybridization as a veiled referent for replacement; I believe because these authors can envision no other model except replacement that incorporates gene flow, they take evidence of gene flow to mean evidence for replacement. The fundamental role that gene flow plays in the continuity model is yet to be appreciated (see Wolpoff et al. 1988).

However, WLH 50 is ignored in Rightmire's (1988) analysis of evolution in the region, an analysis which concluded that 'solid evidence for evolutionary continuity is in fact not readily compiled', although he does not discuss most of the detailed evidence supporting the continuity interpretation in this region that has already been published for Australasia. In view of the lack of focus on what actually has been written, perhaps it is not surprising that what has come to be the popularly quoted 'rebuttal' of the morphological evidence for regional continuity in Australasia (for instance quoted in both Stringer 1986 and Rightmire 1987) is the paper by Kennedy (1984), which refutes claims of morphological continuity if not identity with *Homo erectus* for the postcranial remains of the subfossil Australians that were never made!

4.3 *Misunderstandings*

Perhaps the most telling failure of the recent critical discussions of the Multiregional Evolution model appears in the characterizations of the model they criticize! For instance, the Stringer and Andrews review is inaccurate and confusing in its interpretations of the Multiregional Evolution hypothesis (Wolpoff et al. 1988), and is anything but the 'undogmatic review' described in the accompanying editorial by Roger Lewin (1988), who invited the authors to write it. The number of misunderstandings about the Multiregional Evolution hypothesis in this paper and elsewhere is disturbing.

Habgood (1989) supports the Multiregional model in a paper in which he first denies that there are regional features linking that earlier and later Australasian remains, and then proposes that there are such features if they are considered in combination. He defines regional features as 'those which in combination appear more often in a specific region'. In fact Habgood and several others have pointed to weaknesses in our original formulation of how we identified features reflecting regionalism, and these will be addressed below. Clearly, independent characteristics that appear individually in high frequencies are very likely to appear together more commonly as a consequence of the laws of chance (for instance see Wolpoff [1980:300] for a discussion of how this model could apply to the rapid decrease in unique Neanderthal features in the post-Neanderthal Europeans).

Rightmire (1988) criticizes the description of Australasian regional features as including large faces and teeth, although in discussing this matter Thorne and I were using these observations to describe an evolutionary trend and not for the comparison of individuals. We were interested in the facts that (1) smaller and larger postcanine dentitions characterize respectively north and south Asia throughout the middle and late Pleistocene of these regions even while postcanine tooth size decreases in both areas, and (2) the patterns of dental reduction throughout the east Asian periphery are not the same – in north Asia reduction is continuous through this time span while to the south the bulk of the reduction before the Holocene is with the first appearance of *Homo erectus* and there is stasis in this feature thereafter. One can disregard all of this with the comment that large teeth don't show anything about regionalism since all fossil humans have them, but it seems to me that comments such as these (see Rightmire 1988:255) do little for advancing the understanding of either the details or the causes of human evolution.

In the Stringer and Andrews 'review' (1988) the Multiregional hypothesis is incorrectly characterized a number of times, and often implications are attributed to it which in many cases do not follow from it, and in some instances are directly contradicted by publications on Multiregional evolution. These invariably provide the 'predictions' then said to be disproved. Stringer & Andrews assert (1988) 'under the Multiregional model, the marked morphological and metrical similarities of ... geographically distinct crania are explained as the result of parallel evolution'. This may be the most misleading of the misunderstandings in their review. Coon's (1962) discredited 'parallel evolution' interpretation is rejected in every publication on Multiregional Evolution and even in Stringer and Andrews' own summary of the hypothesis. What these authors miss, or misunderstand, is the role attributed to gene flow in the Multiregional model. In fact, Stringer and Andrews do not view gene flow as an important evolutionary mechanism. For instance, they assert (1988:1265):

it seems improbable that (gene flow) could prevent increasing differentiation in a very widely distributed species over long periods of time.

But such a role for gene flow is far from improbable (Slatkin 1987). In fact, if Stringer and Andrews' assessment were correct, polytypic species such as *Canis familiaris* could not persist, but in fact they are quite common and have been common in the past (for instance, see Mayr [1963, Chapter 12]). The common presence of widespread gene flow is recognized and its consequences are rigorously analyzed in the 50 years of clinal theory following from Huxley's (1939) classic work. Slatkin's (1987) recent book reviews the current understanding of the roles played by gene flow, in his emphasis on the complex interrelationships of all the evolutionary forces. It would be useful if there were a greater appreciation of the fact that commonly there are opposing evolutionary forces of differing magnitudes. These produce morphological gradients and it is unreasonable to expect that when there are opposing forces, selection or drift can 'overwhelm' opposing effects of gene flow, any more than gene flow can 'swamp out' the opposing effects of selection or drift (Barton & Charlesworth 1984, Wolpoff et al. 1984). The published misunderstandings of these critical aspects of gene flow and its role in the evolutionary process leads to many of the incorrect 'predictions' made for the Multiregional model (see Wolpoff et al. 1988). Some of the miss-statements about what this model is (incorrectly) said to predict includes the following: (1) the claim that living peripheral populations are supposed to be more divergent than central ones; (2) the assertion that the Multiregional model predicts 'no universal patterns of *Homo sapiens* origins'; (3) the contention that the model predicts Neanderthal features are expected in the far east because morphological clines supported by gene flow stretch between there and Europe; (4) the assertion that gene flow is supposed to lead to an 'earlier establishment of combinations of *Homo sapiens* characters' at the center of the human range; and (5) the claim that 'transitional (mosaic) fossils should be common' (why more common than other fossils?). None of these so-called 'predictions' were predicted in the development of the Multiregional model, nor in my opinion do they logically follow from it. Their incorporation in a paper claiming to objectively test competing models is inexplicable.

Admission that there are transitional fossils, and their interpretation, is a major source of the problems which have arisen in critical discussions of the 'Garden of Eden' model. Stringer & Andrews (1988) claim, without providing argument of evidence, that Chinese fossils such as Jinniushan (Yinkou) and Dali are not regionally transitional because they exhibit a great resemblance to their European contemporaries while contrasting with their 'supposed local ancestors'. Contradictory evidence is reviewed here, and can be found in every Chinese publication on the specimens (including those quoted by Stringer and Andrews). It is difficult to understand why authors unfamiliar with the materials continue to make assertions such as these in the face of easily available reviews (Pope 1988) and indeed whole volumes to the contrary (for instance the review of paleoanthropology in China edited by Wu & Olsen [1985] with paper after paper by workers having direct experience with the specimens). From Weidenreich on, east Asian workers have drawn the same inferences from these data. Most recently Pope (1988:70) concludes:

Transitional forms between Homo erectus and Homo sapiens seem to be present in both Java and China... In China the evidence for 'regional continuity' is more convincing. The Hexian, Dali, and Jinniushan crania all evince craniofacial characters that occur relatively frequently in Mongoloid populations. The presence of shovel-shaped incisors and agenesis of M3 in the Chenjiawou mandible also support this interpretation.

While many turn to China as the region with the best evidence for transitional specimens, some of the transitional Australasian specimens seem to be more acceptable to Stringer and Andrews themselves, as discussed above, and the question of which part of east Asia has the better case for regional continuity and transitional forms linking 'modern *Homo sapiens*' populations and their predecessors is taking on the aspects of a Miller Lite commercial: 'great taste', 'no, less filling', 'no, great taste', 'no, less filling'. But the real issue is not which area of east Asia has better evidence for a transition. Authors who accept the 'Garden of Eden' model need to resolve the acceptance of any transitional specimen out of Africa with the predictions that follow from the assumptions of their model.

5 THE EARLIEST MODERN POPULATIONS

Another point at which predictions of the models contradict each other is over the question of where the earliest modern populations will be found. According to the 'Garden of Eden' explanation for modern populational origins, and the 'African Origins' model on which it is based, this place must be sub-Saharan Africa. The bases for this supposition are mainly in the early dates said to be associated with the first modern humans from these regions (Bräuer 1984b). However, problems abound with every single one of these African dates. There are some very good reasons to question the South African dates, in particular, but the dates of the so-called early modern humans from the Kibish Formation, Omo, are actually no better established. Neither Border Cave nor Klasies River Mouth Cave, the two sites thought to be especially early, has an absolute radiometric date, and in the case of Border Cave it is not even clear that most of the adult specimens have a provenience (Rightmire 1979:25)!

As far as the Klasies River Mouth Cave specimens are concerned, the most serious question is the validity of their perceived 'modernity' – a perception that in my opinion is a consequence of their fragmentation, the lack of either publications about or research access to the recently discovered very archaic appearing specimens in the Archaeology Department of Stellenbosch University, and the prior lack of systematic comparisons between these materials and recent or modern individuals from the Southern Cape. In fact, such comparisons reveal the fragmentary collection to be anything but modern (Klein 1989, Caspari & Wolpoff 1990, Wolpoff & Caspari 1990). Of course, this assessment depends on what one means by 'modern'. Delson (1988) regards the Djebel Irhoud crania as 'early anatomically modern *Homo sapiens*' and Rightmire (1981, 1984) has come to regard Ngaloba as 'modern *Homo sapiens*'. With definitions as broad as these the Klasies remains could, for the most part, be considered 'modern' as well, but it is unclear what this means. If the assessment is meant to reflect the morphological ties between archaic specimens and their living or recent counterparts, I would concur with this interpretation. However, if it means that the specimens are thought to be similar in morphological details to recent or living modern populations (for instance with a morphological complex that falls within the modern range, as Rightmire has said of Border Cave), I cannot agree with these characterizations. In the case of Ngaloba, the combination of facial gracilization, the lack of occipital tori, and dental reduction, makes the specimen an ideal intermediary between recent and living east Africans, and earlier individuals such as Broken Hill (male) and Ndutu (female) who show the beginnings of trends in these directions compared with their contemporaries from other regions of the world. Many elements of the Ngaloba morphology make the specimen an ideal precursor for the morphology of the latest Pleistocene east Africans, sharing many detailed features with them (Bräuer 1984a, Rightmire 1984). However, this is not the same as contending that Ngaloba is the same as the recent and living individuals. Similarly, the Klasies cranial and gnathic fragments systematically differ from the recent and living inhabitants of the Southern Cape, in ways that in my opinion make them decidedly more archaic (for instance very large facial remains, marked supracilliary development and broad innerorbital distance, the common absence of a mental eminence).

The earliest populations widely regarded as being 'modern *Homo sapiens*' may not be African at all. At the moment, of all the sites that have been claimed to preserve the earliest 'modern *Homo sapiens*' remains over the years, in my opinion Qafzeh has the best chance of being validly dated to earlier than the widely accepted 35 000-40 000 age for the first appearance of 'modern' humans (Valladas et al. 1988). Many features found in the variable sample from this site are not particularly African at all, although a few characteristics such as the marked alveolar prognathism of the Qafzeh 9 juvenile have been claimed to link it with more southern populations. The early Qafzeh dates were initially heralded as an 'unexpected victory' for the 'Garden of Eden' hypothesis (Stringer & Andrews 1988, Delson 1988). But on reflection is there such a 'victory'? These early Qafzeh dates certainly raise some interesting problems. Not the least of these is the designation of these people as 'Proto Cro-Magnons', which is unfortunate not only because the human remains from Cro-Magnon are much younger than the earliest of the post-Neanderthal Europeans, but also from the point of view that especially the male specimens from Cro-Magnon are very much unlike the earliest post-Neanderthal Europeans. Moreover, even if a firm link could be made between the Levantine Mousterians from

Qafzeh and the Aurignacian populations of central Europe, there still would remain the question of what they were doing between their early appearance in western Asia and their late entrance onto the European stage, 'just next door'. Perhaps they were off somewhere else developing what Delson (1988:206) describes as 'Late Paleolithic technology', which he believes then 'spread throughout modern *H. sapiens* populations' (evidently these populations do not include the people of the earliest European Upper Paleolithic, since a Neanderthal is associated with the industry based on this technology). On the other hand, Stringer (1988:566) has come to deny the possibility that there was a special Qafzeh-Cro-Magnon link and states 'it now seems less likely that the Qafzeh sample bears the special relationship to the much younger Cro-Magnons of Europe implied by the name 'Proto-Cro-Magnons'.

Whatever the case, while as a sample Qafzeh may or may not be validly described as 'early modern', the sample as a whole cannot obviously be described as African. In fact, only the so-called 'early modern' specimens from sub-Saharan Africa that clearly have African features are those from both Border Cave, a mostly unprovienced sample. Thus, in spite of the unambiguous predictions of Stringer & Andrews (1988:1263), who assert that the regional African features appear in modern humans before their dispersal (as quoted above), and the Protsch/Bräuer 'African Origins' hypothesis that Africans are to be found at the origin of modern humans, the question remains where are these early modern Africans to be found outside the continent? For all the disparate origins theories for east Asians (for instance comparing Hooton 1947, Thoma 1964, Turner 1987, Wu 1961), the one theory that never seems to appear is 'African origins'. The early 'modern *Homo sapiens*' specimens from the region are not specifically or uniquely similar to Africans (Wolpoff et al. 1984). The early 'moderns' of Australasia have been described as similar to the Ngandong Indonesians, or like specimens from the Asian mainland, depending on which specimens and which authors, but the one thing they have not been described as is African! The 'early moderns' from Europe are inexplicably much later in time (if they are thought to be part of an invasion process some 200 000 years ago that was so successful that there was complete replacement). No presentation of the 'Afro-European *sapiens* hypothesis' (Bräuer 1984a, b, 1989) has yet established that there was a single African, or even a uniquely African feature, to be found in the earliest of the post-Neanderthal Europeans.

In sum, it is far from clear just exactly where the earliest modern populations appear. Almost certainly this is at least in part a consequence of the fact that a definition distinguishing these populations has not been (and may never be) determined. Before embracing the 'Garden of Eden' model, Stringer better appreciated the implications of this situation, concluding at one time (Stringer et al. 1984:121):

> *... it is still possible that the modern characters of the late Upper Pleistocene hominids of this area (the far east and Australia) were ultimately derived from an exotic source such as Africa. Such a simplistic model, however, does not take account of possible modern characters in even earlier fossils such as the Dali cranium from China (unless these are plesiomorphies or parallelisms), does not explain the extremely robust nature of some of the late Pleistocene Australian material (unless these are local 'reversions'), nor does it accord with a record of local cultural continuity ...*

6 REGIONAL OR RACIAL CHARACTERISTICS: THE CENTER AND THE EDGE

The origin, antiquity, and initial geographic distribution of certain racial features that confuse the terminal *Homo erectus* boundary and makes modern humans difficult to define are found much earlier, in the first *Homo erectus* migrants from Africa according to the Multiregional Evolution model. However, in the 'Garden of Eden' model, race origins are very much more recent, linked to the time and place of origin of the species *Homo sapiens* through the mechanism of dispersion – migration with complete replacement, and this hypothesis provides the prediction that the terminal boundary will be easy to delineate since it was marked by a punctuational event. Multiregional Evolution theory clearly predicts a virtually opposite pattern, the late appearance of regional features at the center of the range and their early appearance in *Homo erectus* populations at the geographic or ecological peripheries.

These regional features are to some extent those which serve to identify skeletal variants in the living races. However, there are two differences between skeletal features that might be used to characterize living races and the features that can validly show, or refute, the regional continuity observation. First, for characteristics to validly be used as indications of continuity it is important that they reveal a specific adaptation for functions that can be accomplished in alternative ways. For instance, it is possible that a long distance walking adaptation can only be accomplished by limb lengthening, and that there is only one morphology that provides this adaptation. Long term continuity in limb length is not necessarily supportive of the Multiregional Evolution hypothesis because it could reflect no more than similar adaptations to the same function in genetically unrelated populations. In contrast, consider a changing function such as increased anterior tooth loading. Its functional requirements of the anterior teeth could be met with several alternative adaptations such as differential expansion of the lateral incisors, more pronounced shoveling, hypsodonty, etc. In this case, the persistence of a single morphological alternative in a region would support the explanation of a historical basis for the specific form of the functional morphology expressed, and thereby the Multiregional Evolution hypothesis. Second, to validly reflect regional continuity it is reasonable to expect features to appear in high frequencies in combination (Habgood 1989). There are no unique features for any regions, geographically varying humans after all are not different species, and they are not dramatically different from place to place. This should not be surprising. The small amount of genetic variation underlying human morphological variation is reflected in the fact that only a 1% genetic difference underlies the differences between human and chimpanzee morphology. Consequently, it is the combinations and not the specific characteristics that can be expected to mark different populations over long time spans, and it is in these combinations that distinguishable regional variants can validly be sought.

An early appearance of modern regional features at the center is prevented by a number of factors, many of which are resultant of the mechanisms that promote marked variability there. There is the likelihood that the center is the region of optimal adaptation, perhaps the area where the species evolved. It is likely that a greater range of morphological variation will be tolerated within groups from that region. Moreover, central populations tend to be contiguous and to have relatively higher population densities than peripheral populations. Potential competition between the central populations could increase the range of interpopulational differences, and thus the heterogeneity of the region as a whole. Gene flow through populations will tend to be greatest at a species' geographic center, because gene flow through central populations is multidirectional, and this too promotes heterogeneity. Finally, selection is likely to be relaxed for populations undergoing rapid expansion, which would act to increase the range of phenotypes present, especially at the center of the range of a colonizing species.

In the peripheral (and/or ecologically marginal) populations, events associated with both the initial colonization and the subsequent period of local adaptation help to develop the combination of pronounced difference between allopatric populations and relative homogeneity within them. This is the primary basis for the expectation that in peripheral regions some of the features that mark modern geographic variation can be expected to appear first. The dramatic morphological homogeneity that has been observed to characterize human fossil crania from peripheral sites with large sample sizes such as Zhoukoudian, Sangiran, and Ngandong reflects a reduced number of genetic polymorphisms in the nuclear DNA (and one would expect a marked reduction in the number of mtDNA variants) because the characteristics of the gene pools were initially established by the partial isolation of numerous small populations with histories of founder effect and continuous drift for long time spans. Given the likely pattern of earlier Pleistocene human population movements throughout Eurasia, these are the circumstances under which reduced genetic variation can be expected in the peripheral or marginal regions.

The population structure that probably characterized humans during the process of initial habitation may be similar to that reported for both the Yanomamo Amerinds and also a number of non-human primates. This lies between the panmictic and the isolated island population models, and may be a reasonable prediction for the population structure of Pleistocene hominids for most of their evolution. In the Yanomamo population structure, villages split along lines of kinship so that they are distinguished by marked genetic divergence in spite of the (often considerable) gene flow between them. What is genetically important about this structure is the fact that the variance effective population size will be smaller than the inbreeding effective size. According to Neel (1978), the genetic consequences of this structure are:

1. A considerable portion of the genetic variability is distributed between villages.
2. The loci that are polymorphic within a village have a high level of heterozygosity.
3. However, the overall levels of homozygosity are high.

Templeton (1982) argues that selection will continue to promote heterozygosity in small subdivided populations. A single village drawn out of this net to found a new population would carry with it a reduced number of polymorphisms. Templeton further argues that such a founder population would accumulate inbreeding rapidly because so many of its members would be closely related.

When part of such a village forms a founder population, the founding individuals are almost always more closely related to each other than would be expected in a random sample from the village. Thus, there would also be a reduction in the levels of individual heterozygosities, and one would expect the overall level of genetic variability to be reduced. In his discussion of Neel's data, Templeton concludes that such populations would be less likely to speciate at the periphery than small populations drawn at random from a panmictic species. These are precisely the conditions that seem to have applied to peripheral hominid populations through the middle and late Pleistocene (Wolpoff et al. 1984). Mainly, they vary interregionally, are each monomorphic, and in no recognizable case did they speciate.

As local adaptive responses develop, these initial genetic differences, diverse patterns of selection, and the continued action of drift in the regions with lower population densities would help to retain if not expand the genetic differences initially established (Lande 1976). Moreover, drift and reduced gene flow to the peripheral or ecologically marginal areas might help maintain homogeneity within these populations. This is because at the species' boundaries 'the total amount of gene flow is reduced in peripheral populations and near the periphery gene flow becomes increasingly one-way outward' (Mayr 1970:232). Therefore, it is not surprising to find Mayr's (p. 230) observation that

in polytypic species ... almost invariably ... the degree of polymorphism decreases towards the border of the species and ... many of the peripheral populations are monomorphic.

One factor that might help maintain some of the regional differences at the peripheries of the human range over long time spans at the peripheries would certainly be environmental selection (van Valen 1986). Another less obvious source of selection was suggested by Mark Flinn (Flinn & Wolpoff, in prep.). Criteria for the visual recognition of individuals are largely found in the facial triangle (from the brows to the chin) in primates as well as many other vertebrates. Many, perhaps most, of the regional features distinguishing peripheral hominid popultions are facial, and Flinn suggests that their long term maintenance may be associated with the advantages of recognizing kin in populations undergoing drift because of their small size and immigration of unrelated individuals from other areas. This is an especially seductive idea since it would help account for why facial features initially established through the arbitrary process of drift may have long lasting persistence in regional populations (Flinn & Wolpoff, in prep.).

In east Asia there were two distinct although not geographically discontinuous morphological clades at the north and south ends of the eastern periphery. These morphological clades have been connected by clines of intergradation for as long as there is evidence of geographically intermediate specimens, and while commonly described as 'Mongoloid' and 'Australoid', they could easily and accurately be called northern and southern.

6.1 *Australasia: The Southern End*

From the Australasian perspective, links between the Sangiran *Homo erectus* specimens, the later hominids from Ngandong, and finally the recent inhabitants of Kow Swamp and Coobool Crossing in Australia have been recognized. The case for specific ancestral-descendent relations between hominid samples more than a half million years apart has improved dramatically since Weidenreich (1943b) first proposed it for Sundaland (Thorne & Wolpoff 1981, Pope 1988). Sambungmachan (Jacob 1976, Wolpoff 1980) is an excellent intermediary between the Ngandong hominds and the Sangiran *Homo erectus* remains. While the Sambungmachan male is somewhat smaller than the Ngandong males, it resembles them in the development of an angular trigone at the lateral corner of the supraorbital torus, the reduction in basal pneumatization and the consequently higher position of the maximum cranial breadth, the flattening of the occipital plane of the occiput (and its vertical orientation), the doubled digastric sulci, and the tall, vertical posterior border of the temporal squama. Compared with the male *Homo erectus* fossils from the Kabuh, the Sambungmachan cranial capacity is larger, and the specimen is morphologically more similar to these remains than it is to the Ngandong hominids. Even without Sambungmachan, Weidenreich (1951) was able to link the Kabuh and Ngandong remains. With this specimen, the origin of the Ngandong hominds is firmly established.

The link between the Kabuh hominids and the living Australians has long been evident even without consideration of the intermediate Ngandong sample. Recent discoveries further support this interpretation of evolutionary relationships in south Asia. A new reconstruction of Sangiran 17 (Thorne & Wolpoff 1981), the most complete adult *Homo erectus* cranium and the only complete cranium of an adult male allows comparisons of the face to be made in this region for the first time. This proved to be the anatomical area with the strongest evidence for regional continuity when the Sundaland and the latest Pleistocene and Holocene samples from Australia were compared.

From these comparisons, a number of regional (i.e. intraspecies clade) features were identified. While they cannot be autapomorphic in any geographic area, in combination they characterize Australasia for much (or in some cases all) of human prehistory there, as compared with other regions of the world during equivalent time spans.

Features of the Sangiran 17 face, and of the other more fragmentary Sangiran faces support the contention of a special relation between these and the living samples. Regional facial features include the marked ridge paralleling the zygomaxillary suture, the eversion of the lower border of the zygomatic, the rounding of the inferolateral orbital border, the lack of a distinct line dividing the nasal floor from the subnasal face of the maxilla, the curvature of the posterior alveolar plane of the maxilla that corresponds to the mandibular 'curve of Spee', the posterior position of the minimum frontal breadth, the relatively horizontal orientation of the inferior border of the supraorbital torus, and the marked and dramatic nature of the supraorbital or superciliary expression. Moreover, in my experience Australasian faces fundamentally differ from the faces of other regions in their combination of dramatic size (holding sex constant), lateral orientation of the maxilla, significant development of the superciliary arches (often actually a supraorbital torus), and marked subnasal prognathism. It is clear that the inhabitants of this southern region have had their own distinct morphology for the entire time east Asia has been inhabited (Thorne & Wolpoff 1981, Jacob 1981, Jelínek 1982, Wolpoff et al. 1984). In an independent assessment, Habgood (1989) asserts:

... a number of features (a long flat frontal, posteriorly placed minimum frontal breadth, facial prognathism, eversion of the lower border of the malar, and presence of a zygomaxillary tuberosity) which, when found in combination, appear to document a 'morphological clade' in Australasia.

Later in time, WLH 50 is the most recent addition to the substantial body of evidence linking hominid populations across the Pleistocene of the Sunda subcontinent. The Indonesian *Homo erectus* connection is further strengthened when the earlier discoveries of subfossil remains from Kow Swamp and Coobool Crossing are taken into account, because so many of the evolutionary changes in Australia accelerated during the Holocene, after these populations lived. Because they were earlier these late upper Pleistocene/Holocene specimens retain a much higher frequency of features that are archaic in Sundaland than do the living populations of the continent, and thus reinforce the notion of morphological continuity.

6.2 *North China: The Opposite End*

Since liberation there has been a continuous increase in the fossil record, with new discoveries of *Homo erectus* from Zhoukoudian, Gongwangling, Chenjiawo, Yuanmou, Longgudon, and Hexian. New early (or archaic) *Homo sapiens* specimens include major remains from Dali, Maba, Jinniushan, Changyang, and Xujiayao, and a number of more fragmentary specimens. New individuals that are terminal Pleistocene or perhaps even Holocene in age include Chilinshan, Huanglong, Liujiang, Laibin, Muchienchiao, Tzeyang, and a palate from south China found in a Hong Kong drugstore. These specimens confirm Weidenreich's interpretations, showing both morphological continuity within China and regional distinctions of the Chinese fossils from other areas with reasonably complete fossil records (Aigner 1976, Pope 1988, Wolpoff et al. 1984, Wu & Wu 1982, Wu & Zhang 1978).

Working as he was in north China, Weidenreich (1939, 1943b) was much more detailed in his discussions of morphological continuity for that region, in comparison with his rendering of the Australasian sequence. He described 12 features he felt had particular importance in showing continuity between *Homo erectus* and the modern populations of north China. These were: mid-sagittal torus and parasagittal depression; metopic suture; Inca bones; 'Mongoloid' features of the cheek region; maxillary, ear and mandibular exostoses (the mandibular exostoses forming a mandibular torus); a high degree of platymerism in the femur; a strong deltoid tuberosity in the humerus; shovel-shaped upper lateral incisors; and the horizontal course of the nasofrontal and frontomaxillary sutures. While some of these, especially the postcranial features, are known to be generally characteristic of archaic populations everywhere, most of the others appear to be most frequent in ancient and modern north Asians. Especially in combination they characterize the region for a long period of time.

Over the decades since Weidenreich's death a number of additional observations have been added. Aigner (1976) pointed to: the profile contour of the nasal saddle and of the nasal roof, pronounced frontal orientation of the malar facies and the frontosphenoidal processes of the maxilla, and a rounded infraorbital margin, even with the floor of the orbit. Wu (1981), in his description of the Dali male cranium, was able to add a number of facial features and characteristics of the anterior frontal. These include the frontal boss, large profile angle to the nasal bone, flat face and nasal saddle, angled inferior zygomaxillary junction (145°, approximating the Eskimo mean), and an anterior facing fronto-sphenoidal process of the zygomatic bone. It is the features of the face that have particular importance because the face seems to show more regional distinctions than does the cranial vault in north China, just as is the case in the Sundaland hominids (Thorne & Wolpoff 1981). Indeed, the Dali male (Wu & Wu 1985) and the morphologically similar, newly discovered Jinniushan female (Pope 1988), have added much to an already very convincing case established by Weidenreich and expanded in scope and detail by the discoveries of more fragmentary remains recovered through the end of the last decade (Wolpoff et al. 1984). Besides facial flatness and the specific traits mentioned by Weidenreich and Aigner, we established additional regionally distinct features of these early *Homo sapiens* vaults which include the forehead profile (in particular the development of the frontal boss), the associated distinct angulation in the zygomatic process of the maxilla and anterior orientation of the frontal process of the zygomatic, lack of anterior facial projection and low degree of prognathism, rounded shape of the orbits, minimal nasal projection (the frontonasal and frontomaxillary sutures are virtually level) and a very low nasal profile. Other more fragmentary remains of early *Homo sapiens*, especially those from Maba, Xujiayao, and Dingcun, also show these details when the appropriate parts are preserved. However, Dali and Jinniushan are clearly the centerpieces of the continuity argument in north Asia. The combination of regional features in these specimens so clearly link them specifically with the more recent and modern populations of north China that it is unlikely future discoveries will refute this interpretation. Wu (1961) disapproved of Weidenreich's 'Eskimo, Chinese, Melanesoid' trichotomy for the Upper Cave crania, showing that the three crania fall within the expected range of variation for the area. The early *Homo sapiens* remains have been particularly important in providing firm evidence for a morphological link between the late Pleistocene/early Holocene specimens and the Zhoukoudian *Homo erectus* sample – a link that was lacking in Weidenreich's time, hampering his interpretation (1939) of the Zhoukoudian Upper Cave specimens.

In sum, the basis for claiming regional continuity in China, especially northern China, is at least as good as in Sundaland for three stages of human evolution: *Homo erectus*; archaic *Homo sapiens*; and, the earliest modern *Homo sapiens*. Continuity, however, is not necessarily found in alternative character states of the same traits that are regional in Sundaland. Instead, it is a whole suite of different, unrelated, features that support the continuity argument there through their high frequencies in combination. Regional continuity at both northern and southern ends of the eastern edge of the human range (as it was for most of the Pleistocene) was recognized as long ago as in Weidenreich's time and continually reinforced by both subsequent discoveries and the research of scholars familiar with the fossil remains from both areas (Jacob 1981, Wu & Zhang 1978, Wu & Wu 1982, Wolpoff et al. 1984). It was this evidence that led to Coon's (1962) chapter headings '*Pithecanthropus* and the Australoids' and '*Sinanthropus* and the Mongoloids'.

There is some positive evidence that the transformation of Homo erectus into Homo sapiens was both more or less gradual and occured in different parts of the world... This evidence cannot simply be ignored with impunity. Van Valen (1986:412)

6.3 *Evolutionary sequence within 'anatomically modern' humans outside of Africa*

With regard to the 'Garden of Eden' hypothesis and its 'African Origins' corollary, it is my experience that none of the earlier Australasian, Chinese, or European 'modern *Homo sapiens*' specimens possess features which uniquely resemble archaic or modern Africans. Nor are the earliest 'modern *Homo sapiens*' samples in aggregate more like Africans than their successors.

In Australasia the 'early modern' specimens have been described as similar to virtually anything but Africans in their morphology (Jacob 1967, Thorne 1980, 1984, Wu & Wu 1985, Wu 1987). This contention directly contradicts the assertion that a special relationship exists between WLH 50 and the Ngaloba specimen (see discussion in Delson, 1985), although unique similarities between these particular specimens is clearly predicted by the 'Garden of Eden' model given the near modern (Bräuer 1984a) or modern (Rightmire 1984) status accorded to the Ngaloba cranium. However, this is an assertion of relationship that is demonstrably incorrect. Considering the simplest trichotomy that encompasses the conflicting predictions, involving WLH 50, the earlier sample from the region (Ngandong), and the slightly older potential ancestor from Africa (Ngaloba), WLH 50 and the Ngandong sample were found to share 12 unique features which are not expressed in the Ngaloba specimen. On the other hand, WLH 50 and Ngaloba share virtually nothing uniquely – not also shared with at least some of the Ngandong crania. For instance, while the supraorbitals of both WLH 50 and Ngaloba are divided into medial (superciliary) and very thick lateral toral elements by a weak groove, this is also the case for some of the Ngandong specimens such as cranium 11 although in other crania from the site there is no division. Similarly, a thicker superciliary and thinner lateral toral dimension in these two later specimens is shared with some of the Ngandong remains, but not others. Thus, there are no unique features shared by Ngaloba cranium and WLH 50 that are not also shared with the Ngandong sample. The trichotomy of possible relationships between the three sites is easily resolved.

Among the remains representing the earliest modern populations of China, Liujiang from south China and the Upper Cave specimens from Zhoukoudian have a number of features that are very definitely not characteristically African (Woo 1959, Wu 1961, Thoma 1964, Wolpoff et al. 1984, Wu 1987, Wu & Wu 1985), and therefore show an absence of any tendency for the earlier 'moderns' of the region to appear more African-like than the later populations (as would be expected according to the 'Garden of Eden' hypothesis).

The European evidence fully parallels that of eastern Asia. The earliest post-Neandertal populations of the region do not even seem to be closely related to the Levantine populations said to be more than twice their age (Wolpoff 1989, Stringer 1986), let alone show evidence of unique African features. If anything, there are some limited African characteristics in the much later populations of the Mediterranean (Ferembach 1986), but this is irrelevant to the validity of the 'Garden of Eden' hypothesis.

Table 6. Features linking the Ngandong crania with Willandra Lakes (Australia) Hominid 50 that are not found in the African Ngaloba (Laetoli 18) specimen.

	Shared condition in the Australasian specimens: Ngandong and WLH 50	Condition in the Ngaloba African LH 18
Greatest cranial breadth	At the cranial base, across the supra-mastoid crests	Medium to high across the parietals
Parietal boss	Absent	Distinct
Angular torus	Pronounced	Absent
External mastoid process tri-angular form	Height much greater than basal length	Approximately equilateral
Sagittal keel	Marked on the anterior half of the pa-rietals	Absent
Sagittal region at bregma	Smooth	Slight bulge at bregma with a post-bregmatic depression
Sagittal length of the frontal	Relatively long compared with vault length	Relatively short compared with vault length
Sagittal curvature of the frontal, above supraorbital region	Flatter, less curved	More curved
Crest for temporal line	Extends across the entire frontal	Only encompasses temporal notch
Nuchal torus and supratoral sulcus	Both torus and sulcus present, torus projects markedly	Neither torus nor sulcus present
Inion	Strong inferiorly pointing triangle ex-tending onto the nuchal plane but dis-tinctly above it	No special inion prominence
Occiput along the bottom of the lamboidal suture, at and near the midline	Distinctly bulging parallel to the sutu-re	Flat along the suture

Finally, even in the part of the world closest to Africa, the Levant of western Asia, an 'African link' is difficult to establish for the first 'modern' populations. The frontofacial fragment from Zuttiyeh, dated earlier to 110 000-150 000 years and said to be the earliest of the 'modern *Homo sapiens*' group, has absolutely no resemblance to any Middle or earlier Late Pleistocene African, and shows only specific relations to earlier hominids from eastern Asia (Hublin 1976, 1983, Vandermeersch 1985).

7 CONCLUSIONS

The information presented here and elsewhere unequivocably shows the 'Garden of Eden' hypothesis to be wrong. No data can be found to uniquely support it. Considerable data invalidate it by contradicting its predictions. Many of these data support the alternative, Multiregional

Evolution. In fact, rarely have two conflicting paleoanthropological hypotheses been so clearly contradictory, which probably contributes to how one could be so completely invalidated. Delson's (1988) 'obituary' for Multiregional Evolution, praising the 'valuable legacy' it left behind, may have been premature.

Specific details of regional continuity have been reported both here and elsewhere, describing the particular features that in high frequencies and unique combinations support the interpretation of Multiregional Evolution. Yet, it is unfortunate that the only responses by those who reject this model are to deny that these features actually exist, to confuse regional (clade) features with grade (thereby creating the unwarranted specter of modern 'primitives'), or to assert that the problematic features are due to cultural practices or pathologies – a claim particularly easy to discount. When pathology fails critics of the Multiregional Evolution hypothesis, the only 'explanation' left seems to be that the supporting data do not exist. For instance:

... the interpretation that the archaic appearance of some Australian crania may be due to cultural practices of deformation rather than the persistence of H. erectus features is gaining ground. Delson (1988:206)

Perhaps Australia was a special case where local differentiation, cultural practices, or pathologies led in some cases to apparent evolutionary reversals.
Stringer & Andrews (1988:1267)

There are no clear links in Southeast Asia between Homo erectus and recent humans.
Rightmire (1988:257)

... the claims for clear evidence of regional continuity in this area (east Asia and Australia) appear to be very problematic and hypothetical. Bräuer (1989:148)

Well, the data do exist and it seems to me that the details of regional continuity that have been researched and published by workers familiar with the fossil specimens are sufficient to demand more than unsupported assertions for their valid dismissal. Refutation with new data is the acceptable procedure for validly dismissing observations and their implications. Instead of calling for yet more data to support the Multiregional Evolution model (e.g. Stringer 1989, Bräuer 1989, Habgood 1989), as if there were none, it is time to expect responses to the data that have been published, addressing the details and either bringing new data and observations in refutation or admitting to the validity of the published information.

The analysis of variation in human mtDNA does not present an obstacle to the Multiregional Evolution theory, in fact quite to the contrary. The equilibrium explanation of this variation, balancing the opposing forces of mutation, selection, and random lineage loss, provides an explanation for existing variation that unlinks the mtDNA and nuclear DNA. It is not necessary to posit that widespread but limited mtDNA variation is the result of dramatic population replacements by a new species that did not mix with existing peoples around the world. Quite the opposite, this model relies on widespread interconnected populations (Avise et al. 1988, Wayne et al. 1990). Moreover, the equilibrium model for mtDNA variation implies that the recent dates suggested for modern human origins are unlikely to be valid. It is now recognized that divergence dates based on mtDNA analysis grossly underestimate the actual divergences of species as disparate as eels, catfish, redwinged blackbirds, and black-backed jackals. For late Pleistocene human population divergences there is no valid mitochondrial clock.

When hypotheses have so clearly competing predictions, the refutational approach has a valid role in deciding between them. The paleoanthropological predictions of the 'Garden of Eden' and Multiregional Evolution theories are outlined in Table 2. The question of whether 'modern *Homo sapiens*' is a new species, different from its predecessors once regarded as 'archaic *Homo sapiens* by virtually all workers including Stringer (in Stringer et al. 1979), cannot be easily resolved. However, two facts strongly suggest that the traditional view should prevail and the species hypothesis be rejected. These are the absence of a valid morphological definition for 'modern *Homo sapiens*' and the absence of morphological convergences in the fossil records of various regions toward a similar, presumably African, ancestral morphology. In my opinion, no evidence supports either phenetic or phylogenetic definitions of a separate species for 'modern *Homo sapiens*', and I believe one can validly reject this hypothesis and thereby avoid the need to explain away the numerous problems and inconsistencies following from it.

The distribution of transitional specimens provides additional reasons to reject the 'Garden of Eden' model. Transitional specimens can be found virtually everywhere in the inhabited old world. Indeed, according to Rightmire, whose career has been mainly focused on the African fossil record, one place where they may not be found is sub-Saharan Africa! The dates of these transitional forms do not support the proposal that there were the sweeping migrations and dramatic population replacements required by an 'African Origins' model. For instance, while the earliest so-called 'moderns' may well be from the Levant, the latest are from nearby Europe.

The early appearance of regional features, especially in eastern Asia where very different characteristics (not alternative character states) in the northern and southern samples extend from the earliest inhabitants to the present, provides perhaps the most unambiguous refutation. Only Multiregional Evolution can account for the early appearance of regional features at the peripheries. Again and again, those who are familiar with the Pleistocene fossil records and recent variations at the peripheries find evidence of regional continuities and morphological links between ancient and living populations which are inexplicable by and contradictory to the tenets of the 'Garden of Eden' model.

Finally, the 'African Origins' part of the 'Garden of Eden' model fails. It has never been clear whether or not this Protsch/Bräuer hypothesis involves complete replacement, and although the model rests almost completely on the claim that the earliest modern humans first appear in Africa, these authors do not agree about which are the critically dated African specimens. These ambiguities have made it difficult to test the hypothesis as it has been published other than to quibble about the dates, since it is unclear which specimens are supposed to be the earliest 'moderns' and therefore it is uncertain where refutations based on African materials might lie. However, tests are made possible because of the predictions of the Multiregional Evolution model. Whatever the degree of mixture, or its lack, the 'African Origins' hypothesis posits between the invading Africans and other people, the one unambiguous prediction of this hypothesis must be that Africans or African features should be identifiable in the earliest modern populations outside of Africa, unless the model has no predictions at all. But the prediction that in each region the evolutionary record of 'modern *Homo sapiens*' is expected to appear more African earlier in time, is demonstrably incorrect. Earlier populations of modern humans outside of Africa are more archaic in morphology but absolutely not more African.

Rejection of the 'Garden of Eden' hypothesis does not 'prove' that all of the assumptions and predictions of the Multiregional model are necessarily correct. However, when one of two competing evolutionary models can be rejected, it is time to closely examine the second as the working hypothesis for the Pleistocene human evolutionary sequence, ascertain the details of its predictions and examine all of the relevant data, morphological and biochemical, with the goal of determining whether it can be more accurately modified or whether Multiregional evolution must too be rejected in favor of a new and thus far unestablished alternative. This, in my view, is the situation at present, and I look forward to an interesting decade of publications on this issue.

ACKNOWLEDGEMENTS

I am deeply grateful to G. Bräuer and F.H. Smith for inviting me to participate in their symposium during the International Congress of Anthropological and Ethnological Sciences in Zagreb, and to J. Radovčić and his colleagues at the Croatian Natural History Museum whose interest and enthusiasm were the primary factors that brought paleoanthropologists to Zagreb, and made their stay both informative and interesting. I thank the Eastern European Program of the National Academy of Science for the exchange program grant that allowed me to be in Zagreb and attend the symposium. I am indebted to Jeff Miller, Rachel Caspari, Karen Rosenberg, David W. Frayer, Con Childress, and especially Fred H. Smith and Günter Bräuer for their help in reading and editing this manuscript critically and the many useful suggestions they proposed. This research was supported by a grant from the National Science Foundation (BNS 85-09147), the Committee for Scholarly Exchanges with the Peoples Republic of China (INT 81-17276), the Rackham Graduate School at the University of Michigan, and the National Academy of Sciences Eastern European Program. The concepts and many of the observations in this paper were developed during joint work on the Multiregional Evolution hypothesis with A.G. Thorne and Wu Xinzhi. I am sure they would agree that the good ideas I have presented here are theirs, and the mistakes and bad ideas mine!

REFERENCES

Aigner, J.S. 1976. Chinese Pleistocene cultural and hominid remains: a consideration of their significance in reconstructing the pattern of human bio-culture development. In A.K. Gosh (ed.), Le Paléolithique Inférieur et Moyen en Inde, en Asie Centrale, en Chine et dans le sud-est Asiatique. *UISPP Colloque* VII. Paris: Centre Nationale de la Recherche Sciéntifique, pp. 65-90.

Avise, J.C. 1986. Mitochondrial DNA and the evolutionary genetics of higher animals. *Phil. Trans. Roy. Soc. London, Series B*, 312:325-342.

Avise, J.C., R.M. Ball & J. Arnold 1988. Current versus historical populations sizes in vertebrate species with high gene flow: a comparison based on mitochondrial DNA lineages and inbreeding theory for neutral mutations. *Mol. Biol. Evol.* 5(4): 331-344.

Avise, J.C., J.E. Neigel & J. Arnold 1984. Demographic influences on mitochondrial DNA lineage survivorship in animal populations. *J. Mol. Evol.* 20(2): 99-105.

Barton, N.H. & B. Charlesworth 1984. Genetic revolutions, founder effects, and speciation. *Ann. Rev. Ecol. Syst.* 15: 133-164.

Ben-Itzhak, S., P. Smith & R.A. Bloom 1988. Radiographic study of the humerus in Neandertals and *Homo sapiens sapiens*. *Am. J. Phys. Anthropol.* 77(2): 231-242.

Birdsell, J.B. 1972. The problem of the evolution of human races: classification or clines? *Soc. Biol.* 19: 136-162.

Brace, C.L. 1964. The fate of the 'Classic' Neanderthals: a consideration of hominid catastrophism. *Curr. Anthropol.* 5: 3-43 and 7: 204-214.

Bräuer, G. 1981. New evidence on the transitional period between Neanderthal and modern man. *J. Hum. Evol.* 10: 467-474.

Bräuer, G. 1984a. The 'Afro-European *sapiens* hypothesis' and hominid evolution in East Asia during the late middle and upper Pleistocene. In P. Andrews & J.L. Franzen (eds), The Early Evolution of Man, with Special Emphasis on Southeast Asia and Africa, *Cour. Forsch. Inst. Senckenberg* 69: 145-165.

Bräuer, G. 1984b. A craniological approach to the origin of anatomically modern *Homo sapiens* in Africa and implications for the appearance of modern Europeans. In F. Smith & F. Spencer (eds), *The Origin of Modern Humans: A World Survey of the Fossil Evidence*. New York: Alan R. Liss. pp. 327-410.

Bräuer, G. 1989. The evolution of modern humans: a comparison of the African and non-African evidence. In P. Mellars & C.B. Stringer (eds), *The Human Revolution: Behavioral and Biological Perspectives on the Origins of Modern Humans*. Edinburgh: University of Edinburgh Press. pp. 123-154.

Bridges, P.S. 1985. *A Biomechanical Analysis of Two Prehistoric Amerind Groups: Changes in Habitual Activities and the Division of Labor with the Transition from Hunting-and-Gathering to Agriculture*. Ann Arbor: University Microfilms.

Brose, D.S. & M.H. Wolpoff 1971. Early upper Paleolithic man and late middle Paleolithic tools. *Am. Anthropol.* 73: 1156-1194.

Brown, P. 1981. Artificial cranial deformation: a component in the variation in Pleistocene Australian Aboriginal crania. *Archaeol. Ocean.* 16: 156-167.

Brown, W.M. 1980. Polymorphism in mitochondrial DNA of humans as revealed by restriction endonuclease analysis. *Proc. Nat. Acad. Sciences USA* 77: 3605-3609.

Cann, R.L. 1987. In search of Eve. *The Sciences* 27: 30-37.

Cann, R.L. 1988. DNA and human origins. *Ann. Rev. Anthropol.* 17: 127-143.

Cann, R.L., W.M. Brown & A.C. Wilson 1982. Evolution of human mitochondrial DNA: a preliminary report. In Bonné-Tamir, Cohen & Goodman (eds), *Human Genetics, part A. The Unfolding Genome*. New York: Alan Liss. pp. 157-165.

Cann, R.L., M. Stoneking & A.C. Wilson 1987a. Mitochondrial DNA and human evolution. *Nature* 325: 31-36.

Cann, R.L., M. Stoneking & A.C. Wilson 1987b. Disputed African origin of human populations. *Nature* 329: 111-112.

Carr, S.M., S.W. Ballinger, J.N. Derr, L.H. Blankenship & J.W. Bickham 1986. Mitochondrial DNA analysis of hybridization between sympatric white-tailed deer and mule deer in west Texas. *Proc. Nat. Acad. Sci. USA* 83: 9576-9580.

Caspari, R. & M.H. Wolpoff 1990. The morphological affinities of the Klasies River Mouth skeletal remains (abstract). *Amer. J. Phys. Anthropol.* 81(2): 203.

Charlesworth, B., R. Lande & M. Slatkin 1982. A Neo-Darwinian commentary on macroevolution. *Evol.* 36: 474-498.

Coon, C.S. 1962. *The Origin of Races*. New York: Knopf.

Currey, J.D. & R.McN. Alexander 1985. The thickness of the walls of tubular bones. *J. Zool. (A)* 206: 453-468.

Day, M.H. & C.B. Stringer 1982. A reconsideration of the Omo Kibish remains and the *erectus-sapiens*, transition. *L'Homo erectus et la Place de l'Homme de Tautavel parmi les Hominidés Fossiles*. Nice: Louis-Jean Scientific and Literary Publications, 2: 814-846.

de Lumley, M.A. 1973. Anténéandertaliens et Néandertaliens du bassin Méditerranéen Occidental Europeén. *Etudes Quaternaires* (Université de Provence), Memoire 2.

Delson, E. 1985. Late Pleistocene human fossils and evolutionary relationships. In E. Delson (ed.), *Ancestors: The Hard Evidence*. New York: Alan R. Liss. pp. 296-300.

Delson, E. 1988. One source not many. *Nature* 332: 206.

Dwight, T. 1905. The size of articular surfaces of the long bones as characteristic of sex: an anthropological study. *Am. J. Anat.* 4: 19-32.

Ehrlich, P.R. & P.H. Raven 1969. Differentiation of populations. *Science* 165: 1228-1232.

Endler, J.A. 1977. *Geographic Variation, Speciation, and Clines*. Princeton: Princeton University Press.

Excoffier, L. & A. Langaney 1989. Origin and differentiation of human mitochondrial DNA. *Am. J. Hum. Genet.* 44: 73-85.

Ferembach, D. 1986. Les Hommes de Paléolithique Supérieur autour de Basin Méditerranéen. *L'Anthropol.* 90(3): 579-587.

Ferris, S.D., R.D. Sage, C.M. Huang, J.T. Nielsen, U. Ritte & A.C. Wilson 1983. Flow of mitochondrial DNA across a species boundary. *Proc. Nat. Acad. Sci. USA* 80: 2290-2294.

Flood, J. 1983. *Archaeology of the Dreamtime*. Sydney: Collins.

Frayer, D.W. 1986. Cranial variation at Mladeč and the relationship between Mousterian and Upper Paleolithic hominids. In V.V. Novotný & A. Mizerová (eds), Fossil Man. New Facts – New Ideas. Papers in Honor of Jan Jelínek's Life Anniversary. *Anthropos (Brno)* 23: 243-256.

Frayer, D.W., J. Jelínek, N. Minugh-Purvis, M. Oliva, L. Seitl, F.H. Smith & M.H. Wolpoff 1991. Late Pleistocene Human Remains from Mladeč Cave, Moravia. In prep.

Gould, S.J. 1987. Bushes all the way down. *Nat. Hist.* 87(6): 12-19.

Hale, L.R. & R.S. Singh 1987. Mitochondrial DNA variation and genetic structure in populations of *Drosophila melanogaster. Mol. Biol. Evol.* 4(6): 622-637.

Habgood, P.J. 1989. The evolution of modern humans: evidence from Australasia seen in a global context. In P. Mellars & C.B. Stringer (eds), *The Human Revolution: Behavioral and Biological Perspectives on the Origins of Modern Humans*. Edinburgh: University of Edinburgh Press. pp. 245-273.

Harihara, S, N. Saitou, M. Hirai, T. Gojobori, K.S. Park, S. Misawa, S.B. Ellepola, T. Ishida & K. Omoto 1988. Mitochondrial DNA polymorphism among five Asian populations. *Am. J. Hum. Gen.* 43: 134-143.

Holt, I.J., A.E. Harding & J.A. Morgan-Hughes 1988. Deletions of muscle mitochondrial DNA in patients with mitochondrial myopathies. *Nature* 331: 717-719.

Honeycutt, R.L. & W.C. Wheeler 1987. Mitochondrial DNA: variation in humans and higher primates. In S.K. Dutta & W. Winter (eds), *DNA Systematics: Human and Higher Primates*. Boca Ratin: CRC Press.

Hooton, E.A. 1947. *Up From the Ape*, revised edition. New York: MacMillan.

Horai, S., T. Gojobori & E. Matsunaga 1986. Distinct clustering of mitochondrial DNA types among Japanese, Caucasians and Negros. *Jap. J. Gen.* 61: 271-275.

Howells, W.W. 1973. Cranial variation in man. A study by multivariate analysis of patterns of difference among recent human populations. *Papers of the Peabody Museum of Archaeology and Ethnology* 67: 1-259.

Hrdlička, A. 1927. The Neanderthal phase of man. *J. Roy. Anthropol. Inst.* 57: 249-274.

Hrdlička, A. 1930. The skeletal remains of early man. *Smithsonian Miscellaneous Collections* 83.

Hublin, J.-J. 1976. *L'homme de Galilée. Mémoire de D.E.A. de Paléontologie*. Paris: Université de Paris VI.

Hublin, J.-J. 1978. Anatomie du centre de l'écaille occipitale – Le problème de l'inion. *Cah. d'Anthropol.* 65-83.

Hublin, J.-J. 1983. Que reste-t-il des Présapiens Européens? In M.S. Sakka (ed), *Morphologie Evolutive Morphogenèse du Crâne et Origine de L'Homme*. Paris: Centre National de la Recherche Scientifique, Paris. pp. 171-181.

Hublin, J.-J. 1987. Qui fut l'Ancêtre de l'*Homo sapiens? Pour la Science* 113: 26-35.

Jacob, T. 1967. *Some Problems Pertaining to the Racial History of the Indonesian Region*. Utrecht: Drukkerij Neerlandia.

Jacob, T. 1976. Early populations in the Indonesian region. In R.L. Kirk & A.G. Thorne (eds), *The Origins of the Australians*. Canberra: Australian Institute of Aboriginal Studies. pp. 81-93.

Jacob, T. 1981. Solo man and Peking man. In B.A. Sigmon & J.S. Cybulski (eds), *Homo erectus. Papers in Honor of Davidson Black* (combined bibliography at end of volume). Toronto: University of Toronto Press. pp. 87-104.

Jelínek, J. 1967. Der Fund eines Neandertaler Kiefers (Kůlna 1) aus der Kůlna-Höhle in Mähren. *Anthropologie (Brno)* 5: 3-19.

Jelínek, J. 1969. Neanderthal man and *Homo sapiens* in central and eastern Europe. *Curr. Anthropol.* 10: 475-503.

Jelínek, J. 1982. The east and southeast Asian way of regional evolution. *Anthropologie (Brno)* 20: 195-212.

Jelínek, J. 1983. The Mladeč finds and their evolutionary importance. *Anthropologie (Brno)* 21: 57-64.

Jelínek, J. 1985. The European, Near East and North African finds after Australopithecus and the principal consequences for the picture of human evolution. In P.V. Tobias (ed), Hominid Evolution: Past, Present, and Future. *Proceedings of the Taung Diamond Jubilee International Symposium*. New York: Alan R. Liss Inc. pp. 341-354.

Jones, J.S. 1986. The origin of *Homo sapiens:* the genetic evidence. In B. Wood, L. Martin & P. Andrews (eds), *Major Trends in Primate and Human Evolution*. Cambridge: Cambridge University Press. pp. 317-330.

Kennedy, G.E. 1984. Are the Kow Swamp hominids 'archaic'? *Am. J. Phys. Anthropol.* 65(2): 163-168.

Lande, R. 1976. Natural selection and random genetic drift in phenotypic evolution. *Evol.* 30: 314-334.

Lande, R. 1982. Rapid origin of sexual isolation and character divergence in a cline. *Evol.* 36: 213-233.

Larnach, S.L. & N.W.G. Macintosh 1974. A comparative study of Solo and Australian Aboriginal crania. In A.P. Elkin & N.W.G. Macintosh (eds), *Grafton Elliot Smith: The Man and his Work*. Sydney: Sydney University Press. pp. 95-102.

Latorre, A., A. Moya & F.J. Ayala 1986. Evolution of mitochondrial DNA in *Drosophila subobscura*. *Proc. Nat. Acad. Sci. USA*. 83: 8649-8653.

Leakey, R.E. & R. Lewin 1977. *Origins*. New York: Dutton.

Leroi-Gourhan, A. 1958. Étude des restes humains fossiles prouvenant des Grottes d'Arcy-sur-Cure. *Ann. Paléont*. 44: 87-148.

Lewin, R. 1988. Modern human origins under close scrutiny. *Science* 239: 1240-1241.

Lewin, R. 1990. Molecular clocks run out of time. *New Scientist* 2/10/90: 38-41.

Liškova, M. 1965. The thickness changes of the long bone after experimental stressing during growth. *1st report, Plzensky lekarsky Sbornik* 25: 95-100.

MacRae, A.F. & W.W. Anderson 1988. Evidence for non-neutrality of mitochondrial DNA haplotypes in *Drosophila pseudoobscura*. *Genetics* 120: 485-494.

Matiegka, J. 1934. *Homo Předmostensis: Fosiliní Člověk z Předmostí na Moravě 1. Lebky*. Prague: Česká Akademie Věd i Umění.

May, R.M., J.A. Endler & E. McMurtne 1975. Gene frequency clines in the presence of selection opposed by gene flow. *Am. Nat*. 109: 659-676.

Mayr, E. 1963. *Animal Species and Evolution*. Cambridge: Belknap.

Mayr, E. 1970. *Populations, Species and Evolution*. Cambridge: Belknap.

Neel, J.V. 1978. The population structure of an Amerind tribe, the Yanomami. *Ann. Rev. Gen*. 12: 365-413.

Pope, G.G. 1988. Recent advances in Far Eastern Paleoanthropology. *Ann. Rev. Anthropol*. 17: 43-77.

Protsch, R. 1975. The absolute dating of Upper Pleistocene sub-Saharan fossil hominids and their place in human evolution. *J. Hum. Evol*. 4: 297-322.

Radovčić, J. 1988. Dragutin Gorjanović-Kramberger and Krapina Early Man: The Foundation of Modern Paleoanthropology. Zagreb: Školska knjiga and Hrvatski prirodoslovni muzej.

Radovčić, J., F.H. Smith, E. Trinkaus & M.H. Wolpoff 1988. *The Krapina Hominids: An Illustrated Catalog of the Skeletal Collection*. Zagreb: Mladost and the Croatian Natural History Museum.

Rightmire, G.P. 1979. Implications of the Border Cave skeletal remains for later Pleistocene human evolution. *Curr. Anthropol*. 20: 23-35.

Rightmire, G.P. 1981. Later Pleistocene hominids of eastern and southern Africa. *Anthropologie (Brno)* 19: 15-26.

Rightmire, G.P. 1984. *Homo sapiens* in sub-Saharan Africa. In F.H. Smith & F. Spencer (eds), *The Origins of Modern Humans: A World Survey of the Fossil Evidence*. New York: Alan R. Liss. pp. 295-325.

Rightmire, G.P. 1987. L'Evolution des premiers hominidés en Asie du Sud-Est. *L'Anthropologie* 91: 455-465.

Rightmire, G.P. 1988. *Homo erectus* and later middle Pleistocene humans. *Ann. Rev. Anthropol*. 17: 239-259.

Rightmire, G.P. 1989. Middle Stone Age humans from eastern and southern Africa. In P. Mellars & C.B. Stringer (eds), *The Human Revolution: Behavioural and Biological Perspectives on the Origins of Modern Humans*. Edinburgh: Edinburgh University Press. pp. 109-122.

Saban, R. 1980. The middle meningeal veins system in the fossil men of Czechoslovakia by meaning of the endocranial casts. *Abstracts of the Second European Anthropological Congress, Brno*. pp. 13-14.

Saitou, N. & K. Omoto 1987. Time and place of human origins from mtDNA data. *Nature* 327: 288.

Slatkin, M. 1987. Gene flow and the geographic structure of natural populations. *Science* 236: 787-792.

Smith, F.H. 1978. Some conclusions regarding the morphology and significance of the Krapina Neandertal remains. In M. Malez (ed.), *Krapinski Pračovjek i Evolucija Hominida*. Zagreb: Jugoslavenska Akademija Znanosti i Umjetnosti.

Smith, F.H. 1982. Upper Pleistocene hominid evolution in South-Central Europe: A review of the evidence and analysis of trends. *Curr. Anthropol*. 23: 667-703.

Smith, F.H. 1984. Fossil Hominids from the Upper Pleistocene of central Europe and the origin of modern Europeans. In F.H. Smith & F. Spencer (eds), *The Origins of Modern Humans: A World Survey of the Fossil Evidence*. New York: Alan R. Liss. pp. 137-209.

Smith, F.H. 1985. Continuity and change in the origin of modern *Homo sapiens*. *Z. Morph. Anthropol*. 75(2): 197-222.

Smith, F.H., D.C. Boyd & M. Malez 1985. Additional Upper Pleistocene human remains from Vindija cave, Croatia, Yugoslavia. *Am. J. Phys. Anthropol.* 68(3): 375-383.

Smith, F.H. & G.C. Ranyard 1980. Evolution of the supraorbital region in Upper Pleistocene fossil hominids from South-Central Europe. *Am. J. Phys. Anthropol.* 53: 589-609.

Spuhler, J.N. 1988. Evolution of mitochondrial DNA in monkeys, apes and humans. *Yrbk. Phys. Anthropol.* 31: 15-48.

Stoneking, M. & R.L. Cann 1989. African origin of human mitochondrial DNA. In P. Mellars & C.B. Stringer (eds), *The Human Revolution: Behavioral and Biological Perspectives on the Origins of Modern Humans.* Edinburgh: University of Edinburgh Press. pp. 17-30.

Stringer, C.B. 1982. Towards a solution to the Neanderthal problem. *J. Hum. Evol.* 11: 431-438.

Stringer, C.B. 1986. Documenting the origin of modern humans. ARS Seminar paper, from E. Trinkaus (ed.), *The Origins of Modern Human Adaptations.* Seminar Preprint.

Stringer, C.B. 1988. The dates of Eden. *Nature* 331: 565-566.

Stringer, C.B. 1989. The origins of early modern humans: a comparison of the European and non-European evidence. In P. Mellars & C.B. Stringer (eds), *The Human Revolution: Behavioral and Biological Perspectives on the Origins of Modern Humans.* Edinburgh: University of Edinburgh Press. pp. 232-244.

Stringer, C.B. & P. Andrews 1988. Genetic and fossil evidence for the origin of modern humans. *Science* 239: 1263-1268.

Stringer, C.B., F.C. Howell & J. Melentis 1979. The significance of the fossil hominid skull from Petralona, Greece. *J. Archaeol. Sci.* 6: 235-253.

Stringer, C.B., J.-J. Hublin & B. Vandermeersch 1984. The origin of anatomically modern humans in western Europe. In F.H. Smith & F. Spencer (eds), *The Origins of Modern Humans: A World Survey of the Fossil Evidence.* New York: Alan R. Liss. pp. 51-135.

Tattersall, I. & E. Delsen 1984. *Ancestors. Four Million Years of Humanity.* New York: American Museum of Natural History.

Templeton, A.R. 1982. Genetic architectures of speciation. In C. Barigozzi (ed.), *Mechanisms of Speciation.* New York: Alan R. Liss. pp. 105-121.

Thoma, A. 1964. Die Entstehung der Mongoloiden. *Homo* 15: 1-22.

Thorne, A.G. 1972. Recent discoveries of fossil man in Australia. *Austral. Nat. Hist.* 17: 191-195.

Thorne, A.G. 1976. Morphological contrasts in Pleistocene Australians. In R.L. Kirk & A.G. Thorne (eds), *The origin of the Australians.* Canberra: Australian Institute of Aboriginal Studies. pp. 95-112.

Thorne, A.G. 1980. The longest link: human evolution in Southeast Asia and the settlement of Australia. In J.J. Fox, R.G. Garnaut, P.T. McCawley & J.A.C. Maukie (eds), *Indonesia: Australian Perspectives.* Canberra: Research School of Pacific Studies. pp. 35-43.

Thorne, A.G. 1981. The centre and the edge: The significance of Australian hominids to African paleoanthropology. In R.E. Leakey & B.A. Ogot (eds), *Proceedings of the 8th Panafrican Congress of Prehistory and Quaternary Studies, Nairobi, September 1977.* Nairobi: TILLMIAP. pp. 180-181.

Thorne, A.G. 1984. Australia's human origins – how many sources? *Am. J. Phys. Anthropol.* 63(2): 227.

Thorne, A.G. & S.R. Wilson 1977. Pleistocene and recent Australians: A multivariate comparison. *J. Hum. Evol.* 6: 393-402.

Thorne, A.G. & M.H. Wolpoff 1981. Regional continuity in Australasian Pleistocene hominid evolution. *Am. J. Phys. Anthropol.* 55: 337-349.

Thorpe, R.S. 1985. Character number and the multivariate analysis of simple patterns of geographic variation: categorical or 'stepped clinal' variation. *Syst. Zool.* 34: 127-139.

Trinkaus, E. 1977. A functional interpretation of the axillary border of the Neandertal scapulae. *J. Hum. Evol.* 6: 231-234.

Trinkaus, E. & F.H. Smith 1985. The fate of the Neandertals. In E. Delson (ed.), *Ancestors: The Hard Evidence.* New York: Alan R. Liss. pp. 325-333.

Trinkaus, E. & D.D. Thompson 1987. Femoral diaphyseal histomorphometric age determinations for the Shandiar 3, 4, 5 and 6 Neandertals and Neandertal longevity. *Am. J. Phys. Anthropol.* 72(1): 123-129.

Turner, C.G. 1987. Late Pleistocene and Holocene population history of east Asia based on dental variation. *Am. J. Phys. Anthropol.* 73(3): 305-321.

Valladas, H., J.L. Reyss, J.L. Joron, G. Valladas, O. Bar-Yosef & B. Vandermeersch 1988. Thermoluminescence dating of Mousterian 'Proto-Cro-Magnon' remains from Israel and the origin of modern man. *Nature* 331: 614-616.

Van Valen, L.M. 1986. Speciation and our own species. *Nature* 322: 412.

Vandermeersch, B. 1984. À propos de la découverte du squelette Néandertalien de Saint Césaire. *Bull. Mém. Soc. d'Anthropol. Paris, sér.* 14, 1: 191-196.

Vandermeersch, B. 1985. Le Peuplement d'Israel au Paleolithique: les hommes de Páleolithique inférieur et moyen. *Dossiers histoire et archaeologie No.* 100 (December): 38-44.

Vlcěk, E. 1969. *Neandertaler der Tschechoslowakei.* Prague: Tschechoslowakische Akademie der Wissenschaften.

Weidenreich, F. 1939. Six lectures on *Sinanthropus pekinensis* and related problems. *Bull. Geol. Soc. China* 19: 1-110.

Weidenreich, F. 1941. The extremity bones of *Sinanthropus pekinensis. Palaeont. Sin. New Series* D, 5 (Whole Series 116).

Weidenreich, F. 1943a. The 'Neanderthal Man' and the ancestors of '*Homo sapiens*'. *Am. Anthropol.* 45: 39-48.

Weidenreich, F. 1943b. The skull of *Sinanthropus pekinensis:* A comparative study of a primitve hominid skull. *Palaeont. Sin.*, n.s. D, Number 10.

Weidenreich, F. 1945. Giant early man from Java and south China. *Anthropological Papers of the American Museum of Natural History* 40: 1-134.

Weidenreich, F. 1946. *Apes, Giants and Man.* Chicago: University of Chicago.

Weidenreich, F. 1947. Are human races in the taxonomic sense 'races' or 'species'? *Am. J. Phys. Anthropol.* 5(3): 369-371.

Weidenreich, F. 1951. Morphology of Solo man. *Anthropological Papers of the American Museum of Natural History* 43(3): 205-290.

Weiss, K.M. 1984. On the number of members of the genus *Homo* who have ever lived, and some evolutionary implications. *Hum. Biol.* 56(4): 637-649.

Weiss, K.M. 1988. In search of times past: the roles of gene flow and invasion in the generation of human diversity. In N. Mascie-Taylor & G. Lasker (eds), *Biological Aspects of Human Migration.* London: Cambridge University Press.

Whittam, T.S., A.G. Clark, M. Stoneking, R. Cann & A.C. Wilson 1986. Allelic variation in human mitochondrial genes based on patterns of restriction site polymorphism. *Proc. Nat. Acad. Sci. USA.* 83: 9611-9615.

Wilson, A.C., R.L. Cann, S.M. Carr, M. George, U.B. Gyllensten, K.M. Helm-Bychowski, R.G. Higuchi, S.R. Palumbi, E.M. Prager, R.D. Sage & M. Stoneking 1985. Mitochondrial DNA and two perspectives on evolutionary genetics. *Biol. J. Linn. Soc. London* 26: 375-400.

Wolpoff, M.H. 1975. Some aspects of human mandibular evolution. In J.A. McNamara Jr (ed.), *Determinants of Mandibular Form and Growth.* Ann Arbor: Center of Human Growth and Development. pp. 1-64.

Wolpoff, M.H. 1980. *Paleoanthropology.* New York: Knopf.

Wolpoff, M.H. 1985. Human evolution at the peripheries: the pattern at the eastern edge. In P.V. Tobias (ed.), Hominid Evolution: Past, Present, and Future. *Proceedings of the Taung Diamond Jubilee International Symposium.* New York: Alan R. Liss. pp. 355-365.

Wolpoff, M.H. 1986. Describing anatomically modern *Homo sapiens:* a distinction without a definable difference. In V.V. Novotný & A. Mizerová (eds), Fossil Man. New Facts, New Ideas. Papers in Honor of Jan Jelínek's Life Anniversary. *Anthropos (Brno)* 23: 41-53.

Wolpoff, M.H. 1989. Multiregional evolution: the fossil alternative to Eden. In P. Mellars & C.B. Stringer (eds), *The Human Revolution: Behavioral and Biological Perspectives on the Origins of Modern Humans.* Edinburgh: University of Edinburgh Press. pp. 62-108.

Wolpoff, M.H. & R. Caspari 1990. Metric analysis of the skeletal material from Klasies River Mouth, Republic of South Africa (abstract). *Am. J. Phys. Anthropol.* 81(2): 319.

Wolpoff, M.H., F.H. Smith, M. Malez, J. Radovčić & D. Rukavina 1981. Upper Pleistocene human remains from Vindija Cave, Croatia, Yugoslavia. *Am. J. Phys. Anthropol.* 54: 499-545.

Wolpoff, M.H., J.N. Spuhler, F.H. Smith, J. Radovčić, G. Pope, D.W. Frayer, R. Eckhardt & G. Clark 1988. Modern human origins. *Science* 241: 772-773.

Wolpoff, M.H., Wu Xinzhi & A.G. Thorne 1984. Modern *Homo sapiens* origins: a general theory of hominid evolution involving the fossil evidence from east Asia. In F.H. Smith & F. Spencer (eds), *The Origins of Modern Humans: A World Survey of the Fossil Evidence*. New York: Alan R. Liss. pp. 411-483.

Woo, Jukang (Wu Rukang) 1959. Human fossils found in Liujiang, Kwangsi, China. *Vert. PalAsiatica* 3: 109-118.

Wright, S. 1980. Genic and organismic selection. *Evol.* 34: 825-843.

Wu, Rukang & J.W. Olsen (eds) 1985. *Palaeoanthropology and Paleolithic Archaeology in the People's Republic of China*. New York: Academic Press.

Wu, Rukang & Wu Xinzhi 1982. Comparison of Tautavel man with *Homo erectus* and *Homo sapiens* in China. In *L'Homo erectus et la Place de l'homme de Tautavel parmi les Hominides Fossiles*. Volume 2: 605-616. Nice: Jean Scientific and Literary Publications.

Wu, Xinzhi 1961. Study of the Upper Cave man of Choukoutien. *Vert. PalAsiatica* 3: 202-211.

Wu, Xinzhi 1981. A well-preserved cranium of an archaic type of early *Homo sapiens* from Dali, China. *Sci. Sin.* 24(4): 530-541.

Wu, Xinzhi 1987. Relation between Upper Paleolithic men in China and their southern neighbors in Niah and Tabon. *Acta Anthropol. Sin.* 6(3): 180-183.

Wu, Xinzhi & Wu Maolin 1985. Early *Homo sapiens* in China. In Wu Rukang & J.W. Olsen (eds), *Palaeoanthropology and Paleolithic Archaeology in the People's Republic of China*. New York: Academic Press. pp. 91-106.

Wu, Xinzhi & Zhang Yinyun 1978. Fossil man in China. In IVPP (eds), *Symposium on the Origin of Man*. Beijing: Science Press. pp. 28-42.

A mitochondrial perspective on replacement or continuity in human evolution

Rebecca L. Cann
Department of Human Genetics, University of Hawaii at Manoa, Honolulu, USA

1 INTRODUCTION

Humans, like all other organisms, have a wide variety of mtDNA sequences which appear to be fully functional under standard conditions for a given gene or locus (Whittam et al. 1986). Polymorphisms in human mtDNA have been catalogued since the early 80's, and a stronger evolutionary perspective is now needed to account for the patterns detected. Review papers which enlarge on the details of mitochondrial gene structure, function, and history in a host of animal species (Moritz et al. 1987, Avise et al. 1988, Harrison 1989) help clarify inferences drawn about human mtDNA polymorphism and evolutionary processes. MtDNA can augment more traditional morphological, behavioral, and fossil-based research, and taken with the older cytogenic and isozyme literature, give biologists added confidence in their ability to infer evolutionary process (Harrison 1989). MtDNA can also offer new perspectives into the evolution of a species, as was shown by our suggestion of the evidence for the recent African origin for all human mtDNA (Cann et al. 1987).

Mitochondrial genes occur as multiple copies within the cell, where some copies may accumulate mutations as the DNA sequence changes. Mechanisms which account for this are currently under intense study (Simpson & Shaw 1989, Chang & Clayton 1987). If the mutation occurs in germ cells of the maternal parent, it may be passed on to the offspring. As with nuclear genes, it is likely that most mutations are selectively neutral. It is impossible to infer with certainty the adaptive significance associated with any particular mutation, and when or if natural selection may have operated in the past. However, rapid progress is now being made in defining certain rare, deleterious phenotypes associated with mitochondrial mutations affecting the central nervous system (Wallace et al. 1988). Individuals with health problems showing mitochondrial involvement often present complex phenotypes and may actually be genetic mosaics, possessing some fully functional mitochondrial genomes in addition to defective ones. With a few exceptions, phenotypically 'normal' humans appear to have the same or nearly the same mtDNA sequence common to all mitochondria throughout their bodies (Monnat & Loeb 1985, but see also Laipis et al. 1988). Data derived from the comparison of mitochondrial genomes of different individuals within demes now stimulates the research of both population and molecular biologists.

One particular problem of interest to anthropologists is the extent to which mtDNA sequences might help uncover new information about the behavior of our ancestors, especially regarding mating preferences. Because of the asymmetry of inheritance, mitochondrial genes are useful in exploring female biased activities concerning sexual behaviors and parental care, which may as an end result yield gene phylogenetic trees that are not concordant with population trees as a whole. Moreover, inferences drawn from mtDNA data are not confined simply to female behaviors, as we know from information collected with frog hybrids that document the mating success of satellite males from the 'wrong' species. These males, found in association with calling males, intercept females on their way to mating ponds (Lamb & Avise 1986).

MtDNA has been used to examine the behavior of individuals in many hybrid zones (Barton & Hewitt 1985, Hewitt 1988, 1989), and evolutionary biologists might anticipate that it could also inform our discussions about potential hybrid zones in humans, especially in the Middle East during the last 100 000 years. I now wish to push evolutionary inference even further and ask if we can deduce hybridization between isolated populations, especially between one that is now extinct and the remaining modern ones, if all we have are modern DNA samples? The killer bee phenomenon, reflecting a recent hybrid zone formed in the last 30 years, helps illustrate the limitations of mtDNA in this controversy.

2 BASIC EVOLUTIONARY INSIGHTS USING mtDNA

Assuming genetic neutrality as a first approximation until we know otherwise, mtDNA polymorphisms allow biologists to contribute to organismal level studies by suggesting first a time scale for modern population diversification. The time of genetic diversification in the maternal lineages of modern humans, estimated to be roughly 200 000 years BP, predates by about 100 000 years the critical transition from archaic to anatomically modern morphology, as judged by the fossil evidence (Stringer & Andrews 1988). The inference of 200 000 years ago for a common mother comes from measuring the differences between mtDNA sequences in modern people and calculating the late possible date for a common ancestress from which they could have descended. Underestimating or overestimating the rate of mitochondrial evolution in humans would cause errors in this date.

Second, the patterns of particular mtDNA polymorphisms found in modern populations trace the survival of the oldest lineages to Africa, even though other geographic regions contained archaic populations (Stoneking & Cann 1989). The time scale for the deepest node in the human mtDNA tree, the African only lineages versus the dispersed African lineages along with lineages now found in other geographic regions, could only be placed strictly speaking at 50 000 to 500 000 years ago on the basis of restriction enzyme mapping. It is now estimated confidently at 200 000 years ago from DNA sequences of the most highly variable region (D-loop) from hundreds of authentic sub-Saharan Africans and other human populations using chimpanzees as an outgroup (Vigilant & Wilson 1989, Kocher & Wilson 1989).

Population thinking assures us that at the organismal level, modern humans did not evolve from a single female (even a demi-goddess, such as the straw woman of Excoffier et al. 1989). We also did not necessarily experience a severe crash in population size during the transition from archaic to modern morphologies (Cann 1988), even if only one maternal lineage can be reconstructed. If we use genetic neutrality as a first approximation or the null hypothesis from which to evaluate other hypotheses, neutral polymorphisms in mitochondria provide a data base that can potentially contribute more than just to knowledge of species diversity at the molecular level. They also allow scientists to infer population structure and mating systems.

MtDNA models, like their nuclear counterparts, help scientists estimate effective population size (N e) and rates of dispersal, extinction and migration (Birky et al. 1983, Takahata & Slatkin

1984, Nei 1987). N e in humans has been estimated to be 10 000 using genetic data (Nei & Grauer 1984) and neutral polymorphisms move through subspecies at a rate limited only by the dispersal of individuals carrying them and the number of generations since contact (Endler 1977). Taking 30 000 years as the last date of contact between subspecies in Europe and 20 years per generation as a starting point, 1500 generations have elapsed. Dispersal estimates could range over orders of magnitude, from very high if we use modern hunter-gatherers, to very low if we use swidden agriculturalists. Hybrid zones reflecting contact between two human isolates could therefore theoretically be preserved within the mtDNA of modern humans.

3 MtDNA AND HYBRIDIZATION

Genetic and morphological characters often change in a congruent pattern across a hybrid zone, except where founder events may cause one taxa's mtDNA to be found in another group's, nuclear gene background (Wilson et al. 1985). It is possible that African-based human populations were an example of the house mouse pattern of hybrid zone formation, where the mtDNA of one species of mouse (*Mus domesticus*) has invaded the nuclear background of another species (*Mus musculus*) in Scandanavia during the last 5000 years (Gyllensten & Wilson 1987). Additional examples of this kind point out the danger of blindly assuming that phylogenetic trees derived from one genome or locus will yield branching diagrams consistent with the evolutionary history of the entire population (Felsenstein 1988).

Modern analysis of hybrid zones almost always includes some estimate of gene flow between taxa inferred from protein or DNA markers (Hewitt 1989). While potential or realized exchange of genes is usually the premier criteria for delimiting boundaries between animal taxa, this focus is incomplete for potential human hybrid zones as well as those in other species. Barton (1989) and Templeton (1989) have argued that the current focus on isolating mechanisms emphasized by the biological species concept for maintaining the boundaries between taxa obscures our actual understanding of the product (isolation) and how it relates to one evolutionary process (speciation).

Systematists, judging Neanderthals as subspecific in their degree of differentiation from anatomically modern people, imply that partial genetic isolation may be due only to the geography of dispersal or genetic drift. However, the persistence of two taxa in sympatry, morphologically distinct at the subspecific level for at least 50 000 years, argues for very limited gene flow between these populations, as well as 2 distinct ecological niches (Mayr 1973). Templeton's (1989) terminology for these populations in overlap suggests they are 'demographically exchangable'. In this case, neutral markers undergo genetic drift to either fixation or eventual elimination. A strictly neutral model would predict that only one ancestral mitochondrial genotype should prevail over time, which is consistent with our observations of the maternal ancestry of modern Europeans.

4 MtDNA AND REPLACEMENT

If we imagine two human isolates sharing most (90%) polymorphisms but also differing at other loci, such as infectious disease resistance, there are at least 6 genetically based hypotheses to account for the stable maintenance of two isolates that can hybridize. Hewitt (1989) outlines some of these hypotheses regarding long-term maintenance of a hybrid zone as due to '1) heterozygote disadvantage, 2) differential environmental selection for the two homozygotes, 3) selective equality of homozygotes and heterozygotes, 4) frequency-dependent selection on homozygotes, 5) superiority of one homozygote over another, and 6) superiority of heterozygotes over homozygotes within the zone of overlap.' These genetic hypotheses all invoke natural selection in hybrid zones and are difficult to test even in extant species that can be manipulated in laboratory settings. These hypotheses are important, however, because they suggest limits to the apparent amount of gene flow and the strength of selection necessarily associated with the persistence over thousands of years of 2 morphologically distinct human populations in Europe, as well as other regions of the world.

A focus on the demographic characteristics of small populations suggests that chance often figures in demographic catastrophies (Soule 1987). Results of these natural experiments can sometimes lead to the evolution of new morphologies and adaptations when founders reinvade an empty niche (Carson & Templeton 1984). Human populations, characterized by high rates of local extinction due to infectious disease (Crosby 1986), show some demographic features of mathematically chaotic behavior that emphasize the potential importance of founders. High demic extinction rates driven by infectious disease as population densities build is the expectation, a bit counter intuitive considering the usual focus on inbreeding and fitness depression in small populations. Recolonization from refugia in Africa or later the Middle East, would lead to the result seen in human mtDNA studies suggesting that European and Asian populations were replaced by African founders. I therefore stand in opposition to those biologists who consistently invoke gene flow as the primary force affecting human populations over thousands of years and underestimate the importance of drift in shaping the successful persistence of a lineage, or deny that mtDNA variation could be linked to any phenotype of consequence under natural selection (Wolpoff 1989). Selection for disease resistance would have been a potent force for 2 taxa suddenly experiencing secondary contact, but like gene flow, it is still not the only microevolutionary force operating on these groups.

5 REPLACEMENT OR HYBRIDIZATION? EXPERIMENTS IN PROGRESS

The problem I will now pursue is exploring under what conditions actual and measurable gene flow might look like population replacement, even with DNA data from the known hybrids and the ancestral populations. The best example for consideration comes from current problems in the genetics of honey bees, which as the 'Killer Bee' phenomenon, share a measure of the same popularity in the tabloid press as Eve. Range expansion of African honey bees into the neotropics was initially the result of human action to import a new brood stock for purposes of improving certain characteristics of honey bees in Brazil in 1956. Populations of neotropical American honey bees are showing replacement or take over that is fully documented to have occurred since then (Needham et al. 1988). This phenomenon may help enlighten our understanding of the data necessary for inferring evolutionary processes. In the case of the bees, limited hybridization at the point of release in Brazil has yielded population replacement and not continuity, and appears to be the result of asymmetrical gene flow. Indeed, introgression of European genes into the 'Africanized' honey bees is almost nil (Hall & Muralidharan 1989).

Africanization of the South, Central, and North American honey bee population is an ongoing process with economic and public health consequences that entails hybridization betwen *Apis mellifera scutellata* from Africa and 3 European subspecies *A.m. mellifera, ligustica and carnica* (Smith et al. 1989, Hall & Muralidharan 1989). These subspecies all differ in behavior, genetics and morphology. The bee example is particularly enlightening because some anthropologists suspect that anatomically modern people would differ in these same classes of characters compared to Neanderthal populations (Mellars 1989). In the case of the bees, it is possible to examine the interaction of behavior and gene flow, a process that we can only infer at a distance from our incomplete knowledge of the past with humans (Gargett 1989).

Careful studies of nuclear and mitochondrial genomes of Africanized bees allows us to see that for a large number (N = 70) of feral honey bee swarms analyzed in Mexico and Venezuela, 98.5% (69) had African mtDNA. The bees carrying African genes have moved over 5000 miles in roughly 150 generations, taking about 30 years. Drones retain some European nuclear genes, scored by protein electrophoretic markers. The persistence of these markers is suspect because the European alleles, present in Africanized Brazilian bees, are lost in the Africanized Venezuelan bees (Hall & Muralidharan 1989). European maternal lines are also being gradually lost in the migration north from the release point in Brazil, helping researchers to reconstruct that Africanization does not involve just the overproduction of drones in the African hives, a process which floods the gene pool with African alleles. The increased activity of females carrying African mtDNA in dispersal and mating may lead to the takeover of hives by African swarms and be the actual driving force behind this process (Smith et al. 1989).

Both groups reporting on the mtDNA of Africanized bees suggest that the bees with mtDNA derived from African populations have some selective advantage in comparison to European bees, perhaps related to sustained metabolic activity in flight and earlier development. It is somewhat humerous to note that this is exactly the opposite pattern in humans, where the group eliminated or lost through introgression (Neanderthals) shows evidence of greater sustained metabolic activity and stamina associated with hard physical labor (Trinkaus 1983) compared to anatomically modern people. Based on an estimate of 150 generations, in keeping with the bee example and using a figure of 20 years per human generation, maternal lineages of the group with a strong selective disadvantage would disappear in just 3000 years, even in the face of repeated mating.

6 CONSTRAINTS FROM CALIBRATION POINTS IN THE MITOCHONDRIAL CLOCK

Sudden European genetic and cultural dominance on two continents and hundreds of Pacific Islands in the last 400 years is as difficult to understand from our modern perspective as the total disappearance of the Neanderthal phenotype in Europe, without the historical record to aid in informing us of the role that disease and technology played in displacing indigenous populations. Disallowing a disease hypothesis, regional continuity could be preserved as a viable hypothesis only if dispersal to Europe resulted in the early isolation, around 180 000 years ago, of morphologically distinct (archaic) females from the same African mitochondrial branch as the immigrants who came 90 000 years later. The population carrying this lineage would have to survive 150 000 years in the margins of Europe, be associated with the Neanderthal phenotype, and later be adsorbed by anatomically modern populations through hybridization.

This time constraint is necessary because modern Europeans, like modern Asians and modern Africans, show no evidence of ancient genetic lineages contributed by distant Neanderthal relatives laced with isolated European, African, or Asian *H. erectus* maternal genes. If we hypothesize that Neanderthals had European *H. erectus* maternal ancestors, it is necessary to push the calibration of the mitochondrial DNA clock back to an initial African dispersal of *H.*

erectus. Descendants of these *H. erectus* individuals in Europe become the founders for some Europeans. Stoneking et al. (1986) review how certain geographically isolated populations can be used to internally calibrate the mtDNA clock, and the problems an early dispersal model entails.

In order to justify more ancient genetic roots for Europeans, it is necessary to also hypothesize the founding of New Guinea highland and Australian Aboriginal populations almost 200 000 years ago, along with the migration of North Asians to Alaska no later than 50 000 years ago (Stoneking & Cann 1989). If regional continuity in Europe is a physical reality and not just morphological convergence or poor control of dating, Neanderthal's shallow genetic roots must have traced to Africa, hybrid individuals with Neanderthal mothers must have had poorer relative fitness in the population, and/or Neanderthal females were excluded from the mating structure of the hybrid population. Forced copulation, a measurably frequent human sexual behavior (Brownmiller 1975) increases the probability that Neanderthal females would have been included in the pool of available mates and the primary avenue for gene flow between these two groups. All of these suggestions are untestable speculation.

7 HYBRID ZONE RESEARCH REQUIRES MULTIPLE AND SIMULTANEOUS LINES OF INQUIRY

Finally, it may be informative to discuss the limitations of current knowledge concerning genetic introgression in hybrid zones and how this data contributes to the continuity/replacement debate in humans. Examples from the zoological literature where hybrid zones are actively studied (Hewitt 1989) show that research is pursued along several fronts at once, and studies of mtDNA form just a small part. Any attempt to uncover the true significance of hybrid zones requires the full-time effort of a team of researchers with a variety of skills.

Genetic data include samples from both non-hybrid or parental types as well as hybrids, with nuclear and mitochondrial gene samples prepared from the same individuals, numbering in the hundreds. At least 20 to 30 nuclear loci are studied electrophoretically, while mtDNA is usually mapped at a minimum of 60 restriction sites for statistical significance. Karyotypic data are also necessary if chromosomal rearrangements help define the hybrids. Animals are sampled in well-defined transects that extend into the range of non-hybrid phenotypes. Morphological measurements are taken from these same voucher specimens, and the transects may be censused on a seasonal basis in order to allow for dispersal of juveniles and study the mating success of smaller individuals thought to be less sexually mature. Osteological and dental measurements as well as the pelage or plumage data are included in this research. Behavioral information includes field observations of mating between both hybrids and non-hybrids, with competition and mate removal studies. Foraging efficiency of hybrids as it relates to parental care is also an important element in behavoral research. Additional studies of animals taken into the lab establish the endocrinology, physiology and pathology of hybrids, helping to document developmental shifts, stages of sexual maturity and receptivity, or disease susceptibility.

As I hope will be clear from this description, attempts to establish the evidence for a hybrid zone in the human fossil record using just the morphology of fossils and mitochondrial genetics of modern populations are a fruitless expenditure of time. Unless our ability to recover DNA from well-dated and contextually understood fossils somehow makes a quantum leap in the next 10 years, we may forever be unable to resolve this controversy. For the present, from the DNA standpoint, it is necessary to set potential hybridization aside and consider more concrete if boring problems.

8 SUMMARY AND CONCLUSIONS

In the process of learning about human genetic diversity, we have uncovered a pattern of human population structure with mitochondrial polymorphisms which was only hinted at by studies of nuclear loci. These early protein studies and later DNA work utilizing restriction fragment length polymorphisms (RFLP's) and linguistic trees are consistent with African genetic roots for modern people. MtDNA has some advantages to historical reconstruction due to the fewer number of assumptions which must be made about genetic recombination, and this allows a time scale for our dispersal from Africa to be set. MtDNA sequence data now strengthen this estimate to a split in the population ancestral to modern humans dating to 200 000 years ago, providing the stimulus for geneticists to deal with significant problems in modeling lineage competition/replacement and natural selection or drift. If true hybridization took place between Neanderthal and anatomically modern people, we are about 30 000 years too late to see the persistence of Neanderthal maternal lineages, using the example of the Africanized honey bees. Here, even though gene flow between European and African parental types can be measured, the European nuclear and mitochondrial alleles are lost in only 150 generations. Using the bee example, if replacement of archaic human populations by anatomically modern groups did actually happen, unknown selective differentials between these genomes are implicated. If continuity is the case, drift has removed those Neanderthal mitochondrial lineages from modern humans through demographic exchangability. For the purpose of contributing to this current controversy between continuity/replacement adherents, mtDNA specialists must still sit firmly on the fence until one or the other of the proposed parental types is recovered in a state suitable for DNA extraction. Even then, we might predict the outcome but be unable to say exactly how the transition actually took place. That particular knowledge also requires sampling a few hybrid individuals over well-spaced intervals of time. So far, they appear to be in short supply.

REFERENCES

Avise, J.C., R.M. Ball & J. Arnold 1988. Current versus historical population sizes in vertebrate species with high gene flow: a comparison based on mitochondrial DNA lineages and inbreeding theory for neutral mutations. *Mol. Biol. Evol.* 5: 331-344.

Attardi, G. & G. Schatz 1989. Biogenesis of mitochondria. *Ann. Rev. Cell Biol.* 4: 289-333.

Barton, N.H. 1989. Founder effect speciation. In D. Otte & J.A. Endler (eds), *Speciation and Its Consequences*. Sunderland, Ma.: Sinauer Associates Inc.

Barton, N.H. & G.M. Hewitt 1985. Analysis of hybrid zones. *Ann. Rev. Ecol. Syst.* 16: 113-148.

Birky, C.W., T. Maruyama & P. Fuerst 1983. An approach to population and evolutionary genetic theory for genes in mitochondria and chloroplasts, and some results. *Genetics* 103: 513-527.

Brownmiller, S. 1975. *Against Our Will*. New York, N.Y.: Simon and Shuster.

Cann, R.L. 1988. DNA and human origins. *Ann. Rev. Anthropol.* 17: 127-143.

Cann, R.L., M. Stoneking & A.C. Wilson 1987. Mitochondrial DNA and human evolution. *Nature* 325: 31-36.

Carson, H.L. & A.R. Templeton 1984. Genetic revolutions in relation to speciation phenomena: the founding of new populations. *Ann. Rev. Ecol. Syst.* 15: 97-131.

Chang, D.D. & D.A. Clayton 1987. A mammalian mitochondrial RNA processing activity contains nucleus-encoded RNA. *Science* 235: 1178-1184.

Crosby, A.W. 1986. *Ecological Imperialism, The Biological Expansion of Europe*, pp. 900-1900. Cambridge: Cambridge University Press.

Endler, J.A. 1977. *Geographic Variation, Speciation, and Clines*. Princeton, N.J.: Princeton University Press.

Excoffier L. & A. Langaney 1989. Origin and differentiation of human mitochondrial DNA. *Am. J. Hum. Genet.* 44: 73-85.

Felsenstein, J. 1988. Phylogenies from molecular sequences: inference and reliability. *Ann. Rev. Genet.* 22: 521-565.

Ferris, S.D., R.D. Sage, C.-H. Huang, J.T. Nielsen, U. Ritte & A.C. Wilson 1983. Flow of mitochondrial DNA across a species boundary. *Proc. Natl. Acad. Sci. USA* 80: 2290-2294.

Gargett, R.H. 1989. The evidence for Neandertal burial. *Curr. Anthropol.* 30, 2: 157-189.

Gyllensten, U. & A.C. Wilson 1987. Interspecific mitochondrial DNA transfer and the colonization of Scandinavia by mice. *Genet. Res. Camb.* 49: 25-29.

Hall, G.H. & K. Muralidharan 1989. Evidence from mitochondrial DNA that African honey bees spread as continuous maternal lineages. *Nature* 339: 211-213.

Harrison, R.G. 1989. Animal mitochondrial DNA as a genetic marker in population and evolutionary biology. *TREE* 4, 1: 6-11.

Hewitt, G.M. 1988. Hybrid zones: natural laboratories for evolutionary studies. *TREE* 3, 7: 158-167.

Hewitt, G.M. 1989. The subdivision of species by hybrid zones. In D. Otte & J.A. Endler (eds), *Speciation and its Consequences.* Sunderland, Ma.: Sinauer Associates Inc.

Kocher, T.D. & A.C. Wilson 1989. Mitochondrial DNA and Human Evolution. *Presentation at ICN/UCLA colloquium 'Molecular Evolution' 27 Feb-6 March, Lake Tahoe, Ca.*

Laipis, P.J., M.J. van de Walle & W.W. Hauswirth 1988. Unequal partitioning of bovine mitochondrial genotypes among siblings. *Proc. Nat. Acad. Sci. USA* 85: 8107-8110.

Lamb, T. & J.C. Avise 1986. Directional introgression of mitochondrial DNA in a hybrid population of tree frogs: The influence of mating behavior. *Proc. Natl. Acad. Sci. USA* 83: 2526-2530.

Mayr, E. 1973. *Animal Species and Evolution.* Cambridge, Ma.: Belknap Press of Harvard University Press.

Mellars, P. 1989. Major issues in the emergence of modern humans. *Curr. Anthropol.* 30, 3: 349-385.

Monnat, R.J. & L.A.Loeb 1985. Nucleotide sequence preservation of human mitochondrial DNA. *Proc. Natl. Acad. Sci. USA* 82: 2895-2899.

Moritz, C., T.E. Dowling & W.M. Brown 1987. Evolution of animal mitochondrial DNA: relevance for population biology and systematics. *Annu. Rev. Ecol. Syst.* 18: 269-292.

Needham, G.R., R.E. Page Jr, M. Delfinado-Baker & C. Bowman (eds) 1988. *Africanized honey bees and bee mites.* Chincester: Ellis Horwood.

Nei, M. 1987. *Molecular Evolutionary Genetics.* New York, N.Y.: Columbia University Press.

Nei, M. & D. Grauer 1984. Extent of protein polymorphism and neutral mutation theory. *Evol. Biol.* 17: 73-118.

Simpson, L. & J. Shaw 1989. RNA editing and the mitochondrial cryptogenes of kinetoplastid protozoa. *Cell* 57: 355-366.

Smith, D.R., O.R. Taylor & W.M. Brown 1989. Neotropical Africanized honey bees have African mitochondrial DNA. *Nature* 339: 213-215.

Soule, M.E. (ed.) 1987. *Viable Populations for Conservation.* Cambridge: Cambridge University Press.

Stoneking, M. & R.L. Cann 1989. African origin of human mitochondrial DNA. In P.A. Mellars & C.B. Stringer (eds), *The Human Revolution.* Edinburgh: Edinburgh University Press.

Stoneking, M., K. Bhatia & A.C. Wilson 1986. Rate of sequence divergence estimated from restriction maps of mitochondrial DNAs from Papua New Guinea. *Cold Spring Harbor Symposia on Quantitative Biology* 51: 433-439.

Stringer, C.B. & P. Andrews 1988. Genetic and fossil evidence for the origin of modern humans. *Science* 239: 1263-1268.

Takahata, N. & M. Slatkin 1984. Mitochondrial gene flow. *Proc. Natl. Acad. Sci. USA* 81: 1764-1767.

Templeton, A.R. 1989. The meaning of species and speciation: a genetic perspective. In D. Otte & J.A. Endler (eds), *Speciation and its Consequences.* Sunderland, Ma.: Sinauer Associates Inc.

Trinkaus, E. 1983. *The Shanidar Neandertals.* New York, N.Y.: Academic Press.

Vigilant, L. & A.C. Wilson 1989. Human mitochondrial DNA sequence polymorphism in Africa: the origin of Pygmies. *J. Cell. Biochem. Sup.* 13C, 1989: 126.

Wallace, D.C., X. Zheng, M.T. Lott, J.M. Shoffner, J.A. Hodge, R.I. Kelly, C.M. Epstein & L.C. Hopkins 1988. Familial mitochondrial encephalomyopathy (MERRF): genetic, pathophysiological, and biochemical characterization of a mitochondrial DNA disease. *Cell* 55: 601-610.

Whittam, T.S., A.G. Clark, M. Stoneking, R.L. Cann & A.C. Wilson 1986. Allelic variation in human mitochondrial genes based on patterns of restriction site polymorphism. *Proc. Natl. Acad. Sci. USA* 83: 9611-9615.

Wilson, A.C., R.L. Cann, S.M. Carr, M. George, U. Gyllensten, K.M. Helm-Bychowski, R.G. Higuchi, S.R. Palumbi, E.M. Prager, R.D. Sage & M. Stoneking 1985. Mitochondrial DNA and two perspectives on evolutionary genetics *Biol. J. Linn. Soc.* 26: 375-400.

Wolpoff, M.H. 1989. Multiregional evolution: the fossil alternative to Eden. In P.A. Mellars & C.B. Stringer (eds), *The Human Revolution*. Edinburgh: Edinburgh University Press.

African Pygmies have the more ancestral gene pool when studied for Y-chromosome DNA haplotypes

Gérard Lucotte
Laboratoire d'Anthropologie Physique, Collège de France, Paris, France

1 INTRODUCTION

The human Y-chromosome presents a very great interest for research into human population genetics and concerning the origin of our own species. When mutations appear on the Y-chromosome, they are transmitted through males only, and many of such variants have become fixed in this way, recording the evolutionary story through paternal inheritance. This provides interesting comparisons with results of analyses using maternally-inherited mitochondrial DNA (mtDNA). Examining Y-chromosome DNA differences in aboriginal human populations from different geographical regions thus has great potential for reconstructing recent events in human evolution.

Several cosmidic Y-specific sequences have been used for the detection of restriction fragments length polymorphisms (RFLPs) in a panel of unrelated DNA (Ngo & Lucotte 1986). Clone p49 constitutes a highly polymorphic probe, which detects several RFLPs on Y-chromosomes after digestion of genomic DNA with the restriction enzyme TaqI (Lucotte & Ngo 1985): this probe detects on autoradiograms, after the Southern blotting molecular hybridization procedure, about 16 Y-specific Taq bands, corresponding to a low-copy number sequence family of genes (Ngo et al. 1986a), located in the sub-region Yq11-2 (Quack et al. 1988). Five of these bands, named A, C, D, F and I, each representing a single DNA fragment, can be either present, absent, or variable in length; other rarer variant bands are also described (e.g. Guérin et al. 1988a).

Familial segregation studies in paternal lineages have shown that variations of these five main fragment variants are inherited in a mendelian (but holandric) fashion and are strictly Y-linked (Guérin et al. 1988b). A survey of numerous male individuals indicated that the five variable Taq fragments detected by the p49 probes (sub-clones 49f and 49a) can be considered as five independant allelic series.

These several RFLPs define multiple potential haplotypes on the human Y-chromosome. A preliminary study (Ngo et al. 1986a) had shown that the common haplotypes are present in all three of the main racial groups (Caucasoïd, Negroïd and Asiatic). A detailed analysis of the RFLPs was performed by a survey of about 120 Caucasoïd unrelated males (Europeans, North-Africans and Southwest-Asians); and among the individuals analysed, 16 different combinations of the five RFLPs (sixteen haplotypes numbered I to XVI) have been found (Ngo et al. 1986a). Since all the elements of a combination are syntenics and separated on the chromosome only by several kilobases, it can be considered that each combination defines a true haplotype (in the genetic sense) on the human Y-chromosome.

Probe 49 represents the most powerful tool so far developed for the analysis of patriarchal relationships in human populations; and because of the numerous mutation events the probe can detect, it is also of more special interest in studying polymorphisms in isolates (small ethnic and residual groups). For example, the Taq polymorphism of probe 49 was studied in some Papuas from New-Guinea (the Baruya tribe, living in the Wonenara valley). Results indicate (Breuil et al. 1987) that all the individuals screened are identical at the variable loci A, C, F and I, but fixed at a specific allele (Db) for the D locus; the deduced haplotype, number XVII (A2, Co, Db, F1, I1), is 'Baruya-specific'.

2 GENEALOGY OF THE HUMAN Y-CHROMOSOME

TaqI polymorphism can be considered as one aspect of CpG suppression in mammals, in which the high level of DNA methylation accounts for the CpG deficiency. This implies that, in parallel to the loss of CpGs, there should be a corresponding loss of TaqI sites (and therefore an increase in the size of TaqI restriction fragments) during evolution. As a corollary, restoration of TaqI sites must occur rarely.

Hence, in Taq RFLPs, the smallest allele should represent the original form, and this observation illustrates the possibility of establishing evolutionary relationships among the various haplotypes (all the observed haplotypes could then be derived from another by single mutations). We believe that, in most cases concerning our allelic series, the smaller allelic forms of the Taq fragments actually represent the original ones. So, simulations have been used to construct a plausible genealogy of the 17 initially described Y haplotypes, using the parsimony principle, and on the basic of a limited set of basic hypotheses concerning polymorphic allele evolution in human ancestry (Hazout & Lucotte 1986). The possibility to establish such a genealogy is based on the observations of some incompatibility rules between alleles (Hazout & Lucotte 1989), permitting the construction of evolutionary relationships among the various haplotypes observed. Four solution networks are finally retained (Hazout & Lucotte 1986), with the ancestral haplotype deduced as being probably haplotype XIII (A3, Co, D1, F1, I1).

3 CHIMPANZEE RETAINS PRIMITIVE ALLELES

49f is Anthropoid-specific (Ngo et al. 1986b), as established by hybridization studies of DNAs coming from various primate species restricted with BamHI (with BamHI enzyme, all the main Y-specific sequences are located in only one autoradiographed band). In these experimental conditions, and autoradiographic signal is obtained only with chimpanzee and gorilla species. After TaqI restriction, most of the 49f fragments are conserved between chimpanzees and humans (Abbas et al. 1988). For polymorphic bands in humans, chimpanzees have A3, Co, F1 and I1. Chimpanzees are also polymorphic for various bands, and one individual in the sample studied is Ao, D1. Consequently, four of the primitive alleles postulated previously by the model are fixed in chimpanzees (and the chimp is also polymorphic for D1). All the allelic components of the predicted ancestral haplotype XIII are in fact already present in the chimpanzee species.

4 HAPLOTYPE XIII IS THE PREDOMINANT ONE IN AFRICAN PYGMIES

Pygmies represent the most typical African gene pool and, for some classical genetic markers, they are considered as the most direct descendants of 'proto-Africans' (Cavalli-Sforza 1986). In our initial study (Lucotte et al. 1989), we dealt with two indigenous African groups: Bantus and Pygmies. Blood samples were collected from these two populations. The non-Pygmy black Africans are unrelated male natives of Bangui (Central African Republic = CFA). The Pygmies, also unrelated, are natives of CFA (Aka tribe). Only 12 of the 16 haplotypes (I to III, VI to VIII and XI to XVI) already present in Caucasian populations (Ngo et al. 1986a, and new data that we obtained since from 63 unrelated Caucasians) are also found in the two African groups studied (Table 1).

Haplotype distributions are very similar in Bantus and Caucasians. Haplotype XV is the most frequent one, and haplotypes XII, XIV and XIII are also well represented. The frequencies of other haplotypes are low.

Haplotype distribution in Pygmies is quite unique, with the most frequent one being XIII (about 30% of all the haplotypes). Failure to identify in Pygmies several intermediate haplotypes that are found in Bantus may possibly reflect a founder effect that decreased the number of surviving haplotypes, but haplotypic diversity seems present in this population (at least 8 haplotypes are found, in the limited sample studied).

Two new haplotypes were observed in Pygmies only (Fig. 1). In haplotype XVIII, the A allele is A4, the C allele is absent (Co) and other variable alleles are D1, F1 and I1. In haplotype XIX Taq bands A, C, F and I are absent, and the D allele is D1.

Only one unique optimal solution, represented by the oriented networks in Figure 2, is obtained from Pygmies when all the haplotypes are considered together. This calculated solution between the different haplotypes is optimal (Hazout & Lucotte 1987) first because there are no intermediates (haplotypes not observed, but created by the algorithmic procedure used) in the corresponding network, and second because the number of operations to transform this network into such a tree is minimal (number of operations = 10). In these trees, haplotype XVIII is selected as the ancestral one.

Table 1. Numbers of individuals and percentages of Y-haplotypes for each African population.

Haplotype numbers	Haplotypes	No (%) of Bantus N = 56	No (%) of Pygmies N = 37
I	Ao, Co, Do, F1, I1	1 (1.7)	0
II	Ao, Co, D1, F1, I1	0	3 (8.1)
III	A1, Co, Do, F1, Io	2 (3.6)	0
VI	A2, Co, D1, Fo, I1	2 (3.6)	0
VII	A2, Co, D1, F1, Io	2 (3.6)	0
VIII	A2, Co, D1, F1, I1	2 (3.6)	0
XI	A3, Co, Do, F1, I1	4 (7.1)	0
XII	A3, Co, D1, F1, Io	10 (17.9)	4 (10.8)
XIII	A3, Co, D1, F1, I1	8 (14.3)	11 (29.7)
XIV	A3, C1, D1, F1, I1	8 (14.3)	7 (18.9)
XV	A3, C1, D2, F1, I1	15 (26.8)	6 (16.2)
XVI	A4, Co, D1, F1, Io	2 (3.6)	1 (2.7)
XVIII	A4, Co, D1, F1, I1	0	2 (5.4)
XIX	Ao, Co, D1, Fo, Io	0	3 (8.1)

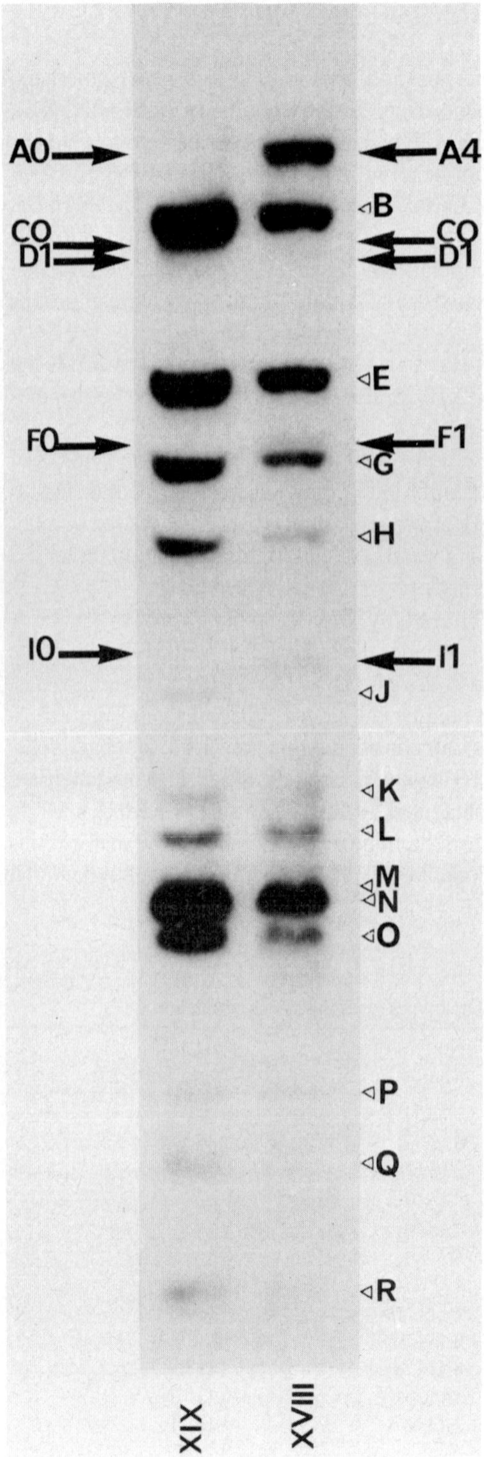

Figure 1. Autoradiograms of the two new haplotypes detected in the Pygmies. Haplotype XVIII is A4, Co, D1, F1, I1 and haplotype XIX is Ao, Co, D1, Fo, Io.

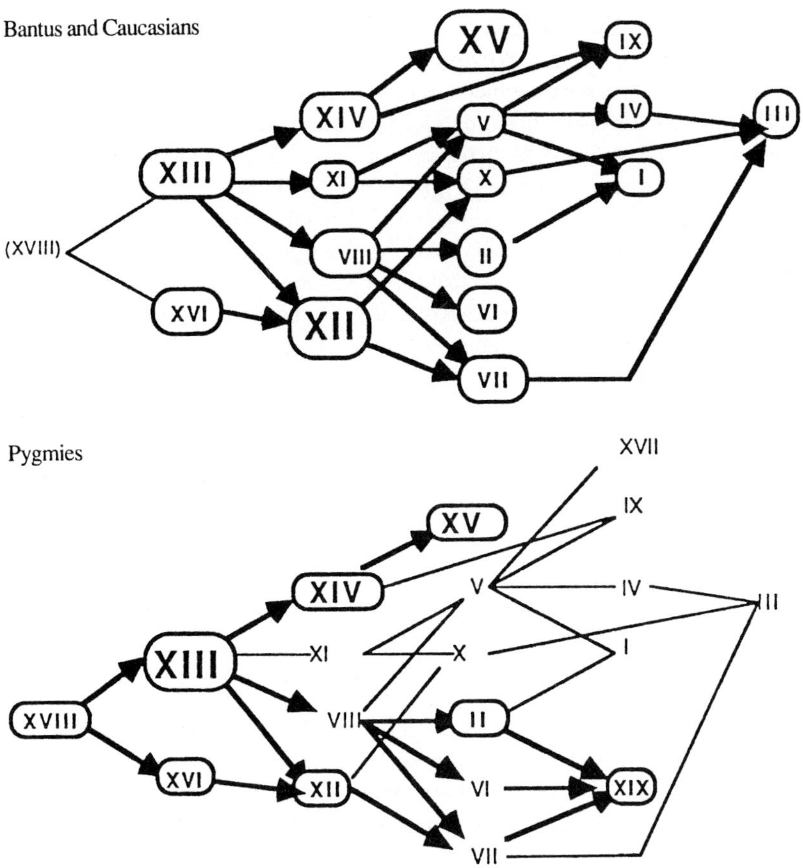

Figure 2. Optimal filiations between haplotypes obtained in the two African populations studied. Only real haplotypes are found in this represented optimal solution. Caucasians and Bantus are lumped together because of similarities in haplotype distribution. In the two African groups studied, present haplotypes are circled according to their representations in percentage. Vertical columns indicate, when possible, the different successive mutational steps (from the left to the right part of the figure) concerning filiations postulated between the different haplotypes.

5 DISCUSSION

Molecular genetics constitutes an important source of objective and quantitative information about the evolutionary history of our own species. Recently published results on nuclear (Wainscoat et al. 1986) and mitochondrial (Cann et al. 1987) RFLPs have led molecular geneticists to adopt a new hypothesis about human phylogeny (Bräuer 1984) – i.e. that all modern human populations are derived from an African gene pool. Mitochondrial data rely on a strict female-to-female genetic transmission, and the type of work presented here has now commenced (with the use of a highly polymorphic Y-chromosome specific probe) on the construction of our paternal ancestry.

The present study shows that, for African Pygmies (but not for Bantus), haplotype XIII (the ancestral one previously postulated, and often found in chimpanzee) is the most common haplotype. In the Pygmy population studied, two new haplotypes (XVIII and XIX) are also present. Haplotype XVIII is similar to haplotype XIII (A3 → A4) and is illustrated here as the ancestral one. As noted above it has been argued recently that, at the molecular genetic level, the evolution of modern humans took place in Africa. Our data concerning the use of a polymorphic Y-specific probe system is consistent with such a scheme for our paternal inheritance, but this old contribution can be seen mainly in African Pygmies, not in Bantus.

Only by accumulating data over a large number of independant systems will an accurate reconstruction of phylogenies be possible. Samples of two aboriginal Pygmy populations have so far been studied in detail: the Bi-aka from the village of Bagandu (in the southwest corner of CFA) and the Mbuti from Ituri forest (in northeastern Zaïre). The comparison between Pygmies and Caucasians shows that the greatest differences between Africans and Caucasians are for loci like genes coding for alcohol dehydrogenase 3, antithrombin III, and the anonymous locus D4S14 (Cavalli-Sforza et al. 1986). In a more recent study (Bowcock et al. 1987), 47 DNA markers from 30 genes, or chromosomal regions, were investigated in five populations (Bi-aka and Mbuti Pygmies, Melanesians, Chinese and Caucasoïds). The calculated distance between all populations considered in pairs and averaged over all loci also favors a primary split between Eurasia and Africa, but this conclusion is not statistically significant. Condensing the 47 markers into 30 genes (where 10 were treated as haplotypes), it was found that the haplotypes always give higher distances than the separate markers, although similar conclusions can be drawn.

The conclusions are that the data accumulated so far favor an early split between African and Eurasian populations, and that the African Pygmy populations are the most primitive. The differences observed, however, suggest that more probes and more populations than those studied to date are necessary to reach a statistically significant solution. The combined use of autosomal nuclear DNA, of mtDNA, and of Y-specific DNA markers has a great potential for the genetic analysis of human populations and reconstruction of human phylogeny.

REFERENCES

Abbas, N., P. Guérin, P. Bessières, J. Ruffié & G. Lucotte 1988. Séquences Y-chromosome humain spécifiques présentes chez les singes Anthropoïdes. *Biochem. Syst. Ecol.*, 16: 105-109.

Bowcock, A.M., C. Bucci, J.M. Hebert, J.R. Kidd, K.K. Kidd, J.S. Friedlander & L.L. Cavalli-Sforza 1987. Study of 47 DNA markers in five populations from four continents. *Gene Geogr.* 1: 47-64.

Bräuer, G. 1984. A craniological approach of the origin of anatomically modern *Homo sapiens* in Africa, and implications for the appearance of modern Europeans. In F.H. Smith & F. Spencer (eds), *The Origins of Modern Humans*, New York: Alan R. Liss. pp. 327-410.

Breuil, S., L. Hallé, J. Ruffié & G. Lucotte 1987. Polymorphisme d'une sonde Y-spécifique nommée p49 chez les Papous Baruyas de Nouvelle-Guinée. *Ann. Génét.*, 30: 209-212.

Cann, R.L., M. Stoneking & A.C. Wilson 1987. Mitochondrial DNA and human evolution. *Nature*, 325: 31-36.

Cavalli-Sforza, L.L. 1986. *African Pygmies*, Orlando: Academic Press.

Cavalli-Sforza, L.L., J.R. Kidd, J.K. Kidd, C. Bucci, A.M. Bowcock, B.S. Hewlett & J.S. Friedlander 1986. DNA markers and genetic variation in the human species. In *Molecular biology of Homo sapiens*, Cold Spring Harbor Laboratory, pp. 411-417.

Guérin, P., S. Berriche, S. Intrator, P. Rouger, D. Salmon, C. Salmon & G. Lucotte 1988b. Les techniques de polymorphisme de l'ADN de sondes du chromosome Y appliquées au problème des filiations père-fils dans les analyses de paternité. *Rev. Franç. Transf. Immuno-Hématol.*, 31: 513-520.

Guérin, P., P. Rouger & G. Lucotte 1988a. A new TaqI 80 variant detected with the p49 probe on the human Y-chromosome. *Nucl. Acids Res.*, 16: 7 759.

Hazout, S. & G. Lucotte 1986. Vers une généalogie du chromosome Y. *Ann. Génét.*, 29: 246-252.

Hazout, S. & G. Lucotte 1989. Linkage disequilibrium between haplotypes and allelic incompatibilities detected with p49f, a probe located on the Y-chromosome. *Int. J. Anthropol.* (in press).

Lucotte, G., P. Guérin, L. Hallé, F. Loirat & S. Hazout 1989. Y-chromosome DNA polymorphisms in two African Populations. *Am. J. Hum. Genet.*, 45: 16-20.

Ngo, K.Y. & G. Lucotte 1986. Stratégies de recherche des polymorphismes de restriction des séquences du chromosome Y. *Ann. Génét.*, 29: 88-92.

Ngo, K.Y., J. Ruffié, & G. Lucotte 1986b. Comparaison des fragments de restriction spécifiques du chromosome Y chez l'homme et le chimpanzé, ainsi que chez d'autres espèces de Primates. *Biochem. Syst. Ecol.*, 14: 141-148.

Ngo, K.Y., G. Vergnaud, C. Johnsson, G. Lucotte & J. Weissembach 1986a. A DNA probe detecting multiple haplotypes on the human Y-chromosome. *Am. J. Hum. Genet.*, 38: 407-418.

Quack, B., P. Guérin, J. Ruffié, & G. Lucotte 1988. Mapping of probe p49f to the proximal part of human Y-chromosome long arm. *Cytogenet. Cell. Genet.*, 47: 232.

Wainscoat, J.S., A.V.S. Hill, A.L. Boyce, J. Flint, M. Hernandez, S. Thein, J.M. Old, J.R. Lynch, A.G. Falusi, D.J. Weatherall & J.B. Clegg 1986. Evolutionary relationships of human population from an analysis of nuclear DNA polymorphism. *Nature*, 319: 491-493.

Africa's place in the evolution of *Homo sapiens*

Günter Bräuer
Institut für Humanbiologie, Universität Hamburg, Hamburg, Germany

1 INTRODUCTION

In a recent review article on the origin of modern humans, Stringer & Andrews (1988a) made an attempt to test whether the genetic and fossil evidence is in greater agreement with a model of 'Multiregional Evolution' or with one of 'Recent African Evolution'. Because it was more accessible to testing, the extreme cladistic position, which limits the species *Homo sapiens* to modern humans, was used to test the African Evolution model. Stringer & Andrews (1988a: 1263) justified this procedure by arguing that a test of these extreme models also makes it more likely to recognize the predominant mode of *Homo sapiens* evolution than would be possible with models which lie between the extremes.

Although the extreme Recent African Evolution model is thus a test model which, in this form, is adovated by only a few workers, it immediately became a favorite target of the proponents of multiregional evolution (Wolpoff et al. 1988, this volume, Frayer, this volume). Smith (this volume) has termed this the 'Speciation/Replacement model' and holds along with others that 'it is only necessary to show that there is some degree of regional continuity across the archaic/modern boundary' outside of Africa to refute this model. Wolpoff (this volume) is convinced that such evidence is present and thus concludes that there is considerable support for the alternative Multiregional Evolution. This conclusion is, however, not the whole story, as there are other less extreme replacement alternatives which also allow for a certain degree of continuity. Moreover, multiregional evolution implies more than merely some degree of continuity, namely 'worldwide changes in selection acting on already differentiated local populations of archaic *Homo sapiens*' (Wolpoff 1980:298-299). According to Wolpoff (1989:88), gene flow here would not usually have had a 'stimulating' effect on human evolution and would not likely swamp the effects of selection. Smith's (1982:685) interpretation of the south-central European fossil record as evidence of an indigenous archaic-to-modern *H. sapiens* transition is also an expression of such a regional evolution, and implies more than just a certain degree of continuity. At the time, Smith saw no 'reason to believe that gene flow into this region from the Near East or anywhere else was any more responsible for the archaic-to-modern *H. sapiens* transition in south-central Europe than gene flow from south-central Europe was for the same transition in the Near East or elsewhere'. Concerning western Europe, he stated (1982:686) that 'the continuum is less distinct, but new discoveries appear to be closing the gap'. Since the time of this remark, however, almost all workers have abandoned hope in the correctness of a regional transition in western Europe. In other regions as well, the fossil record today shows increasingly clearly that the conception of numerous indigenous evolutionary lines leading to modern humans is not well supported by the present evidence (Bräuer 1984b, 1989a, Trinkaus 1986, Stringer & Andrews 1988a, Habgood, this volume).

Because of the lack of convincing transitional samples, perhaps the only possibility remaining to the proponents of this model appears to involve the search for evidence of 'at least some degree of continuity' in individual anatomical details (see Frayer, this volume). And thus, it is understandable that the proponents of multiregional evolution consider Stringer's and Andrews' extreme test model as the most suitable alternative, for this model implies 'total replacement' and offers a chance – if indeed one exists at all – of calling the overall validity of this extreme model into question. Nevertheless, it is incorrect for the proponents of the Multiregional Evolution model to believe that demonstrating specific indicators of a certain degree of continuity will provide support for their model while simultaneously helping to refute the replacement view. A less extreme replacement model, such as one which would allow for variable degrees of gene flow between modern and archaic populations, could provide a better explanation for a certain degree of continuity than the Multiregional Evolution model.

2 AFRICAN REPLACEMENT MODELS

2.1 *'Test model' (after Stringer & Andrews 1988a)*

As the authors themselves noted, the extreme total replacement test model used by Stringer & Andrews (1988a:1267) was able to explain most but not all of the present evidence. For example, they consider the Australian fossil evidence as posing a serious problem: 'some of the fossil samples display supposed features of regional continuity, while others do not'. Concerning the Neanderthal/modern transition, they see little or no continuity of genuine regional features. On the basis of certain Neanderthal-like features among the early moderns, Stringer (1988:273), however, considers it possible that there was some gene flow between the two forms in Europe (see also Stringer, this volume). For this reason, it seems to me that Wolpoff's conclusion (this volume) that the restriction of the name *Homo sapiens* to modern humans (as in the test model) necessarily means that modern humans represent a new biological species that did (or could) not 'hybridize' with archaic humans is extreme. Delson (1988:206) has also remarked that the assumption of additional species between *Homo erectus* and modern humans 'clouds the apparent continuity of all varieties of *Homo* in the later Pleistocene and may lead to recognition of greater taxonomic distinction among living humans than is warranted'. Both these and other remarks (e.g. Stringer 1987, Stringer & Andrews 1988b:774) clearly indicate that the test model was not intended as a globally valid model of evolution, but rather as a model which should merely aid in recognizing the predominant mode of evolution. For this reason, there is no justification for using evidence of some gene flow or continuity across the archaic/modern boundary outside of Africa to refute the view that replacement was the major factor in the origin of modern humans. Equally unjustified, of course, is the further conclusion that such evidence provides support for the model of Multiregional Evolution.

2.2 *'(African) Hybridization and Replacement Model' (after Bräuer 1984a)*

As the following sections will show, a certain type of replacement theory currently provides a reasonable interpretation for the appearance of modern humans in the fossil record of Europe and Asia, since unequivocally transitional fossils or samples have not yet been discovered in these regions (Bräuer 1984a, b, 1989a). This notwithstanding, evidence from Europe, China and Australasia does suggest that certain features of regional archaic precursors are present in some of the early moderns of these areas. In some regions (e.g. Australia), more archaic genes appear to have been assimilated by the modern gene pool than in other regions (e.g. Europe).

Concerning the Neanderthals and the early modern humans, I have always supported a replacement view which assumes gene flow or 'hybridization' (Bräuer 1980, 1981). The 'Afro-European *sapiens* hypothesis' (Bräuer 1982, 1984a, b) clearly represents such a 'hybridization and replacement' model (Fig. 1): 'a.m. humans seem to have 'spread' further, via the Near East, into Europe, increasingly absorbing and replacing the Neanderthals that were living there. This period of replacement probably lasted for several thousand years, and it can also be assumed that various degrees of hybridization between Neanderthal and modern populations occurred during this time. This would at least partially explain the numerous Upper Paleolithic finds that appear to feature Neanderthalid reminiscences' (Bräuer 1984b:395). Howells (1988:234) also assumes that 'gene exchange with newly arriving anatomical moderns should have been substantial'.

With respect to the Far East, I still think that the fossil evidence can also be seen in quite good agreement with the model known as the '(African) Hybridization and Replacement model' (Bräuer 1984a). 'This process, ... was certainly very complex, multicausal, different in various regions, and hardly rapid or complete. It does not exclude a certain degree of regional continuity in the fossil record – although this may often be interpreted as the result of gene flow or hybridization as well' (Bräuer 1989a:125-126). To make it quite clear, this 'African Hybridization and Replacement model' proposes that replacement was the predominant factor or mode of origin of modern humans in Europe and Asia, but does not assume that modern humans represent a biological species distinct from the archaic *sapiens* forms.

Wolpoff (1989:77) has misinterpreted the current evidence of the mtDNA studies by saying: '... I find it curious to observe both Cann and Stoneking, in various publications, each quote Bräuer's palaeoanthropological interpretations in support of the 'Garden of Eden' Hypothesis when in fact if this interpretation was correct it would actually invalidate the hypothesis. Admixture between invading and indigenous populations would invalidate the interpretation of the genetic evidence ...' Yet it is by no means a compelling conclusion of the genetic studies that the radiating modern humans necessarily involved a new biospecies. Wilson et al. (1987: 162-163), for example, have posed the question: 'Supposing that they (modern humans) came out of Africa..., we can ask what the expected result of mixing with the anatomically archaic population already resident in Eurasia might be'. Yet since no extremely divergent mtDNA lineages have been found among the Asians which have been studied to date, Wilson et al. (1987:167) have cautiously concluded that it is unlikely 'that Asian *Homo erectus* contributed much to the gene pool of anatomically modern *Homo sapiens*'. A future finding of only a very few more greatly deviant mtDNAs could change this interpretation in favor of more gene flow. The results of genetic studies, in other words, do not by any means indicate that only an extreme 'Speciation/Replacement' model appears possible. Cann (pers. comm., 1988, this volume) can hardly imagine that there was a reproductive isolation between the archaic and the radiating modern *sapiens* populations. To explain the current genetic data she favors the view that there must have been a lot of extinction. Cavalli-Sforza (1989:410), too, does not see any reason for a 'complete replacement or elimination of the local pre-modern groups; they may well have been absorbed genetically, if they were cross fertile with *H. sapiens sapiens*, as is possible, or even likely'.

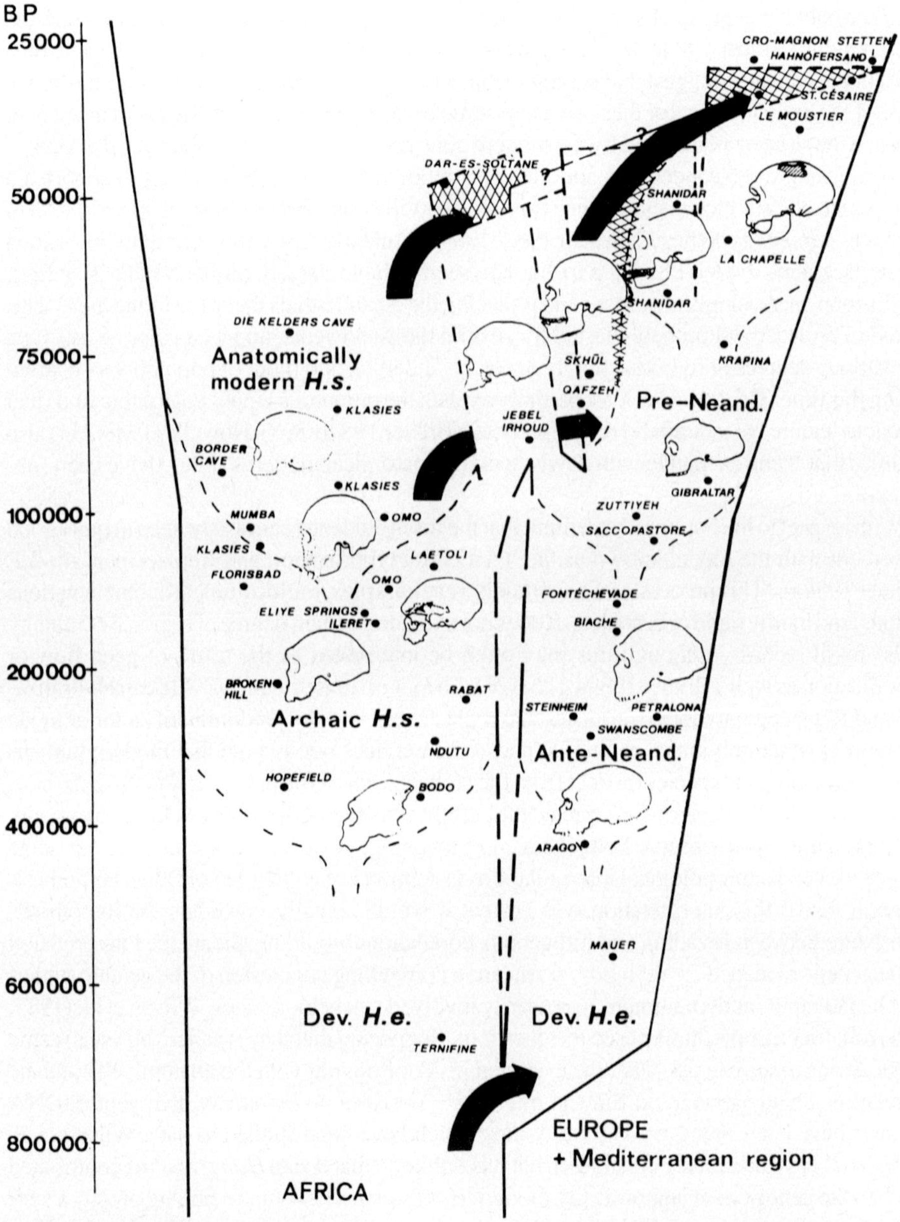

Figure 1. A revised version of the 'Afro-European *sapiens* hypothesis', including especially recent evidence from Northern Africa and South-West Asia.

The 'African Hybridization and Replacement model' is not an extreme test model intended to clarify the predominant mode of *sapiens* evolution, but rather a model which intends to explain the fossil evidence within Africa and outside of Africa. For this reason, it is in comparison to such a model that the 'Multiregional Evolution model' should be shown to offer a more adequate interpretation.

3 REGIONAL EVOLUTION IN AFRICA

Although Wolpoff (1989:76) continues to stress that 'the morphological evidence for continuity between archaic and more modern populations is poorest' in sub-Saharan Africa as compared to other parts of the world, the evolutionary sequence in Africa is actually the best documented and, in the mean time, the least controversial for both the proponents of a replacement model and of a multiregional evolution view. Wolpoff's evaluation of the African evidence thus clouds the facts just as much as his earlier assessment that the St. Césaire Neanderthal provided important evidence for regional continuity to modern humans in western Europe (Wolpoff 1981:823).

Already at the beginning of the 1980s, a new analysis of the African hominid finds from the Middle and Upper Pleistocene resulted in a quite conclusive picture of sub-Saharan evolution from early archaic to the clearly more modern late archaic *Homo sapiens* and then to anatomically modern humans, whereby the transition to the latter was most likely completed by more than 100 000 years ago (Bräuer 1982, 1984a, b). While little doubt was raised about this proposed morphological continuum of *sapiens* evolution, its conclusiveness was largely dependent upon the quite vague dating of various hominids. New hominid finds and numerous new dates during the last decade have, however, continued to provide support for the course of evolution that I proposed, with the result that the presently available evolutionary framework (Fig. 2) rests upon a considerable number of finds as well as quite defendable dates.

Thus, there are no longer any serious doubts that the main representatives of early archaic *Homo sapiens* (Bodo, Eyasi, Ndutu, Broken Hill 1, and Hopefield) are between 200 000 and 400 000 years old (Clark 1989). Earlier dating of the Eyasi hominids to a mere 30 000 years (Protsch 1981) are now no longer tenable. New excavations at Lake Eyasi and new absolute dates instead point to a date around 200 000 years or even older (Mehlman 1987).

There is also important new evidence concerning late archaic *Homo sapiens*. Clarke's re-excavations at Florisbad have demonstrated that this hominid, which stands close to the threshold to modern humans, cannot be dated to 39 000 years (Protsch 1974), but rather to between 130 000 and 200 000 years (Clarke 1985). Recently, the important specimen Laetoli H18, which also stands almost at the threshold to modern humans (Rightmire 1986, Bräuer 1989a), has been dated by the Th-230 method to 129 000 ± 4000 years, by the Pa-231 method to 108 000 ± 30 000 years BP (Hay 1987), and by the isoleucine epimerization method to 100 000 to 200 000 years BP (Bada 1987). All of these new dates thus confirm that a late archaic grade of *Homo sapiens* existed in sub-Saharan Africa between ca. 200 000 and 100 000 years ago (Fig. 2), and that certain of its exponents may already be classified as almost modern.

Two further hominids may also be added to this grade: ES-11693 from Eliye Springs/West Turkana and ER-3884 from Ileret/East Turkana. Although the heavily fossilized and almost completely preserved cranium ES-11693 has not yet been precisely dated, it exhibits a mosaic of archaic and modern features which would most likely belong to the spectrum of late archaic *Homo sapiens* (Bräuer & Leakey 1986a, b, Bräuer 1989b). Stratigraphic provenience indicates that an age of at least 100 000 years may probably be assumed for the similarly important cranial remains ER-3884 (Feibel et al. 1989). The first analyses of this hominid show that the cranium is anatomically modern with respect to many features yet also possesses certain archaic features, especially in the supraorbital region (Bräuer et al., this volume).

Finally, the last years have also provided us with significant new knowledge about the early moderns. While Wolpoff (1989) still considers all of the dates for these early moderns uncertain, he has been unable to prevent their increasingly wide acceptance.

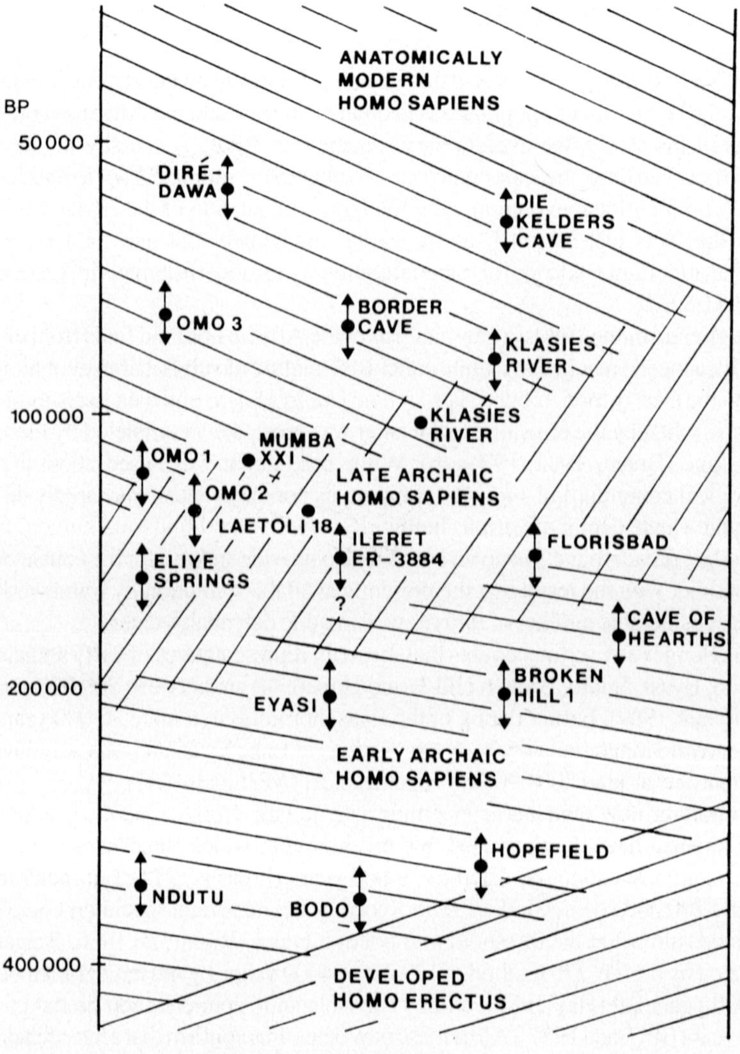

Figure 2. The evolutionary continuity in sub-Saharan Africa as demonstrated by the most important hominids.

In fact, the age between 80 000 and 125 000 years BP which Singer & Wymer (1982), Butzer (1982), and Shackleton (1982) have suggested for the MSA human remains from Klasies River has, with only slight modifications, been essentially confirmed by the new excavations and analyses of Deacon (1989, this volume). The revised stratigraphy indicates that only the 41815 mandible may be a mere 90 000 to 100 000 years of age, as compared to the previously assumed age of ca. 125 000 years. On the other hand, the two anatomically modern maxillary fragments (personal study) found near the base of the sequence (LBS member) during the new excavations may possess an age greater than 120 000 years. In general, Morris (this volume) has noted that today there is an overall consensus that the important MSA II remains (which also include the modern frontonasal fragment 16425 and the very gracile mandible fragment 16424) date to between 80 000 and 100 000 years. In his analysis, Morris (this volume) also concluded that while the Klasies remains are clearly anatomically modern, they still reveal no clear affinities to the Holocene populations of this region. In contrast to earlier assumptions, recent multivariate analyses by Van Vark (1986, Van Vark et al. 1989) have also revealed that Border Cave 1 clearly differs from recent African and non-African populations, although the cranium is basically anatomically modern (Bräuer 1984b). On the basis of all the available evidence, Butzer (pers. comm., 1987) assumes that this cranium is ca. 90 000 years old (see Bräuer 1989a). These new results for Klasies River and Border Cave indicate (Morris, this volume) that the early moderns of southern Africa did not yet unequivocally exhibit any morphology which would allow one to recognize the direct affinities to the present-day populations of sub-Saharan Africa.

This is even clearer with respect to the Omo 1 cranium (Bräuer 1984b, 1989a). It possesses a robust modern form which exhibits affinities to such North African samples as Afalou or Taforalt (Bräuer 1984b) as well as to the early Cro-Magnoids of Europe (Bräuer & Rimbach 1990). There is thus nothing about the morphology of this find which would exclude its belonging to the spectrum of ancestors of the early moderns outside of Africa as well. According to my replacement model, no specialized Africans are considered to be ancestors of the non-African moderns; it instead postulates that a less differentiated anatomically modern form was. For this reason, it would hardly be possible to provide concrete evidence of typically African features among early modern humans in Europe and Asia, as Wolpoff (pers. comm., 1988, this volume) has demanded for an 'Out of Africa' model. Wolpoff's (1989) claim that an age of 40 000 years for Omo 1 is just as sound as a higher age also contradicts the factual situation. A number of ^{14}C dates for the Omo Kibish Members III and IV (Butzer 1969) demonstrated that most if not all of Member III lies beyond the range of conventional Radiocarbon dating (Butzer, pers. comm., 1987). The find horizon of the Omo I skeleton in Member I lies much deeper stratigraphically (see Day & Stringer 1982) than the oldest ^{14}C date of >37 000 years BP in upper Member III. Thus, none of the available evidence contradicts the controversial U/Th date of 130 000 BP for Member I, and there are unpublished supporting dates from the upper Shungura deposits, at Omo (Butzer, written communication to Cambridge Meeting 1987).

Besides Klasies River, Border Cave and Omo 1, there are additional finds which provide evidence of the early presence of modern humans in sub-Saharan Africa. These especially include the earliest Upper Pleistocene teeth found at the Mumba Rock Shelter/Tanzania (Bräuer & Mehlman 1988) and the dental remains from Die Kelders Cave on the south coast, which have been dated to approximately 60 000 to 80 000 BP (Klein 1989). There are uncertainties in the datings of several other finds (Omo 3, Equus Cave, Singa, Kanjera, Diré-Dawa). If all of the evidence for *sapiens* evolution in sub-Saharan Africa is considered together, then the evolutionary sequence (Fig. 2) is well documented from early archaic to modern *Homo sapiens*, for we now have finds for all grades which may be considered to be fairly well dated and morphologically clearly classifiable.

Up to now, it has been rather difficult to describe the course of evolution in northern Africa and to relate this to the sequence in sub-Saharan Africa. This was mainly due to the few primarily mandibular remains from the Middle Pleistocene, the uncertainties in the evaluation of the possibly pathological Salé hominid, the lack of finds for the period between the Middle Pleistocene finds and the Jebel Irhoud hominids (generally dated to ca. 40 000 to 60 000 years), and the disputed influences of the Eurasian Neanderthals upon these Upper Pleistocene northern Africans (Bräuer 1984b). Most recently, however, some evidence has been accumulating which may possibly suggest that the developments in northern and sub-Saharan Africa may have been more closely linked than was previously assumed.

Recent ESR dates indicate that the deposits from which the Jebel Irhoud remains derived

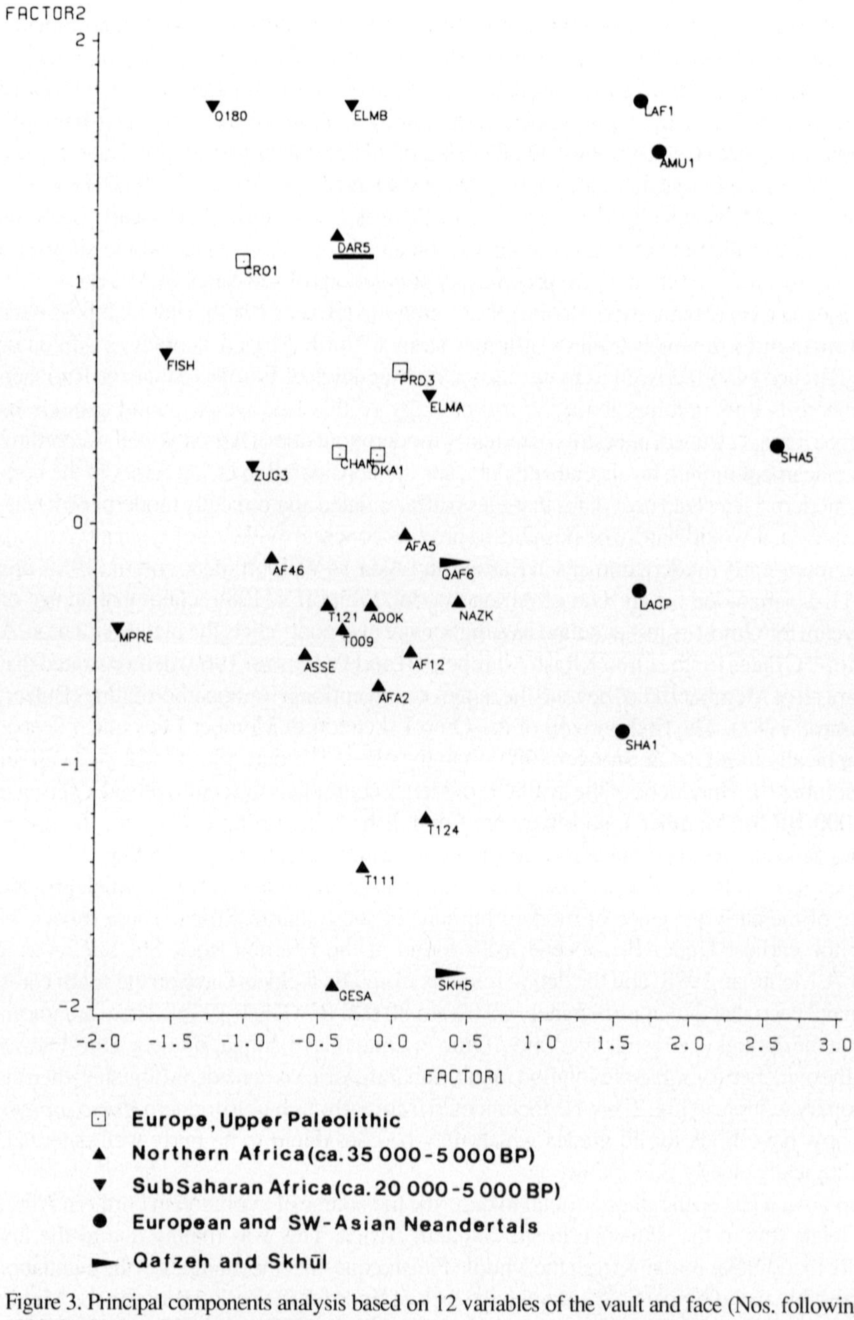

Figure 3. Principal components analysis based on 12 variables of the vault and face (Nos. following Bräuer, 1988: 26, 29, 29b, 32(5), 43a, 43b, 48, 51, 52, 54, 55, 77a). Factor 1 represents 45.62% and Factor 2 20.07% of the total variance (from Bräuer & Rimbach 1990). CRO1 = Cro-Magnon 1; PRD3 = Předmost 3; CHAN = Chancelade 1; OKA1 = Oberkassel 1; AFA5, AF46, AFA2, AF12 = Afalou-bou-Rhummel; T121, T009, T124, T111 = Taforalt; ADOK = Ain Dokhara; ASSE = Asselar 1; NAZK = Nazlet Khater; DARS = Dar-es-Soltane 5; GESA = Gebel Sahaba 18685; O180 = Oakhurst 180; ELMA, ELMB = Elmenteita; FISH = Fish Hoek; ZUG3 = Zuurberg 3; MPRE = Matjes River Pre-Bushman; LAF1 = La Ferrassie 1; LACP = La Chapelle; SHA1, SHA5 = Shanidar; AMU1 = Amud; QAF6 = Qafzeh 6; SKH5 = Skhūl 5.

possibly belong to oxygene isotope stage 5, and might thus be older than 75 000 years (Hublin 1989). Hublin (pers. comm., 1988) considers an age of around 90 000 years or even more most likely. If this early age can be further substantiated, two different interpretations appear possible to me (see Fig. 1).

1. Jebel Irhoud belongs to the same late archaic *Homo sapiens* grade as does LH 18, as there are some morphological similarities between these hominids.

2. The morphological spectrum of late archaic *Homo sapiens* in northern Africa resulted from both late archaic (sub-Saharan) African or even early modern African and Neanderthal elements. This would explain the presence of some Neanderthal influence in the North African material as well as the affinities to sub-Saharan *Homo sapiens*. In any case, it now appears quite possible that the evolution in northern Africa was not as separated from that in sub-Saharan Africa as has been assumed.

With regard to the early modern humans of northern Africa that are associated with the Aterian, there are also recent indications of a greater age than has been hitherto suggested. Thus, it now seems possible that the modern remains from Dar-es-Soltane II and Témara are at least 40 000 years old, although they may even be 50 000 to 70 000 years old (Clark 1989, Hublin 1989, Wendorf et al. 1990). A recent Principal components analysis of the Dar-es-Soltane 5 cranium (Fig. 3), based on 12 measurements of the vault and face, showed strong affinities to European Cro-Magnoids and thus appears to provide additional evidence that a less differentiated early modern African cranial form may also be found outside of Africa (for further details about these comparative analyses see Bräuer & Rimbach 1990). These recent results concerning northern Africa thus provide further evidence of evolutionary continuity in Africa and its special role in *sapiens* evolution.

4 REGIONAL EVOLUTION OR REPLACEMENT OUTSIDE AFRICA

As far as western Asia, Europe and eastern Asia are concerned, the proponents of multiregional evolution have suggested that various regional evolutionary lines leading to modern humans have been connected with one another through gene flow (e.g. Wolpoff 1980, 1985, 1989, Wolpoff et al. 1988). The supporters of the replacement view however, have been unable to find any unequivocal evidence in any of these areas for a regional evolution of archaic into modern *sapiens* forms (e.g. Bräuer 1984a, b, 1989a, Stringer & Andrews 1988a). And indeed, one should have fundamental doubts about palaeoanthropological findings when, in spite of the large number of available finds (in particular from Europe) both models appear to be equally supported by the fossil record. Yet this is quite obviously not the case. Stringer & Andrews (1988a) were even able to risk choosing a very extreme total replacement model as a test model. Their review of the fossil evidence nevertheless clearly demonstrated that most of the evidence supports replacement as the predominant mode of the origin of modern humans outside of Africa. Yet how is it possible that Stringer's and Andrews' conclusions from the fossil record could have been so vehemently disputed by Wolpoff et al. (1988)?

One main reason for this concerns the changed understanding of multiregional evolution which some of the proponents of this model have arrived at during the past years. For Smith (this volume) it is not even necessarily a claim 'that regional archaic human populations were invariably the major contributors to modern human gene pools in their respective geographic regions. At its most basic level, continuity is only an argument that archaic humans played some role in the origin of regional modern human populations...'. In short, this view thus implies that it is also possible to speak of regional evolution when it can be shown that there was only a slight amount of gene flow between the archaic and the subsequent modern humans through the common existence of certain morphological features. Wolpoff's (1981, 1985, 1989) frequent claims of clear evidence for regional evolution nearly everywhere in the Old World may thus in fact be at least partially based upon the presence of only a very few anatomical details. Fundamental differences in cranial form, such as those observed between the Neanderthals and early moderns, are thus ignored by major proponents of multiregional evolution, to whom they no longer appear to present any problem. Yet when regional evolution is understood in this manner, then this concept is used inadequately. For it does not then necessarily reflect regional evolutionary changes caused by local selection and/or drift and gene flow (Wolpoff 1989), but merely the common presence of certain features which could just as well be explained by gene flow between archaic and immigrant modern populations ('Hybridization and Replacement model'). Smith (this volume) recognizes this problem and is willing to accept that it is currently not possible to establish the relative importance of replacement and continuity in peripheral areas such as Europe.

The following considerations focus on the currently most discussed arguments which are held to support the idea of regional evolution outside of Africa. Concerning Southwest Asia, the new dates for the Qafzeh hominids to ca. 90 000 years (Valladas et al. 1988, Bar-Yosef, this volume) suggest that all of the claims for regional evolution between Neanderthals and modern humans ('we have a picture of the Near Eastern Neanderthals 'caught in the act' of evolving into modern *sapiens*', Wolpoff 1980:309) have become more than merely questionable. The morphological differences between the Near Eastern Neanderthals and the hominids from Qafzeh and Skhul are also well illustrated in Figure 3.

With regard to western Europe, Wolpoff (1981:823) has remarked that the late Neanderthal from St. Césaire 'specifically resembles the late Neanderthals from Vindija cave and thus provides important support for the observed trends within the Neanderthal sample'. Vandermeersch (1984:195), however, was unable to observe any transitional characters. If Wolpoff still adheres to regional evolution in western Europe ('Such changes could happen fairly quickly' 1981:823, but see Wolpoff et al. 1988:772), then he may be increasingly isolated in this position. Howells (1988:234) is 'not aware of any present writer who sustains a real local and special evolution of Neanderthals into Europeans, especially in the west'. Such completely modern specimens as Stetten 1 (south-west Germany) are just as old as the late Neanderthals of western Europe, while the latter do not exhibit any clear evolutionary trends to the modern Europeans (see also Fig. 3). There are very good reasons behind the broad acceptance of a replacement model for western Europe, which is now supported by proponents of multiregional evolution as well (Smith & Trinkaus 1991).

In the western portion of the Old World, only the fossil record of south-central Europe is still believed to demonstrate regional evolution across the Neanderthal/modern boundary. Yet here the facts are also far from demonstrating any regional evolution to modern humans. All of the continued claims for regional continuity are based largely upon two lines of evidence: 1. The demonstration of morphological trends from the Krapina to the Vindija Neanderthal sample in the direction of modern humans. 2. The identification of single features among the early moderns which are considered characteristic of the Neanderthals.

With respect to 1, Smith (this volume) considers the Vindija Neanderthals 'an excellent transitional sample' between earlier south-central European Neanderthals and early modern humans from the same region. Yet since sex, age, and individual variation in the sample (which consists only of very fragmentary pieces) cannot be reliably determined, even Frayer (this volume) has stated that it is difficult here to ascertain what is and what is not a transitional fossil.

In his attempts to demonstrate this transitional character, Smith (this volume) uses two measurements ('alveolar height' and 'nasal breadth') which unquestionably possess a great variability, especially with respect to age and sex; moreover, the Vindija sample consists of N = 2. Because of such extremely weak evidence, it is no surprise that many workers have greater doubts about a transitional pattern than about morphological evidence for continuity in sub-Saharan Africa. Even Smith (this volume) admits that it is possible that the causes for the assumed transitional pattern seen in the Vindija sample may also reflect the effects of gene flow from modern human populations further east, like the Qafzeh people, rather than a local, independent development of the modern human anatomical form. Although this is indeed a possibility, especially if the modern find from Crete is indeed 50 000 years old (Facchini & Giusberti, this volume), at the present I do not consider this explanation to be any more likely than the assumption that these are primarily sample problems.

Yet even when one or another feature is present among some of the specimens of the Vindija Neanderthals, as, e.g. when the supraorbital pattern appears to be more modern among some individuals than others, such individual finds can at best support assumptions concerning certain changes within the Yugoslavian Neanderthals; they are not convincing enough to provide evidence for an evolution to modern humans in this region (Bräuer 1984b). What is more, the pattern of the brow ridges of modern Europeans is not a regional characteristic for Europe, but may also be found outside Europe, e.g. in both northern Africa and the Near East. Stringer's & Andrews' (1988a:1266) claim of an absence of Neanderthal-modern transitional fossils in Europe thus appears to be largely justified (see Bräuer 1984a, 1989a).

The second line of evidence for continuity in south-central Europe is seen in several features, in particular occipital bunning, which may be frequently observed among the Neanderthals and in several of the early moderns (Smith & Trinkaus 1991). Although most of these features can also be found in archaic and modern *sapiens* specimens from other continents (e.g. some kind of bunning may be found in the Jinniushan and Liujiang specimens from China, pers observ.), some degree of gene flow between Neanderthals and moderns appears quite likely (Bräuer 1984b, 1989a). Yet in view of the large and generally recognized morphological gap between the Neanderthals and the early moderns (Bräuer 1989a) and the fact that the Vindija specimens are unquestionably Neanderthals (Smith & Trinkaus 1991), it would be wrong to consider this slight evidence for gene flow as a sign of regional continuity in the sense of any model of Multiregional Evolution. With regard to central Europe, Smith & Trinkaus (1991) assume that a substantial external influence was also involved in the emergence of modern Europeans, for the numerous changes from the Neanderthals to early moderns would be difficult to explain without recourse to any extra-European influence. While this interpretation differs considerably from previous ones and estimates the external influence to have been much greater, in my opinion the facts accord better with a model which lays even greater stress upon external influences, namely as some form of replacement like that which Smith & Trinkaus (1991) now also accept for western Europe. In my eyes, a 'Hybridization and Replacement model' such as the 'Afro-European *sapiens* hypothesis' (Fig. 1) provides a reasonable explanation for the numerous differences between the Neanderthals and the modern humans which were their direct successors, for the lack of clear transitional fossils, and for the evidence of some gene flow between the Neanderthals and the early modern Europeans. The implication of the test model, namely, that these moderns represent a new biospecies, thus does not appear very likely.

Turning to the Far East, we find that here an extreme total replacement model also appears to be difficult to reconcile with the facts. Yet the same holds for the model of Multiregional Evolution as well. The major problem concerning this part of the world is the temporal and morphological gap in the fossil record between the latest archaic *Homo sapiens* specimens and the earliest modern humans (Bräuer 1989a). Bartstra et al.'s (1988) new investigations, moreover, indicate that the Ngandong finds are probably only some 80 000 to 100 000 years old. As a result, these archaic *sapiens* finds, which many have considered as *Homo erectus* or *Pithecanthropus* (Jacob 1981, Stringer 1984), would be shifted closer to the oldest modern finds from Australasia. The morphological gap between the Ngandong sample and the up to 30 000 year old and quite gracile Mungo finds from Australia thus remains very large and without a link. Similarly, no continuity between Ngandong and the oldest moderns from Java and Borneo (Wadjak and Niah) can be recognized (Santa Luca 1980). However, more recent finds from Australia (WLH 50, Kow Swamp, and others) possess – although anatomically modern – several archaic features which Wolpoff (1989) has seen as evidence of a clear regional continuity. Rightmire (1987), however, was unable to find any clear evidence for such continuity. In any case, this model provides no explanation for the unbridged gap between Ngandong and the oldest moderns from Indonesia and Australia. On the other hand, it is difficult to attempt to explain the fossil record of Australasia by means of a model of total replacement.

Although Groves (1989) and Habgood (1989a, b) have greatly questioned the value of every one of the regional features listed by Thorne & Wolpoff (1981) and Wolpoff et al. (1984) (many of the proposed Australasian regional features may be frequently found in the Mesolithic Wadi Halfa sample from Nubia!), it appears likely that a certain degree of gene flow between the moderns expanding into the south and the Ngandong populations or their successors is necessary to explain the spectrum of morphology found in Australia. For this reason, a total replacement model may be less likely, while the 'Hybridization and Replacement model' can provide a satisfactory explanation for both the large morphological gap between the late archaic and the early modern humans and the presence of some archaic features among several of the modern specimens (Bräuer 1989a).

With respect to the finds from China, this model, which assumes slow immigrations from outside as well as a certain degree of gene flow between archaic and modern populations, i.e. a certain amount of continuity, appears also capable of explaining the observed variation among the early modern finds (Bräuer 1989a, Habgood, this volume). For the claims for continuity (Wolpoff 1989) in this region are similarly not based upon the presence of transitional samples, but rather upon the presence of features which are believed to occur more frequently in this region than in others. Groves (1989) and Habgood (this volume), however, were able to demonstrate that both the distribution and the phylogenetic relevance of most of these features is rather questionable. Here, further studies are necessary.

Thus, we can see that while in both Europe and the Far East there is a gap in the fossil record between the last archaic *sapiens* forms and the first modern humans, the latter exhibit an astonishing similarity to one another in the various regions of the globe, a similarity which can hardly be attributed to multiregional evolution (Stringer, this volume, Bräuer 1989a). Convincing transitional groups such as those from Africa are lacking in the other parts of the Old World. Thus, the current fossil record reduces the continuous claims for multiregional evolution to a certain and variable degree of gene flow between indigenous archaic and incoming modern populations. The various kinds of genetic data indicate as well that this 'polycentric theory seems almost refuted today' (Excoffier et al. 1987:179, see also Cavalli-Sforza et al. 1988).

5 BETWEEN THE EXTREMES

Stringer's and Andrews' comparison of the extreme models of total replacement and multiregional evolution has shown that the fossil record is in greater agreement with the replacement mode, even though their test model was not able to explain all of the finds. The indications for continuity which Wolpoff, Frayer, Smith and others have proposed are thus most likely to raise questions about such an extreme replacement model. As the discussions in this paper have demonstrated, however, these do not provide support for multiregional evolution. For many of the arguments which are currently brought forward to support this model involve questionable frequencies of features whose relevancies are themselves frequently questionable. Yet since – with the exception of Africa – there are distinct gaps in the regional sequences at the transitions to modern humans in all regions of the globe, the observed 'regional' features among early modern specimens may most likely be explained by hybridization or gene flow between radiating modern groups and indigenous regional archaic populations. These radiations should here be understood as 'territorial change of groups or tribes, primarily due to ecological factors' (Howells 1976, Bräuer 1984b:334).

The available evidence of regional continuity may thus be most parsimoniously and adequately explained through the 'African Hybridization and Replacement model'. While Neanderthals and archaic Asians were thus probably not completely uninvolved in the composition of early modern gene pools, most likely their contribution was generally so small that replacement may be seen as the decisive process. Some gene flow from regional archaic populations may have taken place beyond the archaic/modern boundary, yet by far the largest part of the modern gene pool may be traceable back to a common ancestral modern stock which arose in Africa. The large amount of evidence for the 'African Hybridization and Replacement model' demonstrates that, according to the present state of our knowledge, Africa played the central role in the origin of us all.

ACKNOWLEDGEMENTS

First of all I would like to thank the Deutsche Forschungsgemeinschaft (German Research Council) for continously supporting my research on *Homo sapiens* evolution for more than 10 years and for assisting my participation at the 12th International Congress of Anthropological and Ethnological Sciences in Zagreb. Furthermore, this paper was written during my tenure as a Heisenberg Fellow of the Deutsche Forschungsgemeinschaft.

I would like to thank Fred Smith for discussions which greatly stimulated some parts of this final version, and Chris Stringer for reading the paper. I would also like to thank Karl Butzer, Rebecca Cann, Luigi Cavalli-Sforza, Fred Wendorf and Milford Wolpoff for various information. Last but not least I am most grateful to John Baker for his support with regard to the English translation.

REFERENCES

Bada, J.L. 1987. Palaeoanthropological applications of amino acid racemization for dating of fossil bones and teeth. *Anthrop. Anz.* 45: 1-8.

Bartstra, G.J., S. Soegondho & A. v.d. Wijk 1988. Ngandong man: age and artifacts. *J. Hum. Evol.* 17: 325-337.

Bräuer, G. 1980. Nouvelles analyses comparatives du frontal Pleistocène supérieur de Hahnöfersand, Allemagne du Nord. *L'Anthropologie* (Paris) 84: 71-80.

Bräuer, G. 1981. New evidence on the transitional period between Neanderthal and Modern Man. *J. Hum. Evol.* 10: 467-474.

Bräuer, G. 1982. Early anatomically modern man in Africa and the replacement of the Mediterranean and European Neanderthals. *I. Intern. Congr. Hum. Palaeont. Nice 1982, Resumés:* 112.

Bräuer, G. 1984a. The 'Afro-European sapiens hypothesis and hominid evolution in East Asia during the late Middle and Upper Pleistocene. *Cour. Forsch-Inst. Senckenberg* 69: 145-165.

Bräuer, G. 1984b. A craniological approach to the origin of anatomically modern *Homo sapiens* in Africa and implications for the appearance of modern Europeans. In F.H. Smith & F. Spencer (eds), *The Origins of Modern Humans: A World Survey of the Fossil Evidence.* New York: Alan R. Liss, pp. 327-409.

Bräuer, G. 1988. Osteometrie. In R. Knussmann (ed.), *Anthropologie. Handbuch der vergleichenden Biologie des Menschen* (4th ed. of Martin's textbook). Vol. I Stuttgart: Fischer: pp. 160-232.

Bräuer, G. 1989a. The evolution of modern humans: a comparison of the African and non-African evidence. In P.A. Mellars & C.B. Stringer (eds), *The Human Revolution: Behavioural and Biological Perspectives on the Origins of Modern Humans.* Edinburgh: Edinburgh Univ. Press, pp. 124-155.

Bräuer, G. 1989b. The ES-11693 hominid from West Turkana and *Homo sapiens* evolution in East Africa. In G. Giacobini (ed.), *Hominidae. Proceedings of the 2nd Intern. Congr. Hum. Palaeont. Torino 1987.* Milano: Jaca Book, pp. 241-245.

Bräuer, G. & R.E. Leakey 1986a. A new archaic *Homo sapiens* cranium from Eliye Springs, West Turkana, Kenya. *Z. Morph. Anthrop.* 76: 245-252.

Bräuer, G. & R.E. Leakey 1986b. The ES-11693 cranium from Eliye Springs, West Turkana, Kenya. *J. Hum. Evol.* 15: 289-312.

Bräuer G. & M.J. Mehlman 1988. Hominid molars from a Middle Stone Age level at the Mumba Rock Shelter, Tanzania. *Am. J. Phys. Anthrop.* 75: 69-76.

Bräuer, G. & K.W. Rimbach 1990. Late archaic and modern *Homo sapiens* from Europe, Africa and Southwest Asia: Craniometric comparisons and phylogenetic implications. *J. Hum. Evol.* 19: 789-807.

Butzer, K.W. 1969. Geological interpretation of two Pleistocene hominid sites in the Lower Omo Basin. *Nature* 222: 1133-1135.

Butzer, K.W. 1982. Geomorphology and sediment stratigraphy. In R. Singer & J. Wymer, *The Middle Stone Age at Klasies River Mouth in South Africa.* Chicago: Univ. Chicago Press, pp. 33-42.

Cavalli-Sforza, L.L. 1989. The last 100 000 years of human evolution: the vantage points of genetics and archaeology. In G. Giacobini (ed.), *Hominidae.* Milano: Jaca Book, pp. 401-413.

Cavalli-Sforza, L.L., A. Piazza, P. Menozzi & J. Mountain 1988. Reconstruction of human evolution: Bringing together genetic, archaeological and linguistic data. *Proc. Natl. Acad. Sci. USA* 85: 6002-6006.

Clark, J.D. 1989. The origins and spread of modern humans: a broad perspective on the African evidence. In P.A. Mellars & C.B. Stringer (eds), *The Human Revolution: Behavioural and Biological Perspectives on the Origins of Modern Humans.* Edinburgh: Edinburgh Univ. Press, pp. 565-588.

Clarke, R.J. 1985. A new reconstruction of the Florisbad cranium, with notes on the site. In E. Delson (ed.), *Ancestors: The hard evidence.* New York: Alan R. Liss, pp. 301-305.

Day, M.H. & C.B. Stringer 1982. A reconsideration of the Omo Kibish remains and the *erectus-sapiens* transition. *I. Intern. Congr. Hum. Palaeont. Nice 1982*, CNRS, pp. 814-846.

Deacon, H.J. 1989. Late Pleistocene palaeoecology and archaeology in the Southern Cape, South Africa. In P.A. Mellars & C.B. Stringer (eds), *The Human Revolution: Behavioural and Biological Perspectives on the Origins of Modern Humans.* Edinburgh: Edinburgh Univ. Press, pp. 547-564.

Delson, E. 1988. One source not many. *Nature* 332: 206.

Excoffier, L., B. Pellegrini, A. Sanchez-Mazas, C. Simon & A. Langaney 1987. Genetics and history of sub-Saharan Africa. *Yrbk. Phys. Anthrop.* 30: 151-194.

Feibel, C.S., F.H. Brown & I. McDougall 1989. Stratigraphic context of fossil hominids from the Omo Group deposits, northern Turkana Basin, Kenya and Ethiopia. *Am. J. Phys. Anthrop.* 78: 595-622.

Groves, C.P. 1989. A regional approach to the problem of the origin of modern humans in Australasia. In P.A. Mellars & C.B. Stringer (eds), *The Human Revolution: Behavioural and Biological Perspectives on the Origins of Modern Humans.* Edinburgh: Edinburgh Univ. Press, pp. 274-285.

Habgood, P.J. 1989a. The origin of anatomically modern humans in Australasia. In P.A. Mellars & C.B. Stringer (eds), *The Human Revolution: Behavioural and Biological Perspective on the Origins of Modern Humans.* Edinburgh: Edinburgh Univ. Press, pp. 245-273.

Habgood, P.J. 1989b. The origin of modern humans in East and Southeast Asia. In G. Giacobini (ed.), *Proceedings of the 2nd Intern. Congr. Hum. Palaeont. Torino 1987*. Milano: Jaca Book, pp. 431-435.

Hay, R.L. 1987. Geology of the Laetoli area. In M.D. Leakey & J.M. Harris (eds), *Results of the Laetoli Expeditions* 1975-1981. Oxford: Oxford Univ. Press: pp. 23-47.

Howells, W.W. 1976. Explaining modern man: Evolutionists versus migrationists. *J. Hum. Evol.* 5: 477-495.

Howells, W.W. 1988. The meaning of the Neanderthals in human evolution. In Fondation Singer-Polignac (ed.), *L'évolution dans sa réalité et ses diverses modalités*. Paris: Masson: pp. 221-239.

Hublin, J.J. 1989. Les Origines de l'Homme Moderne: Europe occidentale et Afrique du Nord. In G. Giacobini (ed.), *Proceedings of the 2nd Intern. Congr. Hum. Palaeont. Torino 1987*. Milano: Jaca Book, pp. 423-430.

Jacob, T. 1981. Solo man and Peking man. In B.A. Sigmon & J.S. Cybulski (eds), *Homo erectus: Papers in Honor of Davidson Black*. Toronto: Univ. Toronto Press: pp. 87-104.

Klein, R.G. 1989. Biological and behavioral perspectives on modern human origins in southern Africa. In P.A. Mellars & C.B. Stringer (eds), *The Human Revolution: Behavioural and Biological Perspectives on the Origins of Modern Humans*. Edinburgh: Edinburgh Univ. Press, pp. 529-546.

Mehlman, M.J. 1987. Provenience, age and associations of archaic *Homo sapiens* crania from Lake Eyasi, Tanzania. *J. Archaeol. Sci.* 14: 133-162.

Protsch, R. 1974. Florisbad. Its palaeoanthropology, chronology and archaeology. *Homo* 25: 68-78.

Protsch, R. 1981. The palaeoanthropological finds of the Pliocene and Pleistocene. Tübinger Monographien zur Urgeschichte. Tübingen 4(3).

Rightmire, G.P. 1986. Africa and the origins of modern humans. In R. Singer & J.K. Lundy (eds), *Variation, Culture and Evolution in African Populations*. Johannesburg: Witwatersrand Univ. Press: pp. 209-220.

Rightmire, G.P. 1987. L'évolution des premiers hominidés en Asie du sud-est. *L'Anthropologie* (Paris) 91: 455-466.

Santa-Luca, A.P. 1980. The Ngandong fossil hominids. *Yale Univ. Publ. in Anthropology* 78: 1-175.

Shackleton, N.J. 1982. Stratigraphy and chronology of the KRM deposits: Oxygen Isotope Evidence. In R. Singer & J. Wymer, *The Middle Stone Age at Klasies River Mouth in South Africa*. Chicago: Univ. Chicago Press: pp. 194-199.

Singer, R. & J. Wymer 1982. *The Middle Stone Age at Klasies River Mouth in South Africa*. Chicago: Univ. Chicago Press.

Smith, F.H. 1982. Upper Pleistocene hominid evolution in south-central Europe: a review of the evidence and analysis of trends. *Curr. Anthrop.* 23: 667-703.

Smith, F.H. & E. Trinkaus 1991. Modern human origins in central Europe: A case of continuity (in French). In J.J. Hublin & A.M. Tillier (eds), *Aux Origines de la Diversité Humaine. Nouvelle Encyclopédie des Sciences et des Techniques*. Paris: Presses Univ. de France.

Stringer, C.B. 1984. The definition of *Homo erectus* and the existence of the species in Africa and Europe. *Cour. Forsch.-Inst. Senckenberg* 69: 131-143.

Stringer, C.B. 1987. A numerical cladistic analysis for the genus *Homo*. *J. Hum. Evol.* 16: 135-146.

Stringer, C.B. 1988. *Homo sapiens*. In I. Tattersall, E. Delson & J. van Couvering (eds), *Encyclopedia of Human Evolution and Prehistory*. New York: Garland Publ.: pp. 267-274.

Stringer, C.B. & P. Andrews 1988a. Genetic and Fossil Evidence for the Origin of Modern Humans. *Science* 239: 1263-1268.

Stringer, C.B. & P. Andrews 1988b. Response to letter by Wolpoff et al. 1988. *Science* 241: 773-774.

Thorne, A.G. & M.H. Wolpoff 1981. Regional continuity in Australasian Pleistocene hominid evolution. *Am. J. Phys. Anthrop.* 55: 337-349.

Trinkaus, E. 1986. Les Néandertaliens. *La Récherche* 180: 1040-1047.

Valladas, H., J.L. Reyss, J.L. Joron, G. Valladas, O. Bar-Yosef & B. Vandermeersch 1988. Thermoluminescence dating of Mousterian 'Proto-Cro-Magnon' remains from Israel and the origin of modern man. *Nature* 331: 614-616.

Van Vark, G.N. 1986. More on the classification of the Border Cave I skull (résumee). *5th Congr. of the European Anthrop. Assoc. Lisboa 1986*.

Van Vark, G.N., A. Bilsborough & J. Dijkema 1989. A further study of the morphological affinities of the Border Cave 1 cranium, with special reference to the origin of modern man. *Anthropologie et Préhistoire* 100: 43-56.

Vandermeersch, B. 1984. A propos de la découverte du squelette Néandertalien de Saint-Césaire. *Bull. et Mém. de la Soc. d'Anthrop. de Paris t.1. série* XIV: 191-196.

Wendorf, F., A.E. Close, R. Schild, A. Gautier, H.P. Schwarcz, G.H. Miller, K. Kowalski, H. Krolik, A. Bluszcz, D. Robins & R. Grün 1990. Le dernier interglaciaire au Sahara oriental. *L'Anthropologie (Paris)* 94: 361-391.

Wilson, A.C., M. Stoneking, R.L. Cann, E.M. Prager, S.O. Ferris, L.A. Wrischnik & R.G. Higuchi 1987. Mitochondrial clans and the age of our common mother. In F. Vogel & K. Sperling (eds), Human Genetics. *Proceedings of the 7th Intern. Congr. of Hum. Gen. Berlin.* Springer: pp. 158-164.

Wolpoff, M.H. 1980. *Palaeoanthropology.* New York: Knopf.

Wolpoff, M.H. 1981. Allez Neanderthal. *Nature* 289: 823.

Wolpoff, M.H. 1985. Human evolution at the peripheries: the pattern at the eastern edge. In P.V. Tobias (ed.), *Hominid Evolution: Past, Present and Future.* New York: Alan R. Liss: pp. 355-365.

Wolpoff, M.H. 1989. Multiregional evolution: the fossil alternative to Eden. In P.A. Mellars & C.B. Stringer (eds), *The Human Revolution: Behavioural and Biological Perspectives on the Origins of Modern Humans.* Edinburgh: Edinburgh Univ. Press, pp. 62-108.

Wolpoff, M.H., Wu Xinzhi & A.G. Thorne 1984. Modern *Homo sapiens* origins: a general theory of hominid evolution involving the fossil evidence from East Asia. In F.H. Smith & F. Spencer (eds), *The Origins of Modern Humans: A World Survey of the Fossil Evidence.* New York: Alan R. Liss: pp. 411-483.

Wolpoff, M.H., J.N. Spuhler, F.H. Smith, J. Radovčić, G. Pope, D.W. Frayer, R. Eckhardt & G. Clark 1988. Modern human origins. *Nature* 241: 772-773.

The Middle Stone Age with reference to Tanzania

Fidel T. Masao
Department of Archaeology, University of Dar-es-Salaam, Tanzania

1 INRODUCTION

The term 'Middle Stone Age' was first introduced in 1928 by A.J.H. Goodwin in an attempt to classify the Stone Age cultures of South Africa (Malan 1957:223). Prior to that the Stone Age cultures were classified into Earlier Stone Age (ESA), with strong Lower Palaeolithic affinities, and the Later Stone Age (LSA), i.e. cultures with strong Upper Palaeolithic affinities. Goodwin pointed out that the Later Stone Age required further subdivision on the basis of different technological attributes and associated hominids. Thus, he regarded the Smithfield and Wilton industries (LSA) as neoanthropic and assigned to San or Bush/Hottentot physical type; the Stillbay and allied industries such as the Fauresmith (Middle Stone Age or MSA) were regarded as earlier and associated with the Boskop physical type. He pointed out the presence of the faceted technique with convergent flaking in the Fauresmith, which then evolved and spread to lay the foundation of the Middle Stone Age (Malan 1957:223).

Despite criticisms by L.S.B. Leakey (1953:138), the term MSA was adopted in Zimbabwe, Zambia, Somalia, Angola and Botswana almost immediately after its introduction and is now accepted all over sub-Saharan Africa.

In North Africa, the term is used differently. As Phillipson (1985:88) has recently pointed out, the MSA industry is known as Mousterian or Levalloiso-Mousterian after its contemporary European counterpart. Views of the time seem to have been so eurocentric that the MSA in North Africa was seen as a Mousteroid copy. As such, terms like Levalloiso-Mousterian and Nubia-Mousterian, in addition to Khormusan, Dabban, etc., were used to connote the MSA of North Africa.

The cultural meaning of the term 'Middle Stone Age' can be defined as connoting a group of cultures, differing from region to region and with different names, but all having a great deal in common with regard to lithic technique and typology. The absence of handaxes and cleavers, the employment of the Levallois and particularly the faceted platform technique, convergent and parallel flaking on flake blades and a variety of flake-tool forms are some of the dominant features of these cultures (Malan 1957:225). The chronological meaning of the MSA period is, however, not yet established and does not always denote the same absolute interval of time when it is used. Overall, a time range roughly between 200 000 and 40 000 years (Clark 1988:236) is conceivable, as will become apparent.

MSA industries seem to fall under what Grahame Clark has probably correctly chosen to call mode 3 stone age technological development, which is defined by the presence of bifacial and

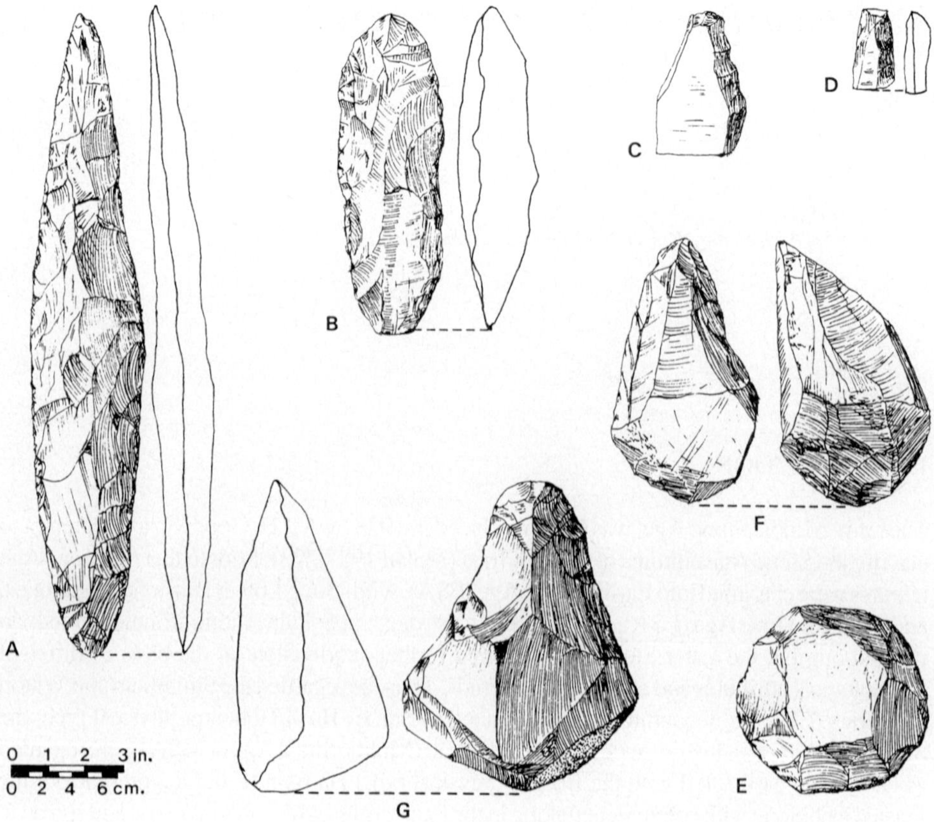

Figure 1. Selected Middle Stone Age artifacts: A) lanceolate; B) core axe; C) end and side scraper; D) double-edged serrated scraper; E) core scraper; F) core axe; G) pick (adapted from Clark 1970:113).

unifacial points and occasionally includes long backed blades (Clark 1973:23). Mode 3 industries gradually grade into those of mode 4 which are dominated by the presence of long punch struck blades (indirect percussion technique with steep retouch). In sub-Saharan African Stone Age technology, mode 4 industrial technology falls in the category of the 'Later Stone Age' which is not within the scope of this survey.

As man's capacity to adapt to varying environmental conditions grew, so did his tool kit. The MSA industrial complexes are generally more diverse and environmentally more specific that the preceeding Acheulean industries. It can also be intimated that the MSA tool kit was more efficient in exploiting the environment. Overall, there are many more MSA sites known in Africa than there are ESA sites, implying, as Clark (1970:72) has suggested, that MSA people spread into parts of the continent not previously occupied. Tool types which are absent in the Acheulean make their appearance in MSA assemblages. These are the points (which are the hallmark of MSA) made by Levallois technique, discs, lanceolates, a variety of scrapers (which are retouched secondarily, either on the side ends or all round), blade cores, etc. (Fig.1). The Levallois technique reaches new levels of refinement in that it becomes a very important way

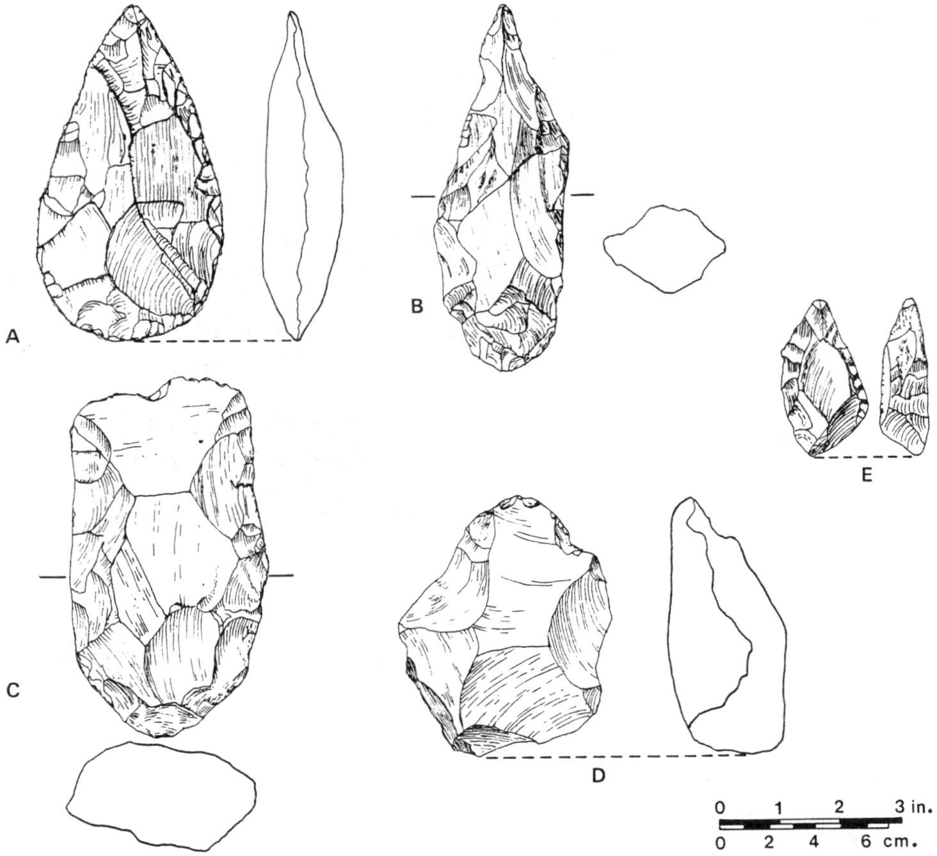

Figure 2. Fauresmith artifacts: A) and B) handaxes; C) cleaver; D) radially prepared core; E) convergent scraper (adapted from Clark 1970:111).

of producing flakes, some of which are long, parallel blades which are non-existent in the Acheulean. The large and heavy tool element is virtually absent from MSA assemblages. Overall, flake production is better controlled than in the ESA, while efficient means of hafting stone tools may have developed during the MSA for tools such as points, burins, discs, and modified and unmodified flakes.

After the end of the Acheulean a number of local and specialized industries arose during what has been known as the 'First Intermediate', a term which is so ill-defined stratigraphically and culturally that it should be discarded altogether and the industries subsumed under it treated as belonging to the MSA (Leakey et al. 1972:328, Cole 1963:341). This would mean, *inter alia*, not only redefining the MSA, but also pushing back its beginning. But since the chronological boundaries of prehistoric cultures are not an issue which can be established without extensive research, the suggestion given here seems appropriate (Leakey et al. 1972, Cole 1963). The industries which are said to belong to the 'First Intermediate' are the Fauresmith, Sangoan, Levalloisian and Proto-Stillbay.

2 FAURESMITH

After about 200 000 years ago the Acheulean large cutting tool tradition underwent more or less technical and tool kit composition modification, depending on specific ecological conditions. The Acheulean gives rise to the Fauresmith industry which has been dated at Rooidam to 174 000 ±20 000 by Uranium Series (Clark 1988:236). Where there was a persistence of ecological conditions to which the Acheulean large cutting tool kit was adapted, the changes were slow and less dramatic, e.g. in East and South Africa (Clark 1970:110). Elsewhere the changes were rapid and, instead of the traditional large cutting tool component, we see a number of retouched scraper forms made on flakes. These forms were often struck from specially prepared cores, producing either thin, broad Levallois flakes or long blade-like forms. There are also small handaxes, cleavers, numerous stone balls and chisels (Cole 1963:184, Clark 1973:110) (Fig. 2).

The industry, a derivative of the Acheulean, was named Fauresmith after the type site in Orange Free State, South Africa. It is said to have been used by communities exploiting the montane grasslands as in Lesotho, the higher plateau and mountains in Kenya, where it was known as the Nanyukian, and in Ethiopia as well (Clark 1970:110, Cole 1963:184).

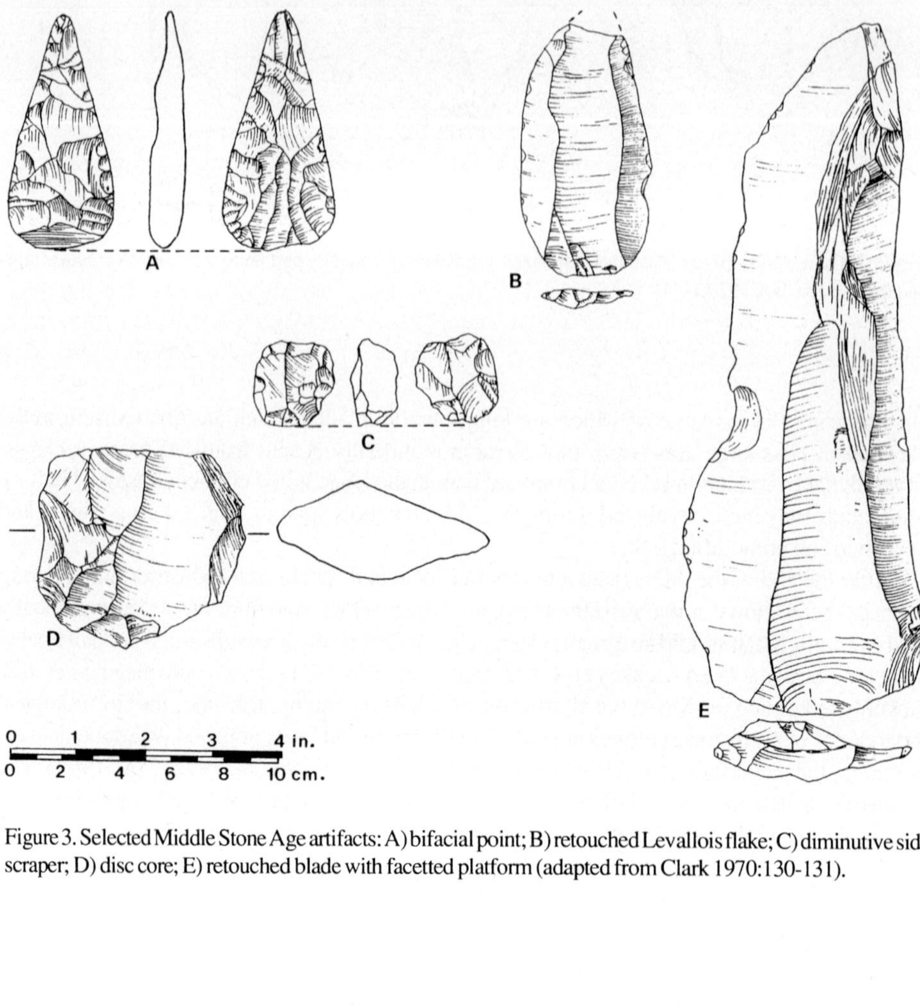

Figure 3. Selected Middle Stone Age artifacts: A) bifacial point; B) retouched Levallois flake; C) diminutive side scraper; D) disc core; E) retouched blade with facetted platform (adapted from Clark 1970:130-131).

3 SANGOAN

The Sangoan, another Acheulean derived industry is contemporary with the Fauresmith but, unlike it, is associated with a more closed vegetation pattern such as savanna woodland and forest, as found around large lakes and rivers. In the Sangoan, the Acheulean large cutting tool tradition is replaced by one composed of several scraping tools, small oblong bifaces, bifacial points, choppers, picks, core axes, axe-like forms, tortoise cores, and flakes made by the facetted platform technique (Cole 1963:185, Clark 1970:172) (Fig.3). The name Sangoan comes from the type site of Sango Bay in Uganda. In Tanzania, isolated finds of Sangoan nature have been found in the Kagera area west of Lake Victoria, Tendaguru and Kilwa Kisiwani. It is also very likely that there are Sangoan sites at Isimila, as suggested by surface finds from the area. The Sangoan at Kalambo Falls is dated to 41 000-38 000 BC, but in the Congo river, Lake Victoria and Lake Tanganyika basins, a new form of tool appears. This is the long lanceolate point made of stone and artistically bifacially worked, which is also common in the later derived Lupemban industry (Clark 1970:112). In Central Africa and much of West Africa, MSA industries are represented by Sangoan type as well as by undated collections from southwestern Mali and central Guinea, which are said to closely resemble the Central African Lupemban (Phillipson 1985:85).

4 LEVALLOISIAN (DEVELOPED LEVALLOISIAN) CULTURES

It must be stated outright that the use of the term 'Levalloisian' to designate specific industries has been questioned for several reasons. First, the same term is more widely known to refer to a special flaking technique. Second, the industries which hitherto have been described as Levalloisian, show such great overlapping with others in technical attributes and temporal range that it is practically impossible to quantify the differences (for example between the so-called 'Levalloisian' and Proto-Stillbay industries). It is therefore proposed that the term should only be used to refer to a flaking technique. At the Pan African Congress of Prehistory in Nairobi in 1947 it was decided to use the term 'facetted platform' instead of the term 'Levallois' to describe this special technique (Cole 1963:180, Clark 1970:84, Mason 1962:227, 298).

François Bordes made statistical investigations on the flaking techniques employed in the Mousterian and concluded that there is no such thing as a 'Levalloisian Culture,' a finding which has been referred to as a 'fundamental contribution to the methodology of Palaeolithic archaeology' (Cole 1963:182). Another flaking technique which has been recognized is the 'Kombewa' method; a term which, like 'Levallois', was used to designate a culture at several sites in the western part of Kenya, but which now refers to a flaking technique. Basically, the cores have rounded butts, though often facetted. The flakes are rounded in outline and show the curious characteristic of double bulb of percussion (Clark 1970:84, Cole 1963:154). Unlike the Levallois technique, which persists to the MSA times, the Kombewa seems to be restricted to the Acheulean industries.

If, as it has been proposed, we abandon the term 'Levalloisian' to describe cultures, then we must find a suitable term for the cultures or industries hitherto referred to as Levalloisian. But, before we do so, let us see how these industries and cultures have been described.

A review of the literature clearly shows that the descriptions are vague and at best qualitative. Levalloisian cultures have been described as possessing flakes made from prepared 'tortoise' cores with facetted striking platforms. They seem to be succeeded by the Proto-Stillbay and

Stillbay industries, but overall, there is considerable overlap, especially in north-eastern Africa, of Levalloisian, Proto-Stillbay and Stillbay industries. These cultures are supposed to have lasted for a very long time (throughout the Later Pleistocene) and were widespread over the whole of the Horn. They are also reported from several sites in Kenya and northern Tanzania. As for the latter, the only site which has been studied and described is Nasera (Apis) rock, although Kohl-Larsen also reported a similar industry from Lake Eyasi.

Basically the 'Levallois' is characterized by points made on flakes with facetted platforms, size scrapers, occasional burins or gravers, and sometimes handaxes made by step-flaking (Cole 1963:132, 171, 175, 180, 195, 207 and 210).

'Levalloisian,' as a culture, should be abandoned and instead we should substitute the term Proto-Stillbay for the earlier industries in East Africa, and the term Stillbay for the industries which hitherto have been referred to as Proto-Stillbay (Cole 1963:196). To facilitate reference and comparison, all these names should be prefixed by the regional name, e.g. Tanzania Proto-Stillbay, Tanzania Stillbay etc. A similar reshuffling has taken place in the Sangoan Culture and, as Cole (1963) has suggested, some of the so-called 'Levalloisian' industries of western Kenya, for example at Muguruk, may in fact be part of the Sangoan.

5 PROTO-STILLBAY

Following from the argument above, the term 'Proto-Stillbay' should now be eschewed in favor of the term Stillbay. The Proto-Stillbay industry as attested by excavations at Broken Hill, succeeds the 'Levalloisian' and later evolves into the Stillbay, henceforth to be referred to as Later Stillbay. In terms of toolkit and temporal range, the Proto-Stillbay is comparable to the Fauresmith and Levalloisian and, as has already been pointed out, there is considerable overlap between the Levalloisian and Proto-Stillbay industries, especially in north-eastern Africa. The Proto-Stillbay and Stillbay industries are centered in South Africa (Natal), Zimbabwe and Zambia and are associated with Kabwe Man, an archaic *H. sapiens* from Broken Hill, Zambia. They are also found at Mambata cave in Zimbabwe. It is most abundant in the Zambezi valley where it overlies Sangoan industries from which it is said to be a direct derivative. In the north, however, it seems to have developed from the Levalloisian. It is characterized by a high frequency of points which are retouched on both surfaces. A Proto-Stillbay industry has been reported from the gorge of Melewa river near Naivasha, Kenya and at Nasera (Apis) rock, Tanzania where it is said to overlie Levalloisian levels and is succeeded by a true Stillbay manifestation (Cole 1963:170, 173, 176, 196, 208 and 341).

What follows now is an attempt to describe the Proto-Stillbay, Stillbay, and Late Stillbay industries in Tanzania. The Sangoan has already been dealt with.

6 THE TANZANIAN PROTO-STILLBAY/STILLBAY/LATE STILLBAY

The Proto-Stillbay industry, formerly known as the Levalloisian, has been described from the two adjacent sites of Nasera (Apis Rock) and Mumba cave near Lake Eyasi. Nasera was first excavated by Leakey in 1932 at which time he recognized a series of occupation layers bearing Proto-Stillbay, Stillbay, Late Stillbay, Magosian and finally Wilton industries. Mumba cave on the other hand was excavated by the Kohl-Larsens in 1936-38. The archaeological material is stored at the University of Tübingen. Subsequently the two sites have been studied by M.J.

Mehlman, who has also examined Kohl-Larsens' collection at the University of Tübingen. According to Mehlman (Mehlman 1977, 1987), there are six beds of deposits at Mumba cave. The top three beds have typical LSA assemblages, while Bed VI contains the MSA industries of Lower Stillbay, Stillbay and Proto-Stillbay. Bed V is half Magosian and half Later Stillbay, each with a varying frequency of microliths (LSA). The main MSA industry of the Mumba sequence is dated by two uranium-series determinations of 109 486 and 131 710 BP (Bräuer & Mehlman 1988). The Eyasi beds with the MSA Nyarasa industry have been shown to antedate the Mumba beds. A date of ca. 173 000 BP has been proposed for the top of the beds (Clark 1988:240). Accordingly, the industry in Bed V has been regarded as being transitional between MSA and LSA.

At Nasera, Mehlman recognized seven beds of cultural deposits and the temporal distribution of the artefacts is comparable to that from Mumba cave. At level 7 the cultural materials are Proto-Stillbay, Stillbay and Late Stillbay, while in Bed VI the deposits were comparable to those of Bed V at Mumba, i.e. transitional between the MSA and LSA. However, the Nasera Bed VI assemblage had a higher frequency of points than at Mumba cave Bed V. Well-made backed pieces, which at Mumba were more numerous than points, were not present at Nasera. This has been interpreted by Mehlman (1977) as a reflection of time difference, for while Nasera level VI is dated to between 22 000 and 23 000 BP, a date for Bed V at Mumba indicates that the transitional industries there could be 8000 to 9000 years earlier.

In early literature, it is reported that the Stillbay gives rise to the Magosian, but at Nasera rock shelter, Mehlman argues that these entities do not exist. The industry seem to have both MSA and LSA features, e.g. the points and some of the cores are in the tradition of the MSA. However, other cores, the backed pieces, the relative microlithic character of the assemblage and the almost exclusive reliance on quartz make it unique. Consequently, it is suggested that the industry be referred to informally as the Naseran. Bone apatite from the Naseran Layer (Mehlman 1977:117) gave a date of 22 250 + 400 BP. If 21 000 BP is accepted as marking the beginning of the LSA (Masao 1979, Leakey et al. 1972, Nelson, unpubl., Deacon 1966), then the Naseran might well be an industry transitional between the MSA and LSA in that part of Tanzania. Thus we must question the use of monothetic terms such as the Stillbay to cover broadly similar industries over the whole continent. Until more research has been done and/or the monothetic terms have been redefined, we should, as suggested before, prefix these names by the regional name of the area where the industries occur. Alternatively, as suggested by Clark (1988), since most of these regional names have not been adequately defined, they should be dropped. This seems to be the consensus as emphasis shifts from assemblage and typological issues to behavioural and environmental problems. However, for comparison purposes usage of such terms has been maintained in this paper.

Lithic finds of Stillbay nature have also been reported from Ilunga, southern Tanzania. A sample was collected by the Chinese crew building the Tanzania-Zambia railway and deposited at the Department of Antiquities, Dar-es-Salaam. On examining the collection I found several struck flakes, Levallois flakes, choppers, diminutive bifaces, bifacial points, crude scrapers, axe-like unifacial implements and radially prepared cores. The typology is obviously suggestive of the Stillbay, but the archaeological context of the sample and the methodology employed in assembling the collection are not known. It would therefore be presumptuous to attach more significance to the sample than to merely acknowledge that it is evidence of Stillbay industry in southern Tanzania. It will be necessary to conduct a thorough archaeological survey of the area, along with sampling and analysis of the assemblage in order to establish the temporal and spatial archaeological context of the sample and its typological composition. Only then can we attempt to describe the industry. The predominant raw materials used are quartz and chert.

At site DKB 11 and 12, in Msia South, Kondoa, I excavated several test pits in which the top 2m. of deposits have an LSA assemblage overlying what seems to be an industry with Stillbay (MSA) as well as LSA features (Masao, unpubl.). The assemblages have not yet been thoroughly analysed, but the Stillbay/LSA occurrence contains the main diagnostic tool types of the Stillbay, i.e. diminutive bifaces, bifacial points, tortoise cores, Levallois flakes, and several crude scrapers. In addition, there are a few microliths which suggest that the industry represents the final phases of the Stillbay just as microliths, a LSA feature, are beginning to appear. The MSA, probably a Stillbay industry, is also known to occur at Olduvai Gorge and West Natron, but has not been described from either area.

At Olduvai Gorge, the MSA is associated with the Ndutu beds whose upper unit is Upper Pleistocene and beyond the range of ^{14}C dating. The overlying Naisiusiu beds, however, have been estimated at an age of 50 000 years with 30 000 to 70 000 as the probable limits of uncertainty (Leakey et al. 1972:334). However, considering the purely microlithic industries (LSA) associated with this horizon, the estimate is most probably on the high side. Other MSA occurrences in Tanzania include exposures along the Olduvai side gorge towards the divide between the Olduvai and Laetoli watersheds, the Ngaloba beds at Laetoli, Loiyangalani River site in Serengeti, Galula in L. Rukwa, the Kilwa, Lindi, Isimila, Nyara river occurrences, Haubi in Kondoa district, Matongo in the Singida district, the Ruaha river basin at Mtera, and the Kagera river in the Lake Victoria basin (Masao, unpubl., Clark 1988).

7 MSA AND MODERN HUMAN ORIGINS

As pointed out already, the MSA is now thought to have begun around 200 000 years BP and persisted until 40 000 BP (Singer & Wymer 1982:149, Clark et al. 1984:37,68, Vogel & Beaumont 1972:51, Rightmire 1978:221). Evidence for this chronology has been documented at sites such as Klasies River Mouth in South Africa, where the MSA occurrence dated to 100 000 BP is associated with hominid remains of *Homo sapiens sapiens*. The first archaic *Homo sapiens rhodesiensis*, or Kabwe man, to be discovered, at Broken Hill in 1921, was previously estimated to be as old at 100 000 years BP (Klein 1973:244), but more recent estimates take it back to at least 200 000 years (Bräuer 1984). Similar MSA occurrences associated with human remains of *Homo sapiens sapiens* have been found at Die Kelders in South Africa and dated to about 80 000 BP (Rightmire 1978:222). In the north, at Porc Epic near Dire-Dawa, MSA occurrences associated with Neanderthaloid remains have been dated by use of the Obsidian hydration method to between 61 000 and 77 500 BP (Clark et al. 1984:37, 63-68). In another site, Gada Motte, also in Ethiopia, MSA assemblages have been dated to between 150 000-180 000 BP (Clark et al. 1984:68). There are more hominid specimens of late archaic and early anatomically modern forms of *Homo sapiens* which are associated with the time period of MSA assemblages. Examples are hominids 1 and 2 at Omo Kibish in Ethiopia, and the LH 18 specimen from Tanzania (Leakey et al. 1972, Rightmire 1983) all of which suggest that development of anatomically modern man may have taken place in Eastern and Southern Africa as argued by Bräuer (1984:160, 1989) in his Afro-European *sapiens* hypothesis. This view has also been supported by Stringer and Andrews (1988:1265) who note, based on the report of Cann et al. (1987), that rates of mitochondrial DNA (mtDNA) evolution give estimates of the origin of the mtDNA of *Homo sapiens* at between 140 000-290 000 years ago, assuming constant rates of mtDNA sequence divergence at 2 to 4% per million years. This estimate falls well within the beginning of the MSA in many sites in Africa, although the early limits of this estimate may predate the development of MSA proper.

The MSA people seem to have had time for leisure and other activities not directly related to procurance of sustenance. Thus there is evidence of the use of paints (rubbed ochre pencils) presumably for body adornment and decorating their surroundings. At Klasies River Mouth for example, presence of paints in the deposits has been reported (Singer & Wymer 1982:117). This may have signaled the beginning of rock art as evidenced by Apollo Cave 11 in South Africa where the deposits date to about 40 000 BP as well as in central Tanzania, where Anati (1983) has estimated the earliest rock paintings to be as old as 40 000 years.

Ancestors of the late archaic *H. sapiens*, i.e. the early archaic group (e.g. Bodo, Broken Hill 1, Hopefield, Ndutu and Eyasi), probably date between 200 000 and 400 000 BP (Bräuer 1989). If this is the case, these hominids could not be responsible for the MSA as they precede the development of the industry. Some of the representatives of anatomically modern man include hominid 1 from Omo Kibish and the Klasies River Mouth specimens. The MSA occurrence at Klasies has been dated to between 60 000 and 125 000 years BP and the hominid remains at the site are associated with the MSA II industry (Bräuer 1989). In another South African site, Border Cave, early modern *H. sapiens* remains are associated with the MSA industry which has been dated to 90 000 years BP and therefore falls within the temporal range of the MSA at the Klasies River Mouth.

In Tanzania, the evidence at hand regarding the makers of the MSA industries seems to complement the story from other parts of Africa. As pointed out earlier, the Eyasi skull excavated by Kohl-Larsen in 1938 was associated with a Stillbay industry. An age of 35 000 BP for the Eyasi hominid has been suggested by amino acid racemization age determinations, but it is believed that these determinations present a minimum age and that the true age is much earlier in time (in the order of 200 000 years). Five other hominids presumably of archaic *Homo sapiens* form were collected by the Japanese expedition in 1986, but these have yet not been published. Another late archaic *Homo sapiens* (LH 18) discovered at Laetoli, Tanzania, has been thoroughly studied by Magori & Day (1983a, b). Its age has now been well established to be 130 000 BP. However, the skull was not found in an archaeological context, and as such we do not know with which MSA industry it would be associated. The three Tanzanian MSA hominid yielding sites, i.e. Ngaloba beds, Mumba and Lake Eyasi are also temporally comparable to findings at Klasies River Mouth and Border Cave in South Africa (Bräuer 1989).

The evidence at hand therefore leaves no doubt as to the makers of the MSA industries. They were archaic *Homo sapiens*, but modern humans (*Homo sapiens sapiens*) continued with the MSA tradition. Most likely some of the former hominids had morphologically developed to the extent that they would be indistinguishable from modern *Homo sapiens sapiens*. We can assume that the makers of the MSA industries were technologically more advanced in terms of adaptability to more varied environments than their predecessors, the makers of the Early Acheulean, *Homo erectus*. Overall, there are many more MSA sites than there are ESA sites implying, as Clark has suggested, that MSA people were technologically able to spread into parts not previously occupied.

The MSA tool kit, though perhaps more efficient than the ESA, shows atavistic atributes. Some enterprising and innovative ESA communities may have found that an industry employing unifacially flaked tools satisfied them best and formed groups where unifacial tool making became traditional and was entrenched in communities evolving towards a *Homo sapiens* level. These communities may then have begun to differentiate among themselves, some developing a culture based almost entirely on unifacial artefacts, like the Pietersburg. Others, like the Stillbay, revived the old bifacial tradition in a new form and defended themselves with delicate but dangerous artefacts such as bifacial points, perhaps an adaptation to hunting in highland/plateau savannas (Mason 1962:232). Still others, like the Sangoan, also maintained the bifacial technique, but modified it so much that instead of producing the traditional biface, a lanceolate, or spear point, was produced as an adaptation to hunting in the forested lands.

This means that there were at least two distinct cultural groups living different life styles in Tanzania. On one hand there was the group of communities whose sites are found in high ground hill tops and rock shelters, the Stillbay. In Tanzania these are represented by such sites as Nasera, Mumba, Olduvai and Msia South. On the other hand, there is the group of hominids whose sites reflect a predilection for river and lake forested areas and the coastal forest belt. These are the Sangoan sites which in Tanzania can be found in the Basins of Kagera river, Lake Tanganyika and Great Ruaha at Mtera, Isimila and in the Kilwa area. These communities made and used stone tools, but there is evidence from a few sites in South Africa that they also made weapons and tools from bone. The extent to which bone was used as a raw material for artefacts cannot be estimated because bone does not preserve very well. However, taphonomic studies are revealing more information about bone at camp sites, although such studies have not been applied to MSA assemblages. MSA communities must also have used materials like leather, wood, bark, horn, and sinew, the last for hafting spears and arrows. They must also have made and used traps, digging sticks and probably wind breaks or shelters (Mason 1962:236). Triangular flakes with ventrally trimmed butts, made in the middle and later stages of the Pietersburg Culture, suggest that the people had learned to haft stone artefacts using gum and sinew, or grass fibre to fix them.

Hunting, and perhaps scavenging, must have been a major activity for sustenance. However, as is the case with extant hunter-gatherer communities, gathering plant food was perhaps just as important. Richly diversified faunal remains of game and other animals including chimpanzees and tortoises have been reported from the MSA Bed V and VI at Mumba and Nasera shelters. Mason (1962:232) has reported 4000 bone fragments and teeth piled on a small area on the edge of an ancient pan, and MSA base. The bones represent one giant extinct buffalo, three hartebeests, five wildebeests, one impala, two waterbucks, one springbuck, one bushbuck, four giant extinct horses, two zebras, eight extinct warthogs, two rhinoceros, one giraffe, two leopards and a lion (Mason 1962:241).

8 SUMMARY STATEMENT

In summary, we can make the following concluding remarks:

1) The MSA in Tanzania is very poorly known and an appeal is made for researchers to do something about this.

2) There is evidence, though not yet documented, of the existence of the two main MSA industrial complexes, the Stillbay and Sangoan industries.

3) The industries which were traditionally considered as Levallois, Proto-Stillbay and Stillbay should now be referred to as Proto-Stillbay, Stillbay and Late Stillbay, respectively.

4) The 'First Intermediate' as a cultural-stratigraphical term is questioned, and it is suggested that the Fauresmith and Sangoan industries, hitherto assumed to belong to the Intermediate, should be viewed as representing the early stages of MSA.

5) The chronological boundaries for the MSA in Tanzania are difficult to establish, but evidence from Olduvai Gorge, Mumba and Nasera would indicate that the MSA industrial complex in Tanzania dates between 130 000 and 20 000 years BP.

6) The makers of the MSA industries were a hominid community of archaic *Homo sapiens*, but more than likely, *Homo sapiens sapiens* had already developed before the last phases of the MSA and therefore may have been responsible for some of the later cultures.

REFERENCES

Anati, E. 1983. The Rock Paintings of Central Tanzania. An unpublished report submitted to the Dept. of Antiquities Dar-es-Salaam.

Bräuer, G. 1984. The Afro-European sapiens hypothesis and hominid evolution in East Asia during the late Middle and Upper Pleistocene. In The Early Evolution of Man with Special Emphasis on S.E. Asia and Africa, *Cour. Forsch.-Inst. Senckenberg* 69: 145-165.

Bräuer, G. 1989. The evolution of modern humans; a comparison of the African and non-African evidence. In P. Mellars & C.B. Stringer (eds), *The Human Evolution. Behavioural and Biological perspectives on the origins of Modern Humans*. Edinburgh: University Press, pp. 123-154.

Bräuer, G. & M.J. Mehlman 1988. Hominid molars from a Middle Stone Age level at Mumba Rock Shelter, Tanzania. *Amer. J. Phys. Anthrop.* 75: 69-76.

Cann, R.L., M. Stoneking, & A.C. Wilson 1987. Mitochondrial DNA and human evolution. *Nature* 325: 31-36.

Clark, G. 1973. *World Prehistory: A New Outline*. London: University Press.

Clark, J.D. 1970. *The Prehistory of Africa*. London: Thames and Hudson.

Clark, J.D. 1988. The Middle Stone Age of East Africa and the beginnings of regional identity. *J. World Prehist.* 2: 237-305.

Clark, J.D., K.W. Williamson, M.J. Michels & C.A. Marean 1984. A Middle Stone Age occupation site at Porc Epic Cave, Dire Dawa (East Central Ethiopia). *Afr. Archaeol. Rev.* 2: 37-71.

Cole, S.M. 1963. *The Prehistory of East Africa*. New York: Weidenfels.

Deacon, J. 1966. An annotated list of Radio Carbon dates for sub-Saharan Africa. *Annals of the Cape Province Museum* 5.

Leakey, L.S.B. 1953. *Adam's Ancestors*. London: Methuen.

Leakey, M.D., R.L. Hay, D.L. Thurber, R. Protsch & R. Berger 1972. Stratigraphy, archaeology and age of the Ndutu and Naisiusiu Beds, Olduvai Gorge, Tanzania. *World Archaeol.* 3: 328-341.

Magori, C.C. & M.H. Day 1983a. Laetoli Hominid 18: an early *Homo sapiens* skull. *J. Hum. Evol.* 12: 747-753.

Magori, C.C. & M.H. Day 1983b. An early *Homo sapiens* skull from the Ngaloba Beds, Laetoli, northern Tanzania. *Anthropos* (Athens) 10: 143-183.

Malan, B.B. 1957. The term 'Middle Stone Age' In J.D. Clark (ed.) *Third Pan African Congress on Prehistory at Livingstone, 1955*: 223-227.

Masao, F.T. 1979. *The Later Stone Age and the Rock Paintings of Central Tanzania*. Wiesbaden: Franz Verlag.

Mason, R. 1962. *Prehistory of the Transvaal*. Johannesburg: University of Witwatersrand Press.

Mehlman, M.J. 1977. Excavations at Nasera Rock Tanzania. *Azania* 12: 111-118.

Mehlman, M.J. 1987. Provenience, age and associations of archaic *Homo sapiens* crania from Lake Eyasi, Tanzania. *J. Archaeol. Sci.* 14: 133-162.

Nelson, C.M. (unpubl.). The Later Stone Age in East Africa. Ph.D. dissertation, U.C. Berkeley.

Phillipson, D.W. 1985. *African Archaeology*. Cambridge: Cambridge University Press.

Rightmire, G.P. 1978. Human skeletal remains from the southern Cape Province and their bearing on the Stone Age prehistory of South Africa. *Quat. Res.* 9: 219-230.

Rightmire, G.P. 1983. The Lake Ndutu Cranium and Early *Homo sapiens* in Africa. *Amer. J. Phys. Anthrop.* 61: 245-254.

Singer, R. & J. Wymer 1982. *The Middle Stone Age at Klasies River Mouth in South Africa*. Chicago: University of Chicago Press.

Stringer, C.B. & P. Andrews 1988. Genetic and fossil evidence for the origin of modern humans. *Science* 239: 1263-1268.

Vogel, J.C. & P.B. Beaumont 1972. Revised radio carbon chronology for the Stone Age in southern Africa. *Nature* 237: 50-51.

A first report on the ER-3884 cranial remains from Ileret/East Turkana, Kenya

Günter Bräuer
Institut für Humanbiologie, Universität Hamburg, Hamburg, Germany

Richard E. Leakey & Emma Mbua
National Museums of Kenya, Nairobi, Kenya

1 DISCOVERY AND STRATIGRAPHIC PROVENIENCE

During a survey in the Ileret area in 1976, Wambua Mangao discovered the partly preserved cranium described here and registered as KNM-ER-3884. Some of the fragments were lying on the surface, the others were excavated from the sediment by Kamoya Kimeu and one of us (R.E.L.) (Fig. 1). All parts were found in an area of less than 1 m^2. The site lies on the western edge of area 5 on the Ileret ridge (Fig. 2), and has formerly been attributed to the 'Guomde Formation' (Feibel et al. 1989). Most of the 'Guomde Formation' which is only exposed on the Ileret ridge, has been subsumed meanwhile into the Chari Member of the Koobi Fora Formation (Brown & Feibel 1986). The hominid is derived from undifferentiated later deposits (Fig. 3), probably representing the time span from 0.5 to 0.1 my ago (Feibel et al. 1989). However, there is some uncertainty as to its age, since the specimen comes from very close to the base of the younger latest Pleistocene and Holocene Galana Boi Formation.

An attempt to get at least a minimum absolute date for the cranium by the ^{14}C accelerator method failed, because the collagen content of the bone was too small (R.A. Housley, pers. comm. 1987). Thus, at present any interpretation of the importance of this hominid must primarily be based on its morphology.

2 THE PRESERVED CRANIAL REMAINS

The excavated remains of ER-3884 consist of a large number of small and well mineralized cranial fragments as well as the complete upper dentition. After removal of the matrix it became evident that many of the fragments showed post-mortem plastic deformation. In spite of this poor state of preservation, Alan C. Walker was able to reconstruct three essential cranial regions: the posterior part of the cranial vault, a frontal part including the supraorbital region, and a maxillary part. In addition to these three reconstructed cranial parts, a larger number of small cranial fragments does exist which could neither be fit together to form larger fragments nor could they be attached to the three main parts. As these small and partly deformed fragments do not provide essential insights into the morphology of the cranium, only the three reconstructed main parts will be dealt with here.

The posterior part of the cranial vault (Fig. 4) consists of most of the occipital and the right parietal. The posterior part of the left parietal adjacent to the lambdoidal suture could also be

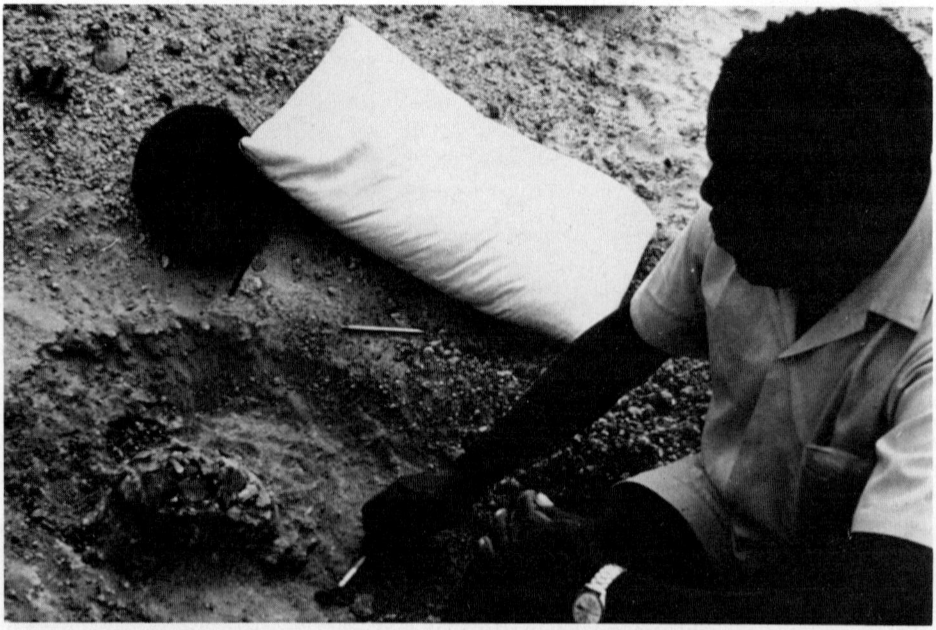

Figure 1. Excavation of the ER-3884 cranium at Ileret by Kamoya Kimeu (photograph: National Museums of Kenya).

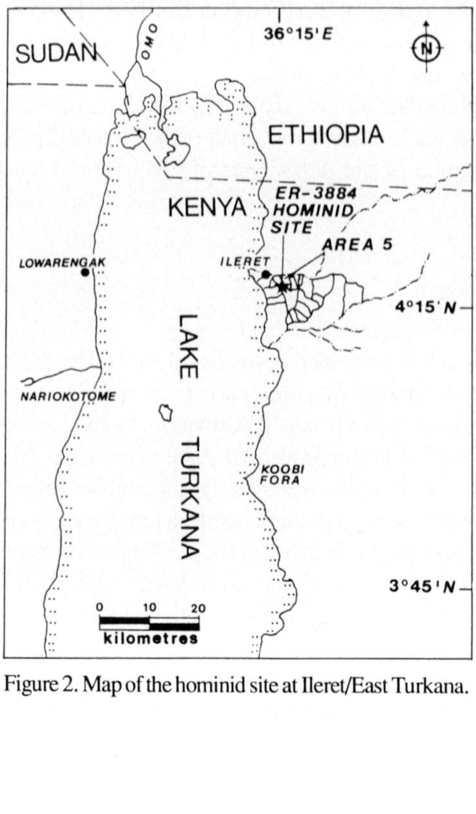

Figure 2. Map of the hominid site at Ileret/East Turkana.

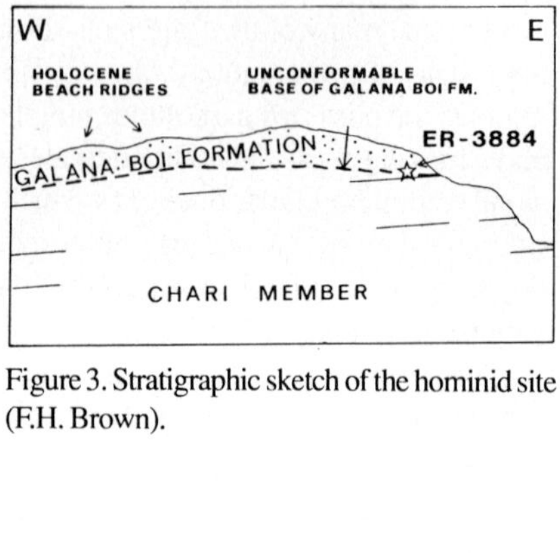

Figure 3. Stratigraphic sketch of the hominid site (F.H. Brown).

reconstructed. Moreover, the right temporal bone is nearly complete, and the left one is largely preserved. Altogether, this major part of the cranial vault does not exhibit significant deformation, with the exception of the basal part around the foramen magnum.

The reconstructed part of the frontal (Fig. 5) consists of four fragments. The lateral pieces of the right and left side include the zygomatic processes and reach medially to a point somewhat beyond mid-orbit (supraorbital notches) on both sides. The other two fragments both are situated more medially and appear not to have a direct contact to the lateral parts. Various attempts to reconstruct the supraorbital area point, however, to the assumption that there is only a very small gap between both fragments on the right side or perhaps even a direct punctual contact, whereas on the left side a somewhat larger gap of a few millimetres is likely. In the glabella area, a segment of probably more than 10 mm is missing, which allows only an approximate reconstruction of the entire supraorbital region (i.e. joining of the right and left portions).

The reconstructed maxillary portion (Fig. 6) includes all teeth, except for the fragmentary crown of the right P^4. On the basis of the preserved parts of the alveolar and palatine processes,

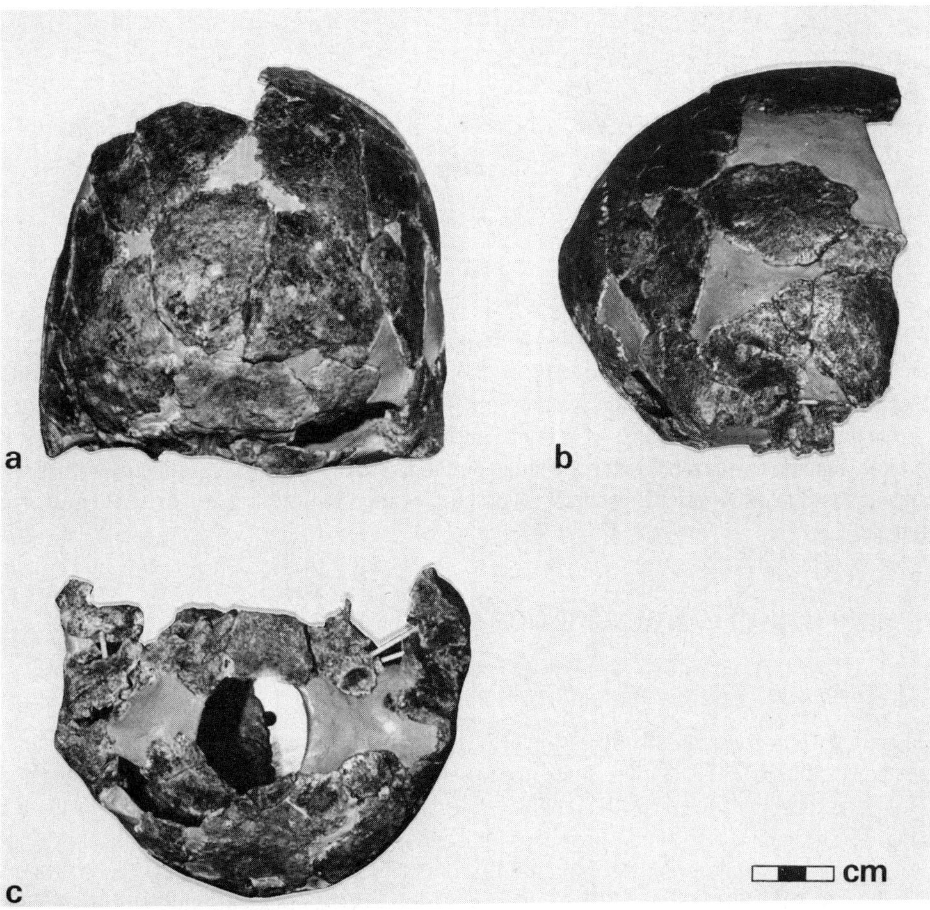

Figure 4. The posterior part of the ER-3884 cranial vault: a) occipital view; b) right lateral view; c) basal view.

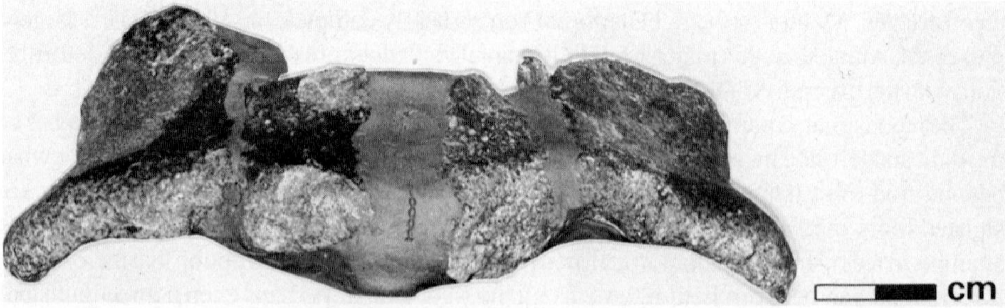

Figure 5. The frontal fragment of ER-3884.

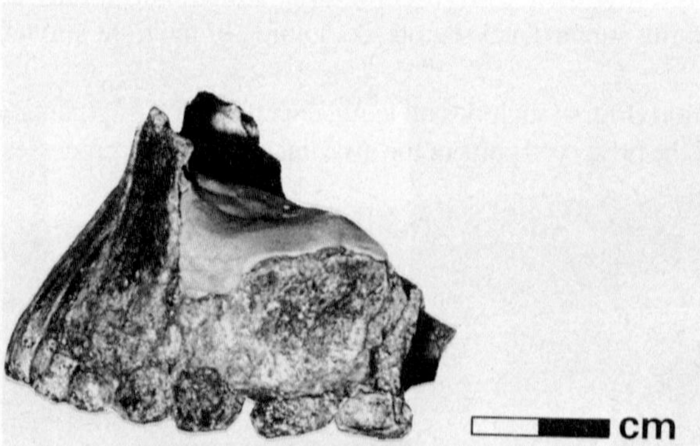

Figure 6. The maxillary fragment of ER-3884 in lateral view.

the dental arcade could be reconstructed with some certainty. Furthermore, parts of the anterior facies on the right and left of the lower part of the nasal aperture, as well as a small part of the zygomatic process on the left side, are present.

On the basis of the relatively slight dental attrition and the condition of the cranial sutures, it is likely that the remains belong to a young adult individual. A reliable sex diagnosis does not appear possible; however, the general size of the specimen is more suggestive of a male than a female.

3 MORPHOLOGICAL DESCRIPTION AND COMPARISONS

3.1 *The posterior part of the vault*

This cranial part gives a good impression of the total size of the vault. With a maximum breadth of about 150 mm, ER-3884 lies in the upper part of the range of late Upper Pleistocene and Holocene specimens from North and East Africa and is, moreover, nearly as broad as Omo 1 (Fig. 7). The vault is also of considerable height and quite similar to that in the male Cro-Magnoid crania from Afalou-bou-Rhummel and Taforalt. With regard to the breadth-height proportion (Fig. 8), ER-3884 falls within the modern as well as in the overlapping ranges of late archaic *Homo sapiens.* In contrast, Broken Hill 1 and also ES-11693 exhibit considerably lower and broader vaults (Bräuer & Leakey 1986).

The large cranial capacity might also point to close relationships to anatomically modern *Homo sapiens*. Regression estimations based on parietal and occipital measurements (after Poissonnet et al. 1978) point to a capacity of about 1400 cc (±). The cranial wall is also relatively thin. At various parts of the parietals it measures only 5 to 6 mm.

In posterior view (Fig. 4a), the lateral walls of the cranium are orientated nearly vertically, so that the total contour does not show any deviation from modern conditions. This also holds in spite of the fact that the maximum breadth is, according to the reconstruction, situated at the level of the weakly developed supramastoid crests. Similar shapes, however, can also be found in late archaic *Homo sapiens* specimens (e.g. Omo 2). ER-3884's minimum biauricular breadth (149 mm) falls completely out of the range of variation of archaic and modern comparative samples. This figure is more than 10 mm greater than that of the largest specimen of the Afalou sample, which is also the largest of all studied African specimens (n = 60) dating from Middle Pleistocene to Holocene (Bräuer, unpubl.). Therefore, it must be considered as possible that this extreme value is partly due to slight deformation. Moreover, it is also possible that this same deformation also affects, to some degree, the considerable maximum cranial breadth.

In lateral view (Fig. 4b), the total profile appears to be rather modern. The dimensions of the occipital support this view. ER-3884's cord/arc index, as well as the occipital angle (Fig. 9), clearly fall into the modern range of variation known from North and East Africa. The slight dorsal protruding of the lower part of the occipital plane is not comparable to the angulation and torus development in Omo 2, but rather is close to the modern condition.

The temporal squama with a height of only 38 mm above the porion is quite low and similar to various archaic *sapiens* specimens such as Omo 2, LH 18, and Ndutu (Fig. 10). When compared to the porion-bregma height, the low temporal squama becomes even more obvious.

The tips of the mastoid processes are broken off on both sides (Fig. 4b). Nevertheless, one gets the impression that the processes were relatively well developed (> 28 mm) and only little shorter than those in Omo 2. They were, however, obviously considerably longer than in LH 18. The tympanic plate extends far laterally and is overhung only slightly by the root of the zygoma. The shape of the meatus is elliptical, and the main axis is inclined obliquely forward. The lateral edge of the tympanic plate is quite well developed anteriorly and inferiorly, but less thick than in LH 18 and Omo 2.

Looking at the temporal from below (Fig. 4c), a deep glenoid fossa and a well developed articular tubercle become evident, especially on the better preserved left half. Tympanic and petrous parts are only slightly angulated with respect to each other; they are orientated in nearly the same axis, which runs obliquely forward to the mid-sagittal. The entoglenoid process is crossed by the sphenotemporal suture. Unfortunately it is no longer possible to determine whether a sphenoid spine was present. The styloid processes are broken off on both sides. The regions medial to the mastoid processes are very fragmentary and do not yield much information. In summary, the temporal exhibits close affinities to the modern range of variation in nearly all features.

The occipital exhibits numerous small gaps and cracks, especially on the nuchal plane; and the surface is somewhat eroded at various parts, so that it is no longer possible to exactly determine the course of the nuchal lines, especially that of the superior nuchal lines. Thus, the position of the inion cannot be securely determined.

3.2 *The frontal part*

Due to the gaps between the four fragments, the minimum frontal breadth is not accurately

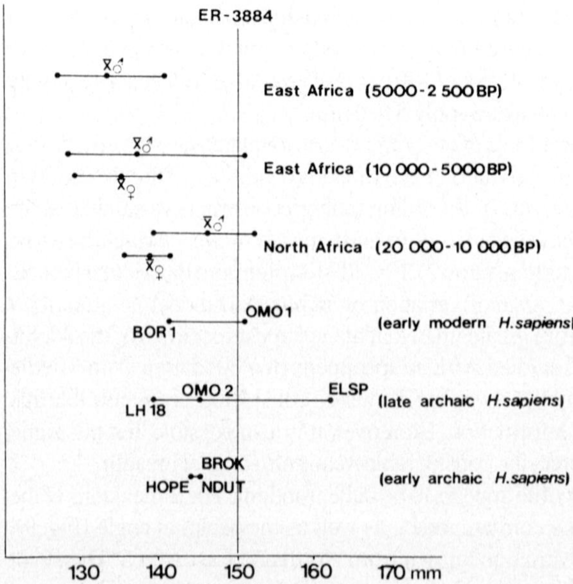

Figure 7. Maximum cranial breadth: ER-3884 compared to African archaic and modern *Homo sapiens* samples (BOR 1 = Border Cave 1; ELSP = ES-11693; HOPE = Hopefield; BROK = Broken Hill 1; NDUT = Ndutu).

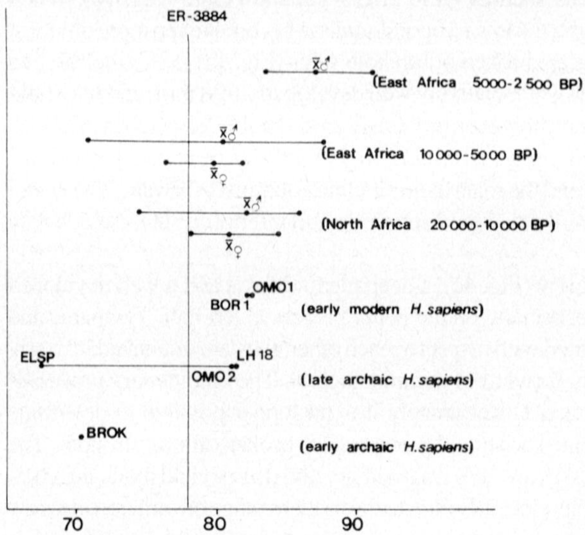

Figure 8. Breadth- Porion/Bregma Height-Index: ER-3884 compared to African archaic and modern *Homo sapiens* samples. Abbreviations: see Fig. 7.

determinable; the value might, however, be relatively large and probably lies above the range of variation of the known early archaic *Homo sapiens* specimens. Similarities in this dimension might probably exist to Omo 2, and to the modern *Homo sapiens* range.

Only vague ideas can be gained about the inclination of the preserved part of the frontal bone. If one tries an anatomically adequate orientation of the orbit roofs, one might conclude that the

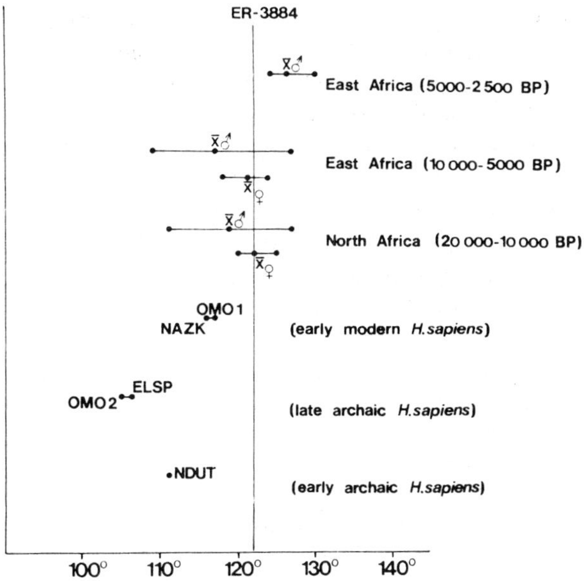

Figure 9. Occipital angle: ER-3884 compared to African archaic and modern *Homo sapiens* samples (NAZK = Nazlet Khater; other abbreviations: see Fig. 7).

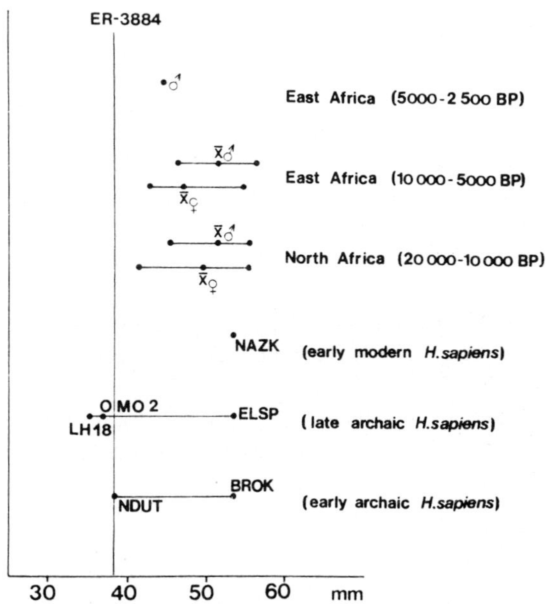

Figure 10. Height of temporal squama: ER-3884 compared to African archaic and modern *Homo sapiens* samples. Abbreviations: see Figures 7 and 9.

profile of the frontal is inclined somewhat rearwards. It is, however, not possible to say whether the degree of inclination falls out of the range of modern humans.

The most striking feature of the frontal fragment is its heavy, torus-like development of the supraorbital region, at least in comparison to modern African conditions. As for the interpretation of this torus-like morphology, some uncertainties remain. For example, the bone surface on the left side in the mid-orbital region is somewhat damaged, and an exact determination of the surface features of the superciliary arch and the transition to the supraorbital trigone is not possible. It is thus not completely clear whether there is a continuous torus-like structure running above the whole orbit up to the lateral process, but there is obviously no sulcus between the arch and the trigone.

The right side also does not allow an absolutely secure assessment of these features, because all of the matrix on the surface between the arch and the lateral process could not be completely removed. In spite of these restrictions, it is evident that at least a torus-like morphology (cf. Bräuer 1984) is present, which runs above each of the orbits in a slightly arched course. Nothing can be determined with regard to the shape of the glabella region.

Comparisons with the supraorbital morphologies of Holocene material from East Africa (e.g. Galana Boi Beds, Lothagam, Hyrax Hill, Naivasha Burial Site, Naivasha Railway Rock Shelter, Njoro River Cave) have shown that ER-3884, with its torus-like, rather thick medial and lateral supraorbital region, is quite different from all of the modern specimens studied. For example, the otherwise quite robust Galana Boi specimens (e.g. ER-5525 A) have considerably thinner supraorbital margins. ER-3884 has certain similarities to the late Upper Pleistocene specimen from Lukenya Hill with regard to shape, thickness and course of the lateral processes. However, above the mid-orbit, Lukenya Hill exhibits a well protruding superciliary arch which is clearly bordered towards the supraorbital trigone by a well developed fossa (Bräuer 1989).

Considering the older hominids, certain similarities might also exist to Omo 1 in the shape and robusticity of the lateral processes, and with regard to the total morphology, also to archaic *sapiens* hominids such as LH 18, Florisbad and even Ndutu with their tori or torus-like structures.

It is quite difficult to decide, whether the differences between ER-3884 and robust modern specimens such as Lukenya Hill and Omo 1 simply reflect a condition of continuous variability, or whether we have here a more fundamental separation between a still archaic and a basically anatomically modern type. Initial comparisons with the material mentioned above make it appear quite probable that ER-3884 still exhibits certain affinities to the non-modern range of variation, which also includes transitions from continuous running tori to more modern conditions.

Above the torus-like supraorbital region, there is a well marked sulcus in ER-3884. More shallow sulci can also be found in ES-11693 and LH 18 while in Florisbad the sulcus is even more weakly developed.

The temporal lines, as far as preserved, are not crest-like but rather weakly marked. With regard to the post-orbital constriction, basically modern conditions do exist as is also the case in the late archaic hominids Omo 2 and Florisbad.

3.3 *The maxillary part*

The maxilla and dental arcade are very large and broad; the respective measurements, nevertheless, still fall within the modern final Upper Pleistocene variation of Africa. With regard to the size, there are also similarities to LH 18 and other archaic *sapiens* specimens.

The preserved parts of the anterior facies of the maxilla next to the nasal aperture are orientated close to the coronal plane (Fig. 6), which might point to a rather shallow canine fossa.

In lateral view, the alveolar portion shows some prognathism. The subnasal part is considerably shorter than that in Broken Hill 1, and is more similar to LH 18's.

4 PRELIMINARY CONCLUSIONS

This first description of the ER-3884 specimen as well as some preliminary comparisons with archaic and modern *Homo sapiens* crania demonstrate that the hominid predominantly exhibits features and dimensions that do not fall outside the range of African anatomically modern *Homo sapiens*. In addition, ER-3884 also exhibits features, especially in the supraorbital region and possibly also in the maxilla, which might point to certain affinities with archaic *Homo sapiens*. Only more detailed analyses including comparisons with a wide spectrum of African fossil and subfossil specimens (Bräuer & Mbua, in prep.) and a better dating will lead to more clarity regarding the phylogenetic relevance of this hominid.

ACKNOWLEDGEMENTS

We thank the Government of Kenya for permission to undertake this study. We are especially grateful to A.C. Walker for the reconstruction of the cranial parts. We would also like to thank F.H. Brown and C.S. Feibel for providing a stratigraphic sketch of the hominid site and R.A. Housley for the microanalysis. This study was supported by the German Research Council (G.B. was a Heisenberg Fellow) and by a grant from the Hamburgische Wissenschaftliche Stiftung (E.M.). We are grateful to these institutions.

REFERENCES

Bräuer, G. 1984. The 'Afro-European *sapiens* hypothesis', and hominid evolution in East Asia during the late Middle and Upper Pleistocene. *Cour. Forsch-Inst. Senckenberg* 69: 145-165.

Bräuer, G. 1989. The evolution of modern humans: a comparison of the African and non-African evidence. In P.A. Mellars & C.B. Stringer (eds), *The Human Revolution: Behavioural and Biological Perspectives on the Origins of Modern Humans*. Edinburgh: Edinburgh University Press, pp. 124-155.

Bräuer, G. & R.E. Leakey 1986. The ES-11693 cranium from Eliye Springs, West Turkana, Kenya. *J. Hum. Evol.* 15: 289-312.

Brown, F.H. & C.S. Feibel 1986. Revision of lithostratigraphic nomenclature in the Koobi Fora region, Kenya. *J. Geol. Soc.* 143: 297-310.

Feibel, C.S., F.H. Brown & I. McDougall 1989. Stratigraphic context of fossil hominids from the Omo Group deposits, northern Turkana Basin, Kenya and Ethiopia. *Amer. J. Phys. Anthrop.* 78: 595-622.

Poissonnet, C.M., G. Olivier & H. Tissier 1978. Estimation de la capacité crânienne à partir d'un os de la voûte. *Bull. et Mém. de la Soc. d'Anthropologie de Paris t.5, sér.* XIII, 217-221.

The origins of modern people: The evidence from Klasies River

Hilary J. Deacon & Ria Shuurman
Department of Archaeology, University of Stellenbosch, Stellenbosch, South Africa

1 INTRODUCTION

Africa has functioned as the biogeographic center in the Neogene for the evolution of the anthropoid apes (Fleagle et al. 1986) and the hominids (Stringer 1984). However, that the evolution of anatomically modern people took place in Africa is a new if not undisputed wisdom (Stringer & Andrews 1988) and only the classic Neanderthals are held to have evolved outside Africa (Tuttle 1988). The evidence for the evolutionary primacy of this continent is largely from palaeontological finds and the African representatives of taxa are consistently of higher antiquity. Evolution in and dispersal out of Africa is a process that has been repeated many times in hominoid and hominid history, most recently in the range expansion and radiation of anatomically modern people.

This contribution looks at the evidence from one locality, Klasies River, at the southernmost tip of the African continent that supports the early dating of the fossil remains of anatomically modern people in Africa. There are a number of human fossils, claimed to date to the later Middle or earlier Late Pleistocene, that have been assessed as transitional between archaic and modern kinds of people or anatomically modern, from sites in sub-Saharan Africa. These would include the finds from Dire-Dawa, Omo, Ndutu, Laetoli, Kabwe, Florisbad and Border Cave as well as Klasies River (Bräuer 1984, Rightmire 1984). Estimated ages in excess of a 100 000 years for such sites are significantly earlier than claims for the antiquity of modern people outside Africa. The exception is the site of Qafzeh (Valladas et al. 1988), on the north-eastern margin of the continent, which is in this age range. The African fossils in general are not well dated and they did not receive due attention until the publication of independent studies in molecular biology (Wainscoat et al. 1986, Cann et al. 1987) that supported an African ancestry for modern people.

The precision with which palaeontological and archaeological materials can be dated within the last 40 000 years by the conventional radiocarbon method has created unrealistic expectations for precision in age determinations in older time ranges. Dating methods, like uranium disequilibrium, amino acid, thermoluminescence and electron spin resonance, which have potential application in measuring the ages of Middle and earlier Late Pleistocene finds, are still at an experimental stage of development and are not routine in application. Such methods can produce consistent results, even if the error bars are large, and the results cannot be discounted because they do not have the precision of radiocarbon dating. It is a reflection of the present stage of research that the best evidence for the early dating of anatomically modern human

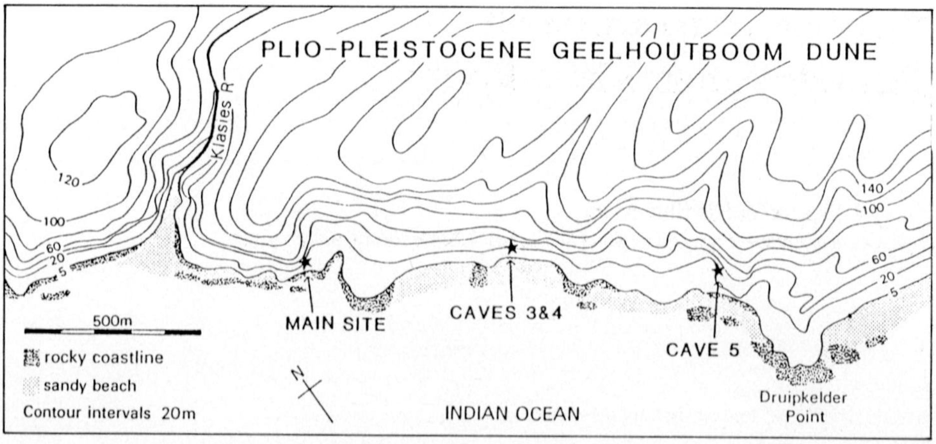

Figure 1. Location of Klasies River main site.

Figure 2. Klasies River main site depository.

remains is from Qafzeh and Klasies River which are the length of the African continent apart. Both sites may sample populations that had already undergone regional differentiation in the late Middle Pleistocene and may date to a time more recent than the primary dispersal of the ancestral population out of the center of origin. The Klasies River evidence in particular is relevant to the ancestry of the southern African endemic form, the San (Deacon 1988) and it supports a specifically sub-Saharan African evolutionary center for anatomically modern people.

2 KLASIES RIVER MAIN SITE

Klasies River, originally 'Kaapsedrif River', refers to a small stream that enters the sea to the west of three discrete cavern systems cut into a cliffed section of the Tsitsikama coast near the town of Humansdorp (Fig. 1). The most westerly of these cavern systems and the only one to produce Pleistocene human fossils is main site, a complex of deep chambers and overhangs that served as a single depository for the accumulation of Late Pleistocene sediments. Different parts of this depository were labelled as caves or shelters 1, 1A, 1B, 1C and 2 (Fig. 2) but with the exception of 1B, strata can be traced from one part to another. The stratigraphic connection with 1B has been removed by erosion. The total thickness of the main site deposit is some 20 meters and the accumulation formed an extensive cone. The cone has been truncated and perhaps only a quarter of the original volume of deposits is preserved. A perched flowstone layer that was deposited on an upper surface prior to erosion dates to the mid-Holocene and the recovery of sea level to between 1 and 2 m above the present in the Holocene is a probable cause of the extensive erosion. This is consistent with the Later Stone Age re-occupation of cave 1 in the last 4000 years after older deposits blocking the entrance had been removed.

A description of the stratigraphy, culture, stratigraphy and the occurrence of human remains at main site, following extensive excavations carried out in 1967-8, has been published by Singer & Wymer (1982). Since 1984 further excavations have been undertaken with the purpose of providing more stratigraphic and contextual detail on the finds, including the human remains, from the site. The current phase of investigation is still in progress and there will be some delay before the analyses of new excavated samples are completed and published.

The stratigraphy of main site has been described as a series of members and these are given here, from the oldest to the youngest, with the Singer & Wymer (1982) equivalents in parentheses; LBS (layer 38), SAS (layers 37, 17-14 in cave 1 and layers 39-23 in cave 1A), RF (layer 22) and Upper (21-1 in cave 1A) members (Fig. 3). A more detailed description is published in Deacon & Geleijnse (1988). The base of the archaeological deposits, the LBS, is a series of sands with discrete intercalations of shell middens and hearths containing artefacts and fauna. The dating of initial human occupation of main site to the Last Interglacial which ended 118 000 years ago (CLIMAP 1984) is not contentious. This age was originally posited by Shackleton (1982) on the basis of oxygen isotope determinations on shell from the LBS middens. His results are confirmed by the oxygen isotope measurements on a more extensive series shell samples through the sequence by Talma (Deacon et al. 1988). The last time the ocean was isotopically as light or lighter than the present was during the Last Interglacial, oxygen isotope stage 5e, and the series of samples from the LBS correlate with 5e and the beginning of 5d. The sediments of the LBS provide additional support for the Last Interglacial age because the sands were derived from a nearby beach source and sea-levels were last as high or higher than the

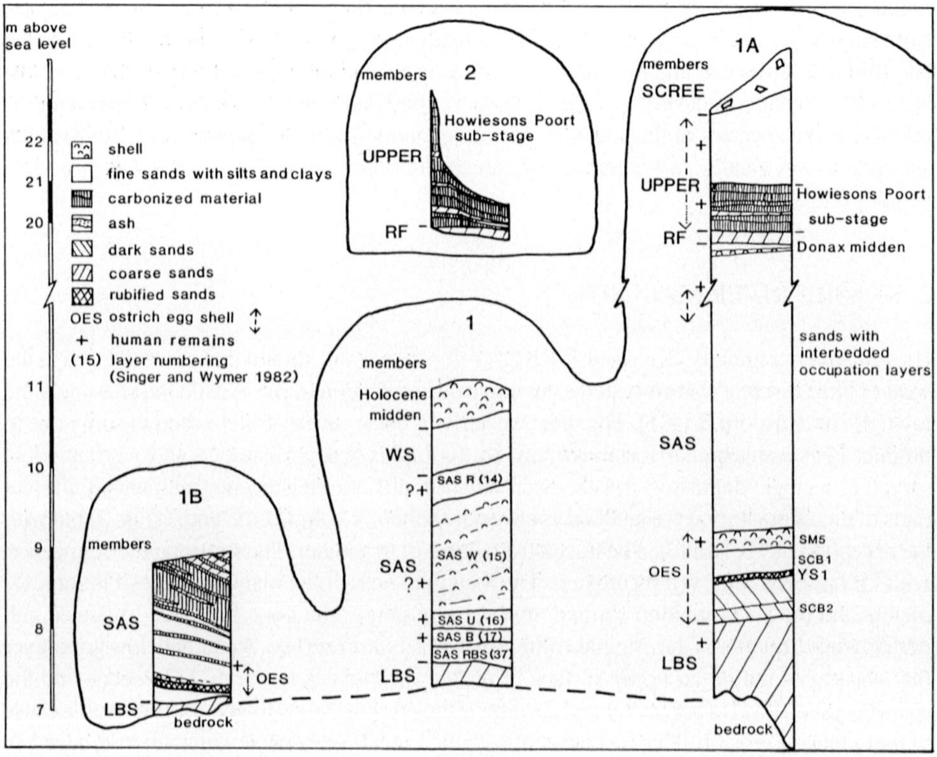

Figure 3. Diagrammatic illustration of the main site stratigraphy showing the correlation of the strata in cave 1B with the rest of main site.

present during stage 5e (Deacon & Geleijnse 1988). A minimum age of 110 000 years for the LBS sediments has been obtained by Vogel (Deacon et al. 1988) which is consistent with the above age estimate, and the aspartic acid dating results of Bada & Deems (1975) indicate this order of age.

There are no diagnostic human remains from the LBS deposits, as defined here, from the Singer & Wymer excavation. The current excavations have produced two maxillary fragments representing two different individuals. This material is from the same cutting and midden horizon and the specimens are essentially penecontemporaneous. The fragmentary nature of all the Late Pleistocene human material from main site as well as evidence for burning and the presence of cut marks on some fragments is consistent with White's (1987) suggestion that cannibalism was practised. One of the maxillary fragments from the LBS is burnt. The two finds which come from a very localized area may be the product of the same episode of cannibalism. There is a marked degree of sexual dimorphism evident in the human remains from Klasies River, a feature noted by Singer & Wymer (1982), and this is obvious even in the small LBS sample.

The deposits immediately overlying the LBS member are again sands with shell and other human occupation interbeds. These SAS deposits began to accumulate in oxygen isotope stage

5c (94 000-105 000 years) (Deacon et al. 1988) after a break in deposition. There are aspartic acid age estimates of the order of 90 000 years which are consistent with such an antiquity. The main sample of human remains has been recovered from the SAS member and reinterpretation of the context of these finds suggests they all come from the lower part of the SAS member and may represent the same taphonomic event, an episode of cannibalism younger in time than that recorded in the underlying LBS member.

It has proved possible to correlate the 1B deposits containing the important mandible, 41815, with the rest of main site (Goede & Hitchman 1987) and resolve any controversy (Binford 1987, Singer & Wymer 1987) about the stratigraphic placement of the 1B sequence. The mandible comes from the lower part of the SAS member and the occurrence of ostrich eggshell in caves 1B and 1A in this zone is a useful marker (Fig. 3). The 1B mandible may not be significantly older than the recent find of the ulna in SM5 in cave 1A and the large number of fragmentary fossils recovered from the SAS member in cave 1. No new excavations have been undertaken in cave 1 but there are grounds for suggesting all the finds made during the 1967-68 excavations and reported as coming from layers 17 through to 14 (Singer & Wymer 1982) relate to the deposition of the lower part of the SAS member. Singer and Wymer assumed that the finds made on one side of the witness baulk in cave 1 were older than the finds made on the other side of the baulk in the West Cutting. The West Cutting is close to the cave wall and the stratigraphy there is unclear because the finer matrix has been washed out of the deposit. Added to this, the younger SAS deposits in cave 1 (layers 15 and 1) are slope deposits and a localized concentration of otherwise rare human remains in a slope deposit is anomalous. It is argued here that all the cave 1 human fossils associated with the Middle Stone Age, the 1B mandible and the 1A ulna, are contemporaneous.

Although fragmentary and mainly dental and cranial parts, the human fossils from the lower SAS member are significant because of their overall modern morphology and because of the degree of sexual dimorphism exhibited. Both robust and gracile individuals are represented in what is a single population sampled by cannibalism and Singer & Wymer (1982) note that the mandible 16424 with three relatively unworn molars in position represents an individual as gracile as a modern San female. These molars and other dental material from the SAS member show much less wear than the teeth from the later Holocene skeleton, 614, at main site and this has implications for differences in diet between the Middle and Later Stone Age populations.

The RF and Upper members on conservative estimates date to more than 60 000 years and are correlated with isotope stages 5a-4 (Deacon et al. 1988). There are no human remains found in the RF deposits which are sands 1 m thick containing a few carbonized partings representing very episodic human occupation. From the Upper member a premolar and an incisor fragment have been recovered in the current research to add to two parietal fragments previously found. The Upper member is between 2 and 4 m thick and the best preserved section is cemented to the west wall of cave 2. Selective preservation of shell due to differences in the pH of ground waters is evident in cave 1A but stalactites that grew from the lower part of the westward sloping roof in cave 2 have had a local buffering effect inhibiting diagenesis. As shell is less well preserved generally in the Upper member the artefact frequencies per unit volume of deposit are very high. Ash lenses and associated carbonized partings are very prominent and these are responsible for banded appearance of the strata. Interest in the Upper member is more archaeological than palaeoanthropological because of the paucity of finds of human remains in spite of the large volume of deposit excavated previously.

3 ARCHAEOLOGY AND PALAEOECOLOGY OF MAIN SITE

Klasies River is the main reference for the cultural stratigraphy of the Middle Stone Age in South Africa. Singer & Wymer (1982) were able to show a sequence of typologically and technologically defined substages with the MSA I represented in the LBS and lower part of the SAS, the MSA II in the middle and upper parts of the SAS and RF and the Howiesons Poort and MSA III in the Upper member. The MSA IV substage is represented by a small sample, in the WS member (Layer 13), in cave 1 but the stratigraphic relationship of this member to the strata in 1A is not satisfactorily established. The limited sample of artefacts from the RF member could be grouped with the Howiesons Poort rather than the MSA II. The definition of these substages periodizes a continuum of temporal changes in artefact size and style. All were made using the same basic prepared core-flake blade technology. The most novel substage is the Howiesons Poort with an increase in the use of non-local raw materials and the production of design types for hafting that were shaped like segments and trapezes with blunting, like on a pen knife blade, on the arc. The distinctive typology of the Howiesons Poort makes it a useful horizon marker and, for example, it is on the basis of its occurrence at both sites that a correlation can be suggested between the Border Cave and Klasies River sequences. Similar Howiesons Poort artefact assemblages, some of which have been called epi-Pietersberg and even Magosian in the older literature, are known from sites south but not north of the Zambezi. The association is with the mosaic of more open southern African habitats and its distribution excludes the woodland habitats of central Africa. The similarity of Middle Stone Age artefact sequences in southern Africa and the lack of distinctive regional variants is noteworthy and the same situation pertains in the Later Stone Age in this region. The implication is that information on the symbolic meaning of artefact styles, and by extension genes, was being exchanged in a network covering the whole subcontinent.

Main site has been facetiously described as the oldest sea food restaurant in the world and indeed it preserves evidence of shell fish collecting on a significant scale. Thick midden accumulations have been built up in a way that indicates patterned waste disposal at living sites as in the Later Stone Age. The marine focus in obtaining food did not extend to fishing and all the fish bone in the site was introduced by birds roosting in the cliff face. Pellets of regurgitated fish bone are still being deposited below the cliff. The lack of evidence for fishing, the absence of the remains of flying birds in the fauna and the fact that shell fish species were not farmed down are among the evidence cited by Klein (1989) to show that the Middle Stone Age people inhabiting Klasies River were not capable of utilizing the resources of the local environment as fully as their Holocene Later Stone Age counterparts. Differences of this kind probably have more to do with the ranking of resources rather than with innate abilities. Later Holocene peoples have a very high archaeological visibility in the landscape, suggesting they were relatively numerous, probably more numerous than at any previous time, and it is not surprising that there is evidence for a greater impact on the resources in the Later Stone Age.

The mammal fauna from main site has provoked a debate on the relative importance of scavenging versus hunting and this has been linked to arguments about food sharing and base camp living. The evidence from main site for food sharing and base camp living is the sheer quantity of shell fish brought to the site to be consumed and the patterned way the empty shells have been disposed of in midden heaps on the living floors. There seems no good reason to question the function of the site as a base for food sharing even if meat was brought back to the site in small packages. Questions about the relative importance of hunting and scavenging have to do with consideration of body part frequencies represented in the mammal faunal sample.

The 1967-8 collection, while large, is selected and only assumed identifiable elements are represented. In spite of any potential sample bias Klein (1976) has argued a case for active hunting while Binford (1984) has stressed the scavenging of larger bovids, a view supported by Blumenschine (1986) for different reasons. The fauna as a whole, like the human component, is very fragmentary which complicates interpretation. A comparative study of the recently excavated faunas from Klasies River and Florisbad (Brink 1987) is underway and this study may indicate the extent to which the body part distributions were biased by selective recovery in the older collections.

While large and particularly large medium bovids are consistently represented in the recently excavated faunal sample so are small bovids especially from the middle part of the SAS member upwards through the sequence. From what is known of the subsistence ecology of Later Stone Age people in the southern Cape at sites like Melkhoutboom (Deacon 1976) the hunting of small, territorial antelope is linked to the use of plant foods and the practice of fire management. The relative frequency of small bovids is indirect evidence for the use of plant foods and there is some support for the contention that plant foods became more important at Klasies River in the well developed carbonized horizons associated particularly with the Upper member. Studies of phytoliths, still at an initial stage, may provide clues as to the classes of plants introduced into the site. Reliance on plant foods should be evident from the wear on the human teeth which is not as marked as in the teeth from late Holocene contexts. Most of the teeth come from the lower part of the sequence and a worn incisor from the Upper member is at best only suggestive of the later importance of plant foods.

There is another level of interest in the changes in subsistence ecology and patterning in the artefact finds from the Klasies River sequence. This is the evidence it provides on the response of human populations to the scale of environmental change in the Late Pleistocene. Main site covers the changes from full interglacial conditions in isotope stage 5e in the LBS member to three quarter glacial conditions in stage 4 in the Upper member. It is in the Upper member times that traces of occupation diminish and there is no record of occupation, specifically of main site, between some 60 000 years and the late Holocene. This is consistent with a general pattern in southern Africa of low archaeological visibility of human populations during an extended period in the mid-Late Pleistocene. In these latitudes it is not glacial cold that is limiting but the aridity that may be associated with cooler oceans and stronger circulation. The scenario that best fits the available data is that Middle Stone Age populations were able to maintain themselves under the environmental conditions of the Last Interglacial *sensu lato*, isotope stage 5, but towards the end of this period, equivalent to isotope stages 5a-4, the populations had to respond to deteriorating environmental conditions and declining resources. This reflects in the record from the site in the increased use of lower ranked foods on the one hand and the intensified use of artefact symbols to promote better social networking on the other. The scenario suggests that in the long term the responses were not sufficient to maintain populations at the former levels of density and with lower rates of recruitment there was a significant population decline not only in the southern Cape but on a subcontinental scale. The study of Howell (1979) on the fertility of !Kung women suggests how finely tuned hunter-gatherers may be to habitat conditions in southern Africa, in this case, through the levels of body fat, and physiological mechanisms of this kind are plausible for the control of Late Pleistocene population levels in relation to the changing effective productivity of the environment. Although there is evidence for a minor peak in archaeological visibility between 30 000 and 40 000 years ago, it is only with the amelioration of environmental conditions after the Last Glacial Maximum, 15 000 years ago, that archaeological visibility increases in southern Africa as a whole, peaking in the last half of the Holocene. It is against this background of significant environmental change and population fluxes that selection for the super gracile San populations operated in the Late Pleistocene in southern Africa.

4 CONCLUDING DISCUSSION

Klasies River main site has yielded the fragmentary remains of a number of anatomically modern individuals. The oldest finds date to some 120 000 years and the bulk of the finds are somewhat younger but still of the order of 100 000 years old. The Late Pleistocene samples represent a single sexually dimorphic population and include notably gracile specimens. The degree of fragmentation is no different from that in the rest of the faunal sample and the primary cause of the fragmentary condition of all the faunal remains is that they were processed as food. Post-depositional weathering and disturbance are lesser factors. These are among the best dated remains of early anatomically modern people and they are as old as or older than any similarly modern remains known from elsewhere.

The inhabitants of Klasies River practised a Middle Stone Age or Middle Palaeolithic technology which simply means they produced standardized flake blanks some of which were hafted. More informative are the stylistic changes through the main site sequence and the novel use of raw materials and backed tools in a transient Howiesons Poort substage of the Middle Stone Age. This non-functional elaboration of artefacts provides some of the best evidence for an ability to switch on and switch off social signalling and there are parallel kinds of behavior shown by Later Stone Age peoples. This suggests these Middle Stone Age people were not only physically but also behaviorally modern. Contrary arguments made on differences in the intensity of resource use between Middle and Later Stone Age people may reflect differences in population densities linked to environmental conditions rather than differences in the capacity for behavior. There is good archaeological evidence from southern Africa for population fluxes in the Late Pleistocene and the Holocene. The timing of these fluxes points to their being controlled by external environmental forcing. The evolution of the San, the southern African endemic form, took place against a background of environmental change and population flux in the Late Pleistocene and the selection for gracility may have much to do with this being an arid subcontinent, and in times past more arid than the present.

ACKNOWLEDGEMENTS

The research at Klasies River has been funded by the Human Sciences Research Council and the University of Stellenbosch. Staff and students in the Department of Archaeology have taken part in the excavations and the task of analysing the finds.

REFERENCES

Bada, J.L. & L. Deems 1975. Accuracy of dates beyond the ^{14}C dating limit using the aspartic acid racemization reaction. *Nature* 255: 218-219.

Binford, L. 1984. *Faunal Remains at Klasies River Mouth*. Orlando: Academic Press.

Binford, L.R. 1986. Reply to 'On Binford on Klasies River Mouth: response of the Excavators'. *Curr. Anthropol.* 27: 57-62.

Blumenschine, R.J. 1986. Carcass consumption sequences and the archaeological distinction of scavenging and hunting. *J. Hum. Evol.* 15: 639-659.

Bräuer, G. 1984. A craniological approach to the origin of anatomically modern *Homo sapiens* in Africa and its implications for the appearance of modern europeans. In F.H. Smith & F. Spencer (eds), *The origins of modern humans: a world survey of the evidence*. New York: Liss, pp. 327-407.

Brink, J.S. 1987. The archaeozoology of Florisbad, Orange Free State. *Mem. nas. Mus., Bloemfontein*, 23: 1-151.

Cann, R.L., M. Stoneking & A.C. Wilson 1987. Mitochondrial DNA and human evolution. *Nature* 325: 31-36.

CLIMAP, Project Members 1984. The Last Interglacial ocean. *Quat. Res.* 21: 123-224.

Deacon, H.J. 1976. Where hunters gathered: a study of Holocene Stone Age people in the eastern Cape. Claremont: S. *Afr. archaeol. Soc. Monograph* 1.

Deacon, H.J. 1988. The origins of anatomically modern people and the South African evidence. *Palaeoecol. Afr.* 19: 193-200.

Deacon, H.J. 1989. Late Pleistocene palaeoecology and archaeology in the southern Cape, South Africa. In P. Mellars & C.B. Stringer (eds), *The human revolution: behavioural and biological perspectives on the origins of modern humans*. Edinburgh: Edinburgh University Press, pp. 547-564.

Deacon, H.J. & V.B. Geleijnse 1988. The stratigraphy and sedimentology of the main site sequence, Klasies River, South Africa. *S. Afr. archaeol. Bull.* 43: 5-14.

Deacon, H.J., A.S. Talma & J.C. Vogel 1988. Biological and cultural development of Pleistocene people in an Old World southern continent. In J.R. Prescott (ed.), Early Man in the southern Hemisphere. *Supplement to Archaeometry: Australasian Studies 1988*. Adelaide: Department of Physics and Mathematical Physics, University of Adelaide, pp. S23-S31.

Fleagle, J.G., T.M. Brown, J.D. Obradovich & E.L. Simons 1986. Age of the earliest African anthropoids. *Science* 234: 1247-1249.

Goede, A., & M.A. Hitchman 1987. Electron spin resonance analysis of marine gastropods from coastal archaeological sites in southern Africa. *Archaeometry* 29: 163-174.

Howell, N. 1979. *Demography of the Dobe !Kung*. New York: Academic Press.

Klein, R.G. 1976. The mammalian fauna of the Klasies River Mouth sites, southern Cape Province, South Africa. *S. Afr. archaeol. Bull.* 31: 74-98.

Klein, R.G. 1989. Biological and behavioural perspectives on modern human origins in southern Africa. In P. Mellars & C.B. Stringer (eds), *The human revolution: behavioural and biological perspectives on the origins of modern humans*. Edinburgh: Edinburgh University Press, pp. 529-546.

Rightmire, G.P. 1984. *Homo sapiens* in sub-Saharan Africa. In F.H. Smith & F. Spencer (eds), *The origins of modern humans: a world survey of the fossil evidence*. New York: Liss, pp. 295-325.

Shackleton, N.J. 1982. Stratigraphy and chronology of the Klasies River Mouth deposits: oxygen isotope evidence. In R. Singer & J. Wymer, *The Middle Stone Age at Klasies River Mouth in South Africa*. Chicago: Chicago University Press, pp. 194-199.

Singer, R. & J. Wymer 1982. *The Middle Stone Age at Klasies River Mouth in South Africa*. Chicago: Chicago University Press.

Singer, R. & J. Wymer 1986. On Binford on Klasies River Mouth: response of the excavators. *Curr. Anthropol.* 27: 56-57.

Stringer, C.B. 1984. Human Evolution and Biological Adaptation in the Pleistocene. In R. Foley (ed.), *Hominid Evolution and Community Ecology*. Orlando: Academic Press, pp. 55-83.

Stringer, C.B. & P. Andrews 1988. Genetic and fossil evidence for the origin of modern humans. *Science* 239: 1263-1268.

Tuttle, R.H. 1988. What's new in African palaeoanthropology? *Ann. Rev. Anthropol.* 17: 391-426.

Valladas, H., J.L. Reyss, J.L. Joron, G. Valladas, O. Bar-Yosef & B. Vandermeersch 1988. Thermoluminescence dating of Mousterian 'Proto-Cro-Magnon' remains from Israel and the origin of modern man. *Nature* 331: 614-616.

Wainscoat, J., A. Hill, A. Boyce, J. Flint, M. Hernandez, S.L. Thein, J.M. Old, J.R. Lynch, Y. Falusi, D.J. Weatherall & J.B. Clegg 1986. Evolutionary relationships of human populations from an analysis of nuclear DNA polymorphisms. *Nature* 319:491-493.

White, T.D. 1987. Cannibals at Klasies? *Sagittarius* 2(2): 6-9.

Biological relationships between Upper Pleistocene and Holocene populations in southern Africa

Alan G. Morris
Department of Anatomy & Cell Biology, University of Cape Town, Medical School, Observatory 7925, South Africa

1 INTRODUCTION

The search for the origin of the earliest anatomically modern *Homo sapiens* has been given a major boost by results of analyses of mitochondrial DNA polymorphism in living human geographic populations (such as Cann et al. 1987). These studies have produced an impressive array of mtDNA data that suggests that all of modern humanity had a common ancestor in Africa some 200 000 years ago. Of particular importance is the linking of the genetic model with current palaeontological models that support an 'Out of Africa' hypothesis (Bräuer 1989, Stringer & Andrews 1988).

Stringer & Andrews's recent review (1988) has outlined the basic view points of the two most important competing models accounting for the origin of modern *Homo sapiens*. They have shown in a convincing manner that the most parsimonious reading of the data favors the model of a relatively recent African origin of a single population of *Homo sapiens* that subsequently replaced all other populations around the world. Central to their African model is the prediction that clearly transitional forms should be apparent solely in Africa. Associated with the transition should be the presence of modern regional characteristics in specimens of relatively great antiquity in Africa but not in other geographic locations. By far the most important palaeontological specimens supporting this argument come from southern Africa in the form of the remains from Klasies River, Border Cave and Equus Cave.

There can be little doubt that the African transitional specimens do provide a sound basis for the transformation of archaic to modern populations (Bräuer 1989), but statements in support of African regional morphology in these early specimens require some critical review. How strong is the evidence for an early African morphology which reflects a continuity with extant groups?

2 KLASIES RIVER: THE LAST PHASE OF THE TRANSITION

The fragmentary human remains from Klasies River are the best provenanced of the Middle Stone Age (MSA) specimens from southern Africa. Although there is still some debate about the age of the site (Parkington 1990, Wolpoff 1989), the general consensus is that the hominids date between some 80 000 and 100 000 years ago. The specimens are distinctive for two reasons; their undoubted modernity (Bräuer 1984, 1989, Rightmire 1984, 1986, 1989) and their extremely fragmentary nature (Deacon 1988b, White 1987). In their descriptions of the specimens from the first excavation, Singer & Wymer (1982:149) stated that the material

tends to bear out the extrapolative inferences from the biological data on modern indigenes by evidencing a mosaic of characters linking the three modern indigenous populations in their ancestral forms.

They recognize the extreme variability in the sample (especially as represented by the mandibulae) and interpret the variation as resulting from the transition from archaic to regionally extant forms. Emphasis is placed on the exceptionally gracile mandibular specimen 16424 as a possible representative individual of a pattern of generalized Negroid (including Khoisan) features. Rightmire (1984, 1986, 1989) has been less assertive in linking the specimens to specific modern populations, but he has indicated that some mandibular and dental features can be found within the range of modern Africans.

Deacon (1988) has stressed the high degree of morphological variability represented by robust and gracile forms in the Klasies sample. The specimens at his disposal include new fragments from the 1984-1986 excavations. He rejects any clear evidence of a temporal trend during the occupation of the Klasies rock shelters, and emphasizes that specimens such as the 'super gracile' 16424 mandible and the robust 13400 mandible are contemporaneous. Deacon is in a speculative mode when he considers the importance of this dimorphism. He suggests that perhaps Klasies was a population under stress with intercommunity violence selecting for large males (Deacon 1989). There certainly is some support for interpersonal violence in the form of the burning and extreme fragmentation of the Klasies hominid remains. White (1987) is strongly supportive of the view that butchering marks on some of the Klasies bones are signs of cannibalism.

Deacon (1989:557) has introduced another idea which may have great importance in relation to the genetic data from living African populations. He suggests on archaeological grounds that the MSA populations of the southern Cape Province were in isolation from other African groups for an extensive period which continued until relatively recent times. In his view, the long isolation of southern Africa might have produced 'a Southern African endemic type ancestral to the San'. This differs from the views of most other current writers who have tended to see sub-Saharan MSA populations as one 'African' source (Bräuer 1989). An echo of this appears in Stringer's (1988) comments on the dating of Qafzeh. He suggests that the early date for the anatomically modern form in the Middle East could indicate an earlier than 90 000 year split in the 'African' anatomically modern stock into a southern African component and a northern group extending from the Omo region into the Near East. This could be interpreted in genetic terms as an extremely early split between Khoisan and Negroid populations.

The actual physical evidence for the linking of the Klasies specimens with any extant regional morphology is actually extremely slim. Only Singer & Wymer (1982) are prepared to make precise statements, but these are on a character for character basis in an often typological manner. Rightmire's more cautious approach is probably more realistic.

3 EQUUS CAVE AND ASSORTED MIDDLE STONE AGE TEETH

At least two southern African sites have produced isolated dental remains from MSA contexts. At Die Kelders, the few teeth recovered are not particularly informative, but they do seem to fall within the range of modern variation (Klein 1989). More hopeful is the larger sample from

Equus Cave. These specimens have been examined in detail by Grine & Klein (1985) who conclude that these Pleistocene specimens fall comfortably within the range of modern variation of southern African populations. More specifically, they feel that Equus Cave presents evidence in support for morphological continuity in southern Africa over the last 100 000 years or so.

As for most of the MSA sites under discussion, dating at Equus Cave is extremely problematical. The original assumption of antiquity at Equus Cave was based on the litho-stratigraphic correlation (Butzer 1974) and the presence of MSA artefacts in the lower units 1B, 2A and 2B. Subsequent fieldwork revealed that the latter items were washed into the cave from an MSA open site further upslope and that the amorphous unabraded *in situ* lithics were most probably of Robberg type (P.B. Beaumont, pers. comm.). That finding is confirmed by work on the pollen (Scott 1987) which suggests that the basal layers may represent the last major cold phase of the Pleistocene and the beginning of the Holocene. Most recently a radiocarbon date of 16 300 ± 160 years before present (Pta 4409) was obtained on collagen from a bone sample just above the base of the sediments in Upper Stratum 2B (P.B. Beaumont, pers. comm., 13 June 1988).

In the light of these new temporal and climatic data, Equus Cave can no longer be considered as a site which bears on the advent of anatomically modern people in Africa.

4 THE ENIGMA OF BORDER CAVE

The hominids from Border Cave have been at the center of an increasingly intense debate both about their morphology and their dating. Much hangs on the association of the fossils with the earlier MSA stratigraphy of the site. Butzer feels that all of the hominid material except the Iron Age specimen (BC4) is best placed between 65 000 and 95 000 years ago (Bräuer 1989), and, for the sake of argument, it is assumed here that this estimate is reasonable.

The debate on the morphology of the Border Cave sample has been specifically about its relationship to living African populations. All studies subsequent to the original description (Cooke et al. 1945) have accepted that the specimens represent an anatomically modern form of humanity.

The first description to use complex statistical methods was that of De Villiers (1973). In this work, she analysed the BC1 cranium by D^2 and Discriminant techniques and concluded that BC1 could be differentiated from all living southern African groups because of the large breadth of the vault. Her more general discussion of the BC2 mandible and the infant skeleton (BC3) recognized many Khoisan and Negro features, but De Villiers felt the specimens could represent a protonegroid stock at a time before differentiation into these modern populations. The detailed study of the BC2 and BC5 mandibles (De Villiers 1976) produced very small D^2 distances between the Border Cave specimens and a spectrum of fossil and modern South African mandibles, and De Villiers recorded that the Border Cave mandibles could not be significantly separated from either Khoisan or Negro samples. BC5 fits closest to the Negro male sample, and BC2 fits closest to the San female sample. The 1978 general statement by Beaumont, Vogel and De Villiers chose not to select Border Cave specimens for membership in any specific modern population, but simply to record their morphology as undoubted anatomically modern with some Khoisan-Negro affinities.

The paper by Rightmire (1979) was the first to place a Border Cave specimen (BC1) into a specific range of modern morphology. Rightmire's detailed study placed BC1 closest to the 'Hottentot' reference centroid and excluded it at the .05 limits from the other samples in his

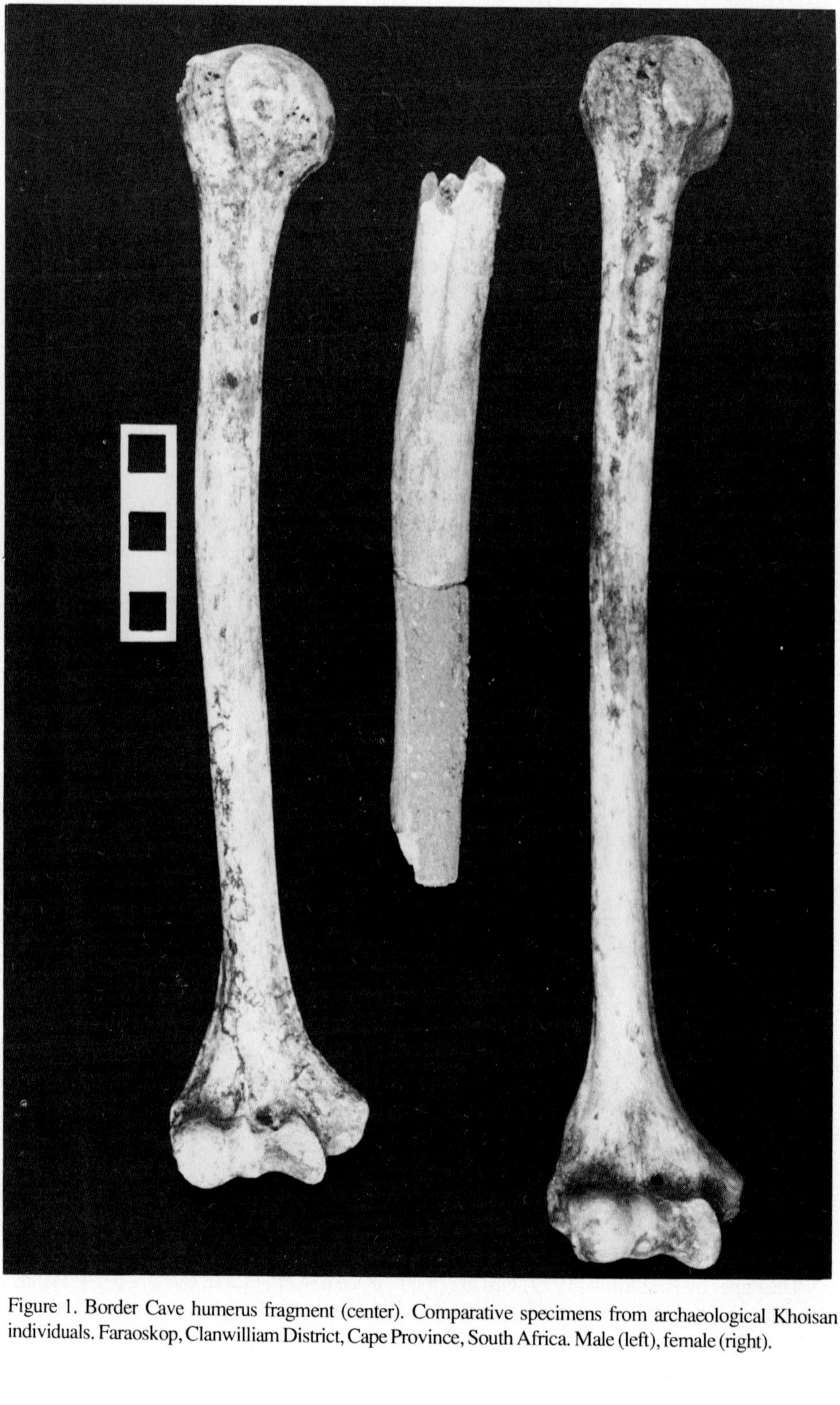

Figure 1. Border Cave humerus fragment (center). Comparative specimens from archaeological Khoisan individuals. Faraoskop, Clanwilliam District, Cape Province, South Africa. Male (left), female (right).

comparative series. His interpretation of this result was that Border Cave represented the remains of large 'Bushman-like' people. The paper sparked lively debate, and Rightmire (1981) responded to criticism with a further analysis that confirmed his original results.

Rightmire's study was followed by a comprehensive review by De Villiers & Fatti (1982) of modern and archaeological Negroid remains from throughout Africa. Using a similar statistical technique, this study clearly identified BC1 as being a member of a population 'with Negro affinities'. More recent studies (Ambergen & Schaafsma 1984, Van Vark 1986) have produced results indicating that BC1 is different from any living population, even though it is still undeniably modern.

Fatti (1986) has reviewed this controversy and has accounted for some of the problems. The largest difficulty appears to be in the choice of measurements. Rightmire's selection of measurements (also used by Ambergen and Schaafsma) is quite different from that of De Villiers and Fatti. With the exception of frontal sagittal chord, these authors have managed to choose entirely different sets of measurements despite the limited size of the vault fragment. Van Vark (1986) reports that he has used other new measurements in his study. Technical differences in the mathematical methodology of Ambergen and Schaafsma (and Van Vark) add to the differences in results, but Fatti defends the technique used by him and De Villiers, and restates their conclusion that the BC1 skull 'is most likely to have been that of a Negro male (or its contemporary counterpart)' (Fatti 1986:33).

The sophistication of the statistical technique and the choice of measurements are not the only problems in the analysis of the Border Cave calvarium. The choice of reference samples for comparison has generated its own difficulties. Fatti (1986) has pointed out variation in the choice of Negroid samples. Rightmire and De Villiers and Fatti restricted their samples of living groups to southern African populations, while Ambergen and Schaafsma included Dogon and Teita sets of crania. Van Vark (1986) has gone farther than this and included a world-wide range of comparative samples. This is of particular importance because the discriminant technique is designed to classify an unknown as a member of a reference sample where it is known that the specimen belongs to one of the groups included in the analysis.

Rightmire (1979) countered this argument by showing that the technique is still valid even if the individual lies somewhat beyond the boundaries of the samples on which the analysis is based.

Far more serious is the problem of the internal consistency of the crucial Khoi and San reference samples used by nearly all of these writers (Morris 1986, 1987). Of Rightmire's 18 Khoi specimens, half are drawn from known-in-life individuals while the remainder are 'Coloured' cadaver crania. Half of De Villiers and Fatti's San sample was typologically selected by L.H. Wells for use in De Villiers 1963 thesis, and the balance is from highly localized prehistoric graves in the eastern Cape Province. These sampling problems result in Rightmire's Khoisan sample representing a range of variation that is possibly larger and quite different from prehistoric Khoisan ranges, whereas De Villiers's San sample is restricted to only a portion of the probable full range of Khoisan variation.

In summary, it must be said that the problems of interpretation surrounding the Border Cave specimens prevent any unambiguous linkage to any geographically local sub-Saharan African populations.

4.1 *New post-cranial specimens from Border Cave*

The human skeletal remains from Klasies River and Border Cave show dramatic differences in

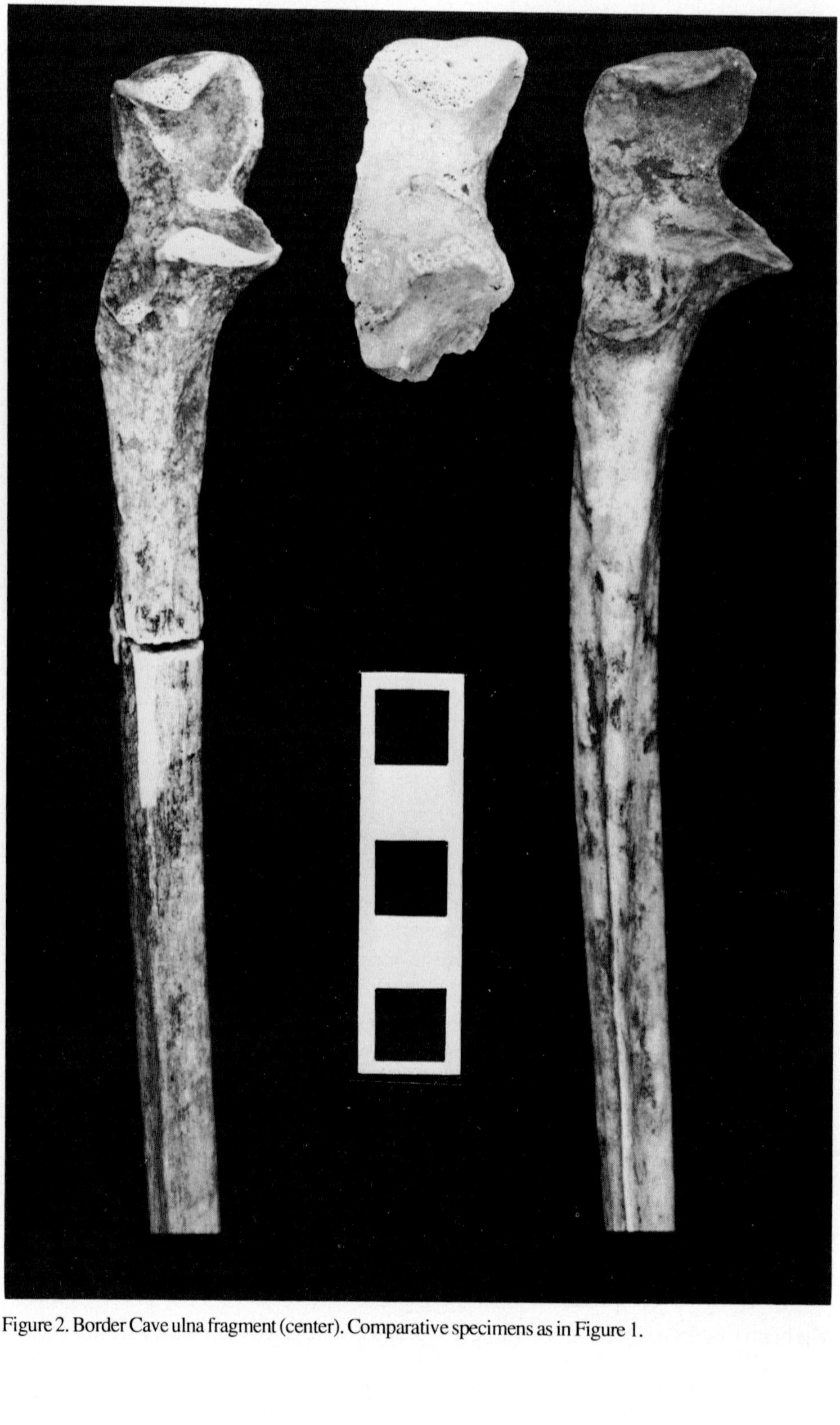

Figure 2. Border Cave ulna fragment (center). Comparative specimens as in Figure 1.

their morphology and taphonomy. The Klasies River material is highly fragmentary and contains remnants of cranial as well as post-cranial body parts. White (1987) has now shown that the fragmentation and burning may very well be the result of cannibalism. Despite the broken nature of the Border Cave specimens, they are far from fragmentary. The two mandibulae are fairly complete, and the BC1 skull consists of most of the braincase. The infant burial is virtually complete and is an intentional interment. Klein (1989) and Parkington (1990) have both remarked on the relatively intact condition of the Border Cave specimens and use this as evidence of a terminal MSA or even an LSA age for the hominids. They suggest that the adult cranial and mandibular material have drifted down into the MSA deposits as isolated bones, despite the data marshalled by Beaumont (1980) to show that the stratigraphy of the deposit is clear and undisturbed and that the hominid bones are contemporaneous with the deeper deposits at the site.

Noticeable in very recent studies of the Border Cave specimens is the lack of post-cranial material. In fact, a number of post-cranial bones were discovered during the disturbance of the site before 1941. Cooke et al. (1945) briefly described a pair of femoral and tibial shafts said to be of the same degree of preservation as the skull, and they also reported on some other finds:

> *A large number of human fragments representing nearly all parts of the skeleton were retrieved from the dump left by Mr. Horton. Some of them, including a few small cranial fragments as well as other skeletal parts, are appreciably different in preservation from the remains originally sent to us by Mr. Drew.* Cooke et al. (1945:12)

Sadly, when De Villiers attempted to study them in 1973, she found that all of this material, including the femora and tibia, had been lost.

During June and July 1987, Peter Beaumont of the McGregor Museum in Kimberley cleaned and sieved all of the sediments that had slumped with time into the 1941-2 excavation in order to plot the stratigraphy within that portion of Border Cave. A number of unprovenanced mammalian bone fragments were recovered during this operation and four of these have been identified as human post-cranial skeletal elements. These specimens are temporarily at the University of Cape Town but will be accessioned in the near future to the collection of the Department of Anatomy of the University of the Witwatersrand in Johannesburg.

Right Humeral shaft

This specimen is composed of two fragments which account for most of the tubular shaft of the right humerus (Fig. 1). Proximally, the distal portion of the intertubercular groove is present but the shaft is broken just below the surgical neck of the bone and the swelling of the lesser tuberosity is not present. At the distal end, the fracture plane of the shaft is above the supracondylar lines. The only anatomical feature which is clearly visible on this specimen is the impression of the deltoid muscle.

The deltoid impression is marked and its massive contour gives a strongly elliptical cross-section to this part of the humeral shaft. The circumference at mid-shaft (estimated to lie at the distal edge of the deltoid tuberosity) is 66 mm.

The preservation of the bone is fair, but it is heavily mineralized and matrix from the deposit adheres to sections of the cortical bone.

Right Proximal Ulnar fragment

The second specimen is a single ulnar fragment comprising the olecranon process, trochlear notch and coronoid process (Fig. 2). The bone is truncated distally below the coronoid process.

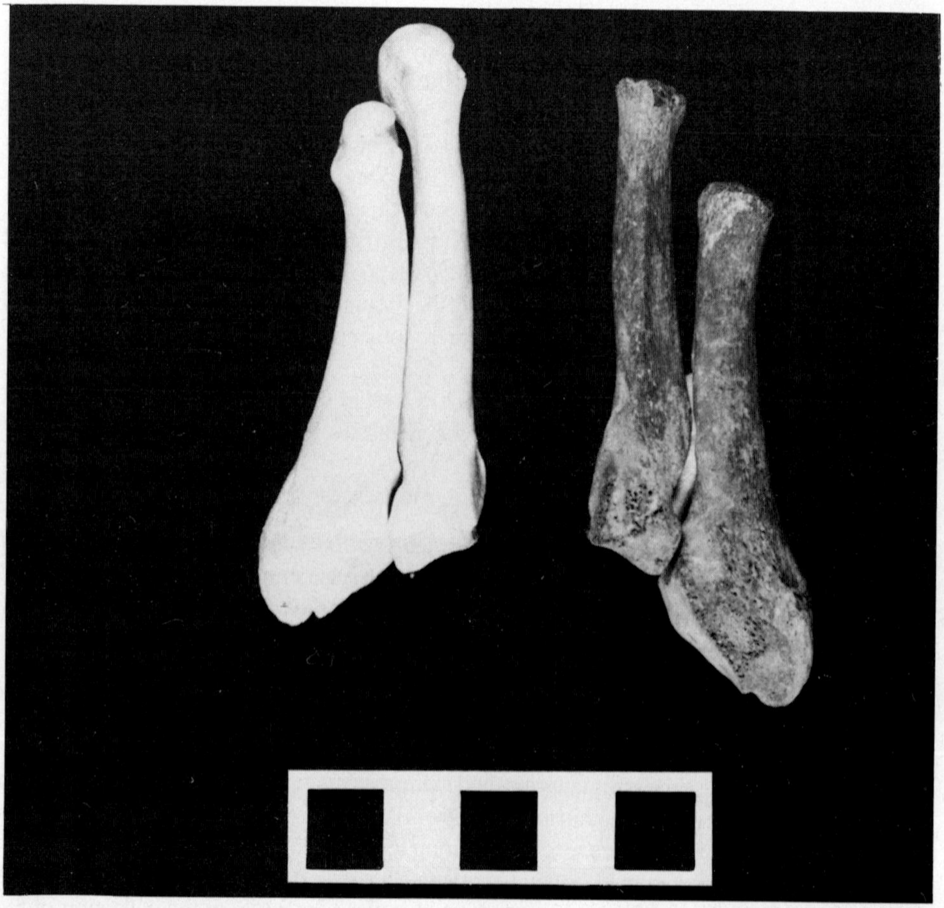

Figure 3. Border Cave metatarsals (right). Comparative specimen from recent University of Cape Town Medical School cadaver.

The olecranon epiphysis is fully united and the ossification state indicates adulthood. The articular surfaces are normal without signs of arthritic changes.

The state of preservation is less mineralized than the humerus fragment, but the distal edge of the fragment and much of the coronoid process is blackened. The blackening is probably evidence of charring as the dark color continues through the whole of the cortex and onto the trabecular bone exposed on the broken surface. The lateral surface of the bone is badly eroded and only a portion of the radial notch remains. There are no signs of cut marks or specific evidence of carnivore damage.

Right Metatarsal IV and V

These two metatarsals are nearly complete, with the exception of the distal epiphyses (Fig. 3). The epiphyses were unfused at the time of death and were not recovered. The interosseous

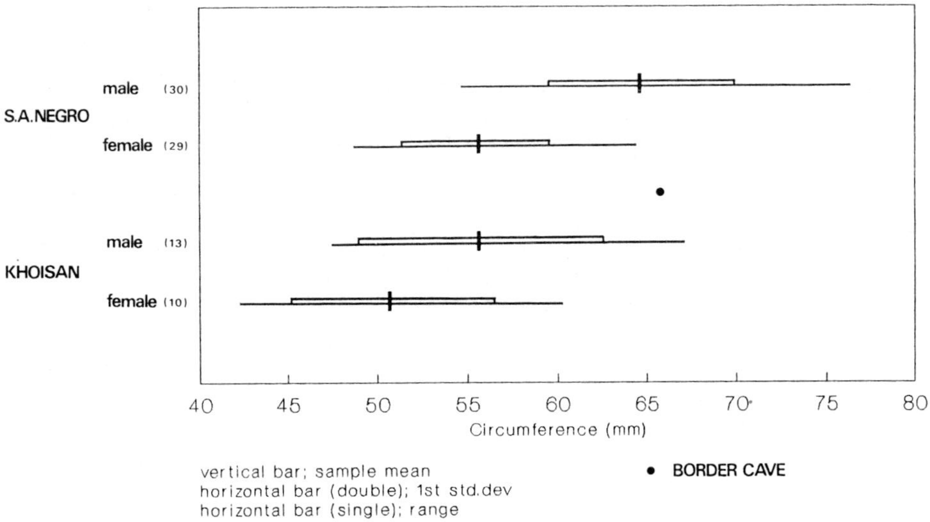

Figure 4. Mid-shaft circumference of humerus.

metatarsal articular surfaces match well, and the bones are evidently of the same individual. Muscle markings are clearly delineated.

The unfused condition of the metatarsal heads indicates that these are from a different individual than the ulna, and probably also the humerus. The timing of the fusion of the distal metatarsal epiphyses is usually by the age of 15 years in the female and slightly later in the male (Krogman & Iscan 1986).

The preservation of the metatarsals is excellent but both bones are very dark in color. The trabeculae exposed by minor surface damage on metatarsal IV, are not excessively darkened and therefore the dark color is restricted to the surface only. This may be evidence for very slight charring but this is not as clear as is visible on the ulnar fragment.

There is relatively little information that can be gleaned from this very small sample of post-cranial material. Its fragmentary nature requires comparison with non-standard measurements which are not readily available in the literature, especially for southern African samples.

The only standard measurement that can be taken on the specimens is the mid-shaft circumference on the humerus shaft. When compared to these measurements for South African Negro (Sotho-Tswana) and Khoisan specimens (Fig. 4), the circumference value exceeds the sample range for both Khoisan and Negro females and is only marginally inside the range for Khoisan males (Morris, unpubl. data). By these standards, the humerus is relatively robust. Interestingly, the now lost lower limb fragments also seem to have been large according to the length as estimated by Cooke, Malan and Wells (Fig. 5). A possibility worth considering is that the humeral, tibial and femoral fragments are all from the same individual, perhaps the same one as represented by the BC1 cranium.

The ulnar fragment is valuable because a similar fragment of a right ulna has been recovered from the recent excavations by Deacon at Klasies River. The Klasies specimen is more complete and is generally better preserved, but the two specimens are very similar in size and degree

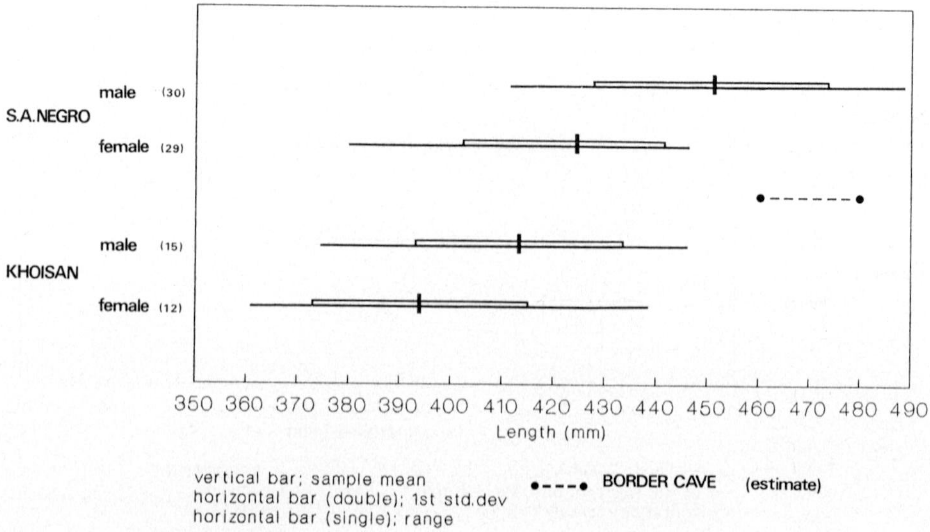

Figure 5. Maximum length of femur.

of massiveness. It might be worthwhile to compare the two specimens formally when Rightmire and Deacon publish the description of the new Klasies material.

As with the other specimens from Border Cave, the dating of these post-cranials is difficult. The material cleared in the 1987 sampling had slumped into the excavation trenches and could therefore not be associated with specific stratigraphic sections. Despite this problem of association, Peter Beaumont feels that, based on the state of preservation, the bone fragments found in the slumped material are almost certainly older than 38 000 years and may very well be as old as the oldest deposits in the site (pers. comm., 8 August 1988). Microchemical analysis is now underway in hopes to confirm the dating by chemical sequencing of these specimens and the Border Cave fauna (A. Sillen et al., in prep.).

5 DISCUSSION

The morphological data from Klasies River and Border Cave do not satisfactorily provide a clear relationship between MSA and Holocene LSA populations in southern Africa. Rightmire's (1984) belief that southern Africa has not witnessed any large scale replacement of its peoples during the last 100 000 years, can either be supported or refuted by the specimens from Klasies River and Border Cave depending on the mathematical technique, choice of measurement, or specimen of interest. The only agreement is that the specimens are anatomically modern.

The genetic picture provided by Cann et al. (1987) and others is compelling evidence that the MSA populations of southern Africa had already taken a specifically 'African' path. Interesting, too, is their diagram showing that the greatest distance between any pair is between a San individual and her neighbouring Negro cluster. Although they do not remark on this particular feature, Jones (1986) has postulated on serogenetic evidence that the San might be a population which split off from the anatomically modern lineage considerably before the time of origin of other Africans. This would fit well with the speculative ideas of Deacon (1989) and Stringer (1988) as summarized above.

The genetic picture is not without problems itself. Excoffier et al. (1987) have shown that the incredibly small sample sizes chosen by Cann et al. (1987) and others probably severely underestimate the mtDNA variability. They feel that in any other classical genetic marker study, these samples would have been rejected. The same paper also warns that mtDNA polymorphisms may be affected by selection and this may skew the perceived relationships between populations. Their conclusion warns that small nonhomogeneous samples and large interindividual differences lead to problematic data interpretation. They present an alternative interpretation to Cann et al. in which the ancestral population of Africa arose outside of the continent.

> *Though Africans are genetically clearly differentiated from other populations of the world, they do not seem to constitute the latter's direct ancestors.* Excoffier et al. (1987:184)

It must be recognized that although the independence of the genetic and palaeontological data can provide an internal check for theoretical models like those outlined by Stringer & Andrews (1988), the data are so different that they may confuse rather than clarify the issue. It is already recognized that mitochondrial and nuclear DNA have vastly different rates, and possibly directions, of evolution. Serogenetic data often do not agree in detail with the DNA differences, and factors of drift and gene flow can mask relationships.

Palaeontological data, on the other hand, identify specific past events. Samples of single crania (such as Border Cave 1) are open to a wide range of interpreation with little hope of clarification unless further individuals from similar contexts are located. The similarity of a single 100 000 year old specimen to an extant population cannot be taken as proof of genetic similarity unless a transforming sequence over time can be demonstrated. This is not the case in southern Africa, and until a larger MSA sample, or a sequence of specimens from the Upper Pleistocene through the Holocene is excavated, proof of the continuity of southern African peoples over the last 100 000 years cannot be demonstrated.

ACKNOWLEDGEMENTS

Access to original specimens was kindly provided by P.B. Beaumont, H.J. Deacon and P.V. Tobias. Especially appreciated has been Peter Beaumont's co-operation in providing unpublished information on both Border Cave and Equus Cave and for his comments on the manuscript. I am also grateful to H. De Villiers, T. Jenkins and J.E. Parkington for their ideas and comments. Elise Fuller provided her expertise for the photographs.

REFERENCES

Ambergen, A.W. & W. Schaafsma 1984. Interval estimates for posterior probabilities, applications to Border Cave. In G.N. van Vark & W.W. Howells (eds), *Multivariate Statistical Methods in Physical Anthropology.* Dordrecht: Reidel, pp. 115-134.

Beaumont, P.B. 1980. On the age of the Border Cave Hominids 1-5. *Palaeont. Afr.* 23: 21-33.

Beaumont, P.B., H. de Villiers & J.C. Vogel 1978. Modern man in sub-Saharan Africa prior to 49 000 years BP: a review and evaluation with particular reference to Border Cave. *S. Afr. J. Science* 74: 409-419.

Bräuer, G. 1984. A craniological approach to the origin of anatomically modern *Homo sapiens* in Africa and implications for the appearance of modern Europeans. In F.H. Smith & F. Spencer (eds), *The Origins of Modern Humans.* New York: A.R. Liss, pp. 327-410.

Bräuer, G. 1989. The evolution of modern humans: a comparison of the African and non-African evidence. In P. Mellars & C.B. Stringer (eds), *The Human Revolution: Behavioural and Biological Perspectives on the Origins of Modern Humans.* Edinburgh: University of Edinburgh Press, pp. 123-154.

Butzer, K.W. 1974. Palaeoecology of South African Australopithecines: Taung revisited. *Curr. Anthropol.* 15: 367-382.

Cann, R.L., M. Stoneking & A.C. Wilson 1987. Mitochondrial DNA and human evolution. *Nature* 325: 31-36.

Cooke, H.B.S., B.D. Malan & L.H. Wells 1945. Fossil man in the Lebombo Mountains, South Africa: The 'Border Cave', Ingwavuma District, Zululand. *Man* 54: 6-13.

De Villiers, H. 1973. Human skeletal remains from Border Cave, Ingwavuma District, Kwazulu, South Africa. *Annals of the Transvaal Museum* 28(13): 229-256.

De Villiers, H. 1976. A second adult human mandible from Border Cave, Ingwavuma District, Kwazulu, South Africa. *S. Afr. Science.* 72: 212-215.

De Villiers, H. & L.P. Fatti 1982. The antiquity of the Negro. *S. Afr. J. Science.* 78: 321-332.

Deacon, H.J. 1988. The origins of anatomically modern people and the South African evidence. *Palaeoecol. Africa.* 19.

Deacon, H.J. 1989. Late Pleistocene palaeoecology and archaeology in the southern Cape. In P. Mellars & C.B. Stringer (eds), *The Human Revolution: Behavioural and Biological Perspective on the Origins of Modern Humans.* Edinburgh: University of Edinburgh Press, pp. 547-564.

Excoffier, L., B. Pellegrini, A. Sanchez-Mazas, C. Simon & A. Langaney 1987. Genetics and History of sub-Saharan Africa. *Yrbk. Phys. Anthropol.* 30: 151-190.

Fatti, L.P. 1986. Discriminant analysis in prehistoric physical anthropology. In R. Singer & J.K. Lundy (eds), *Variation, Culture and Evolution in African Populations: papers in honour of Dr. Hertha de Villiers.* Johannesburg: Witwatersrand University Press, pp. 27-34.

Grine, F.E. & R.G. Klein 1985. Pleistocene and Holocene human remains from Equus Cave, South Africa. *Anthropol.* 8: 55-98.

Jones, J.S. 1986. The Origin of *Homo sapiens*: the genetic evidence. In B. Wood, L. Martin & P. Andrews (eds), *Major Topics in Primate and Human Evolution.* Cambridge: Cambridge University Press, pp. 317-330.

Klein, R.G. 1989. Biological and behavioral perspectives on modern human origins in southern Africa. In P. Mellars & C.B. Stringer (eds), *The Human Revolution: Behavioural and Biological Perspectives on the Origins of Modern Humans.* Edinburgh: University of Edinburgh Press, pp. 529-546.

Krogman, W.M. & M.Y. Iscan 1986. *The Human Skeleton in Forensic Medicine.* Springfield: Thomas.

Morris, A.G. 1986. Khoi and San Craniology: A re-evaluation of the osteological reference samples. In R. Singer & J.K. Lundy (eds), *Variation, Culture and Evolution in African Populations: papers in honour of Dr. Hertha de Villiers.* Johannesburg: Witwatersrand University Press, pp. 1-12.

Morris, A.G. 1987. The Reflection of the Collector: Khoi and San skeletons in museum collections. *S. Afr. Arch. Bull.* 42(145): 12-22.

Parkington, J. 1990. A critique of the consensus view on the age of Howiesons Poort assemblages in South Africa. In P. Mellars (ed.), *The Emergence of Modern Humans.* Edinburgh: University of Edinburgh Press., 34-55.

Rightmire, G.P. 1979. Implications of Border Cave skeletal remains for Later Pleistocene human evolution. *Curr. Anthropol.* 20: 23-35.

Rightmire, G.P. 1981. More on the study of the Border Cave remains. *Curr. Anthropol.* 22: 199-200.

Rightmire, G.P. 1984. Homo sapiens in sub-Saharan Africa. In F.H. Smith & F. Spencer (eds), *The Origins of Modern Humans: A world survey of the fossil evidence.* New York: A.R. Liss, pp. 295-325.

Rightmire, G.P. 1986. Africa and the origins of modern humans. In R. Singer & J.K. Lundy (eds), *Variation, Culture and Evolution in African Populations: papers in honour of Dr. Hertha de Villiers.* Johannesburg: Witwatersrand University Press, pp. 209-220.

Rightmire, G.P. 1989. Middle Stone Age humans from eastern and southern Africa. In P. Mellars & C.B. Stringer (eds), *The Human Revolution: Behavioural and Biological Perspectives on the Origins of Modern Humans.* Edinburgh: University of Edinburgh Press, pp. 109-122.

Scott, L. 1987. Pollen analysis of hyena coprolites and sediments from Equus Cave, Taung, southern Kalahari (South Africa). *Quat. Res.* 28: 144-156.

Singer, R. & J.J. Wymer 1982. *The Middle Stone Age at Klasies River Mouth in South Africa.* Chicago: University of Chicago Press.

Stringer, C.B. 1988. The dates of Eden. *Nature* 331: 565-66.

Stringer, C.B. & P. Andrews 1988. Genetic and fossil evidence for the origin of modern humans. *Science* 239: 1263-1268.

Van Vark, G.N. 1986. More on the classification of the Border Cave 1 skull. *5th Congress of the European Anthropological Association*, Lisbon (abstract).

White, T.D. 1987. Cannibals at Klasies? *Sagittarius* 2: 6-9.

Wolpoff, M.H. 1989. Multiregional Evolution: The fossil alternative to Eden. In P. Mellars & C.B. Stringer (eds), *The Human Revolution: Behavioural and Biological Perspectives on the Origins of Modern Humans.* Edinburgh: University of Edinburgh Press, pp. 62-108.

The role of continuity in modern human origins

Fred H. Smith
Department of Anthropology, Northern Illinois University, De Kalb, USA

1 INTRODUCTION

Recently, Stringer & Andrews (1988) have argued that modern humans arose as the result of cladogenesis in southern Africa, then radiated to Eurasia and ultimately to the rest of the habitated world. Although others had previously suggested southern Africa as the initial point of origin for modern humans (e.g. Beaumont et al. 1978, Bräuer 1984, Rightmire 1984), the Stringer and Andrews Recent African Evolution Model strongly implies that modern humans did not assimilate or 'hybridize' with the archaic humans (e.g. Eurasian Neanderthals or East Asian forms like Dali or Maba) they encountered during their radiation. Rather, the Model asserts that these archaic populations were probably totally replaced throughout Eurasia by a new species, for which Stringer and Andrews would reserve the taxonomic name *Homo sapiens*.. Thus in this scenario, all Eurasian archaic humans became extinct without any contribution to modern human gene pools.

Stringer (this volume) has stated that the Recent African Evolution Model does not require total replacement of archaic human populations by modern ones and, further, that his recognition of species differences between archaic and modern humans is based in an evolutionary (cf. Wiley 1981), not biological, species concept. Nevertheless, the main effect of the Stringer and Andrews Model, regardless of the authors' intent, has clearly been to promote an impression that modern humans were so qualitatively different from Eurasian archaic humans as to preclude the possibility of significant biological connections between them. Consequently, it would seem justified to refer to the Recent African Evolution Model as a speciation/replacement model of modern human origins.

There is a broad spectrum of current thinking regarding the nature of modern human origins (see Smith et al. 1989a for a recent review), including less extreme replacement models such as that of Bräuer (1984, 1989, this volume). Often compared to the Recent African Evolution Model is the Regional Continuity Model (Wolpoff 1985, 1989, this volume, Wolpoff et al. 1984). There are also different perspectives on the nature of continuity in modern human origins (cf. Smith et al. 1989a). However each of these 'continuity' perspectives emphasize that most, if not all, archaic human populations contributed to regional modern human gene pools to some extent; and each of them recognizes that gene flow was critical to the emergence of modern humans throughout the Old World. Thus, regional continuity is no longer a claim for independent, largely isolated emergence of modern humans in several regions of the world, as it was for Coon (1962). Certain perspectives which emphasize the importance of continuity do not even claim that regional archaic human populations were invariably the major contributors to modern human gene pools in their respective geographic regions (Smith et al. 1989a). At its most basic level, continuity is only an argument that archaic humans played some role in the origin of regional modern human populations throughout the Old World and thus cannot be summarily removed from the ancestry of recent people, as speciation/replacement models and their supporters (Delson 1988, Gould 1987, 1988, Lewin 1988) advocate.

Clearly, the last few years have witnessed a quantum surge in the popularity of speciation/replacement models in general and such a model with an African nucleus for modern human origins in particular. The reasons for the popularity of such views are complex and can only be briefly summarized here. In the first place, there has long been a historical pattern of discomfort with the idea that Neanderthals (or similar southwest Asian forms) could be ancestral to early modern people (Spencer 1984, Spencer & Smith 1981, Trinkaus 1982). This has led to numerous attempts to find a way to bypass such hominids in the derivation of modern humans. The various versions of the presapiens and preneanderthal hypotheses formulated since the turn of the century are all examples of such attempts. Second, from a theoretical evolutionary biological perspective, the popularity of explaining all significant macroevolutionary change as resulting from cladogenetic (punctuational) events is increasing in paleoanthropology (see for example the majority of pertinent entries in Tattersall et al. 1988). A speciation/replacement model of modern human origins is viewed as conforming totally with this theoretical perspective (Gould 1987). Third, recent chronological data suggest a much earlier appearance of

Table 1. Theoretical predictions from models of *Homo sapiens* evolution (from Stringer & Andrews 1988:1264).

Aspect	Multiregional evolution	Recent African evolution
1) Geographic patterning of human evolution	Continuity of pattern from middle Pleistocene to present	Continuity of pattern only from late Pleistocene appearance of *H. sapiens* to present
	Interpopulation differences are high, greatest between each peripheral area	Interpopulation differences relatively low, greatest between African and non-African populations
	Intrapopulation variation greatest at center of human range	Intrapopulation variation greatest in African populations
2) Regional continuity and the establishment of *Homo sapiens*	Transitional fossils widespread	Transitional fossils restricted to Africa, population replacement elsewhere
	Modern regional characters of high antiquity at peripheries	Modern regional characters of low antiquity at peripheries (except Africa)
	No consistent temporal pattern of appearance of *Homo sapiens* characters between areas	Phased establishment of *Homo sapiens* suite of characters: (i) Africa, (ii) S. W. Asia, (iii) other areas
3) Selective and behavioral factors involved in the origin of *Homo sapiens*	Factors varied and widespread, perhaps related to technology; local behavioral continuity expected	Factors special and localized in Africa; behavioral discontinuities expected outside Africa

modern humans in southern Africa and western Asia than in other parts of the world (Bräuer 1984, Deacon 1989, Deacon & Geleijnse 1988, Stringer et al. 1989, Valladas et al. 1988). Thus, fully developed modern humans are seen as already existing in Africa and the Near East long before the eventual disappearance of archaic humans in the rest of Eurasia. Fourth, the existence of qualitative differences in behavior, particularly as reflected in the European Middle and early Upper Paleolithic archaeological record, has been increasingly emphasized (e.g. Binford 1985, 1989). Fifth, the basic homogeneity of modern human skeletal form throughout the world (Howells 1973) is argued to reflect a recent common origin for this morphological pattern (Stringer 1989, but see Wolpoff 1986, Kidder et al. this volume). And finally, recent analyses of both mitochondrial (mt)DNA (Cann et al. 1987) and polymorphisms reflecting nuclear DNA (Nei 1985) are interpreted as indicating a relatively recent, most probably African, origin of all living people. It is the combination of these different lines of evidence which, in the minds of supporters of the Recent African Evolution Model, makes it the most parsimonious explanation for the emergence and subsequent radiation of modern humans.

2 PREDICTIONS

Stringer & Andrews (1988:1264) present a series of theoretical predictions of expected patterning in human genetic and paleontological data for both their Recent African Evolution Model and their interpretation of the Regional Continuity Model (Table 1). They stress the importance of both types of data, but particularly note the need to incorporate the genetic data into models of modern human origins. They conclude by stating that '... paleoanthropologists who ignore the increasing wealth of genetic data on human population relationships will do so at their peril' (p. 1268). At the risk of imperiling myself, I will limit discussion in this paper to the paleontological predictions of the Stringer & Andrews model. I would, however, point out that the interpretations of both mtDNA and nuclear DNA variation favored by Stringer & Andrews and accepted by other proponents of this view (e.g. Gould 1987, Lewin 1988) have been called into question on a number of grounds (see Excoffier et al. 1987, Excoffier & Langaney 1989, Smith & Paquette 1989, Spuhler 1988, Wainscoat 1987, Wolpoff 1989). Wolpoff (1989, this volume) and Smith et al. (1989a) in particular have discussed in detail many factors other than a recent African speciation event that logically might have led to the patterns observable in modern human genetic variability. Thus, while there is no questioning the fact that the genetic data is of great importance, there is not unanimity on how these data should be interpreted. Clearly the genetic data do not present the unified, unequivocal support for the Recent African speciation/replacement model often implied by supporters of that model.

Stringer and Andrews make three predictions regarding the nature of the paleontological record that should be met if the Recent African Evolution Model is correct (Table 1). First, transitional fossils between archaic and modern humans should be found only in Africa, since the only transition to modern humans occurred there. Second, regional clade features should be of low antiquity except in Africa. Presumably, they should not predate the earliest modern humans anywhere in Eurasia. Third, modern humans should have a greater antiquity in Africa than anywhere else in the world.

2.1 *Modern human antiquity in Africa*

The case that modern people have substantial antiquity in southern Africa rests primarily on human remains from the Klasies River Mouth Caves (KRM), Border Cave, and the Kibish Formation at Omo (Bräuer 1984, 1989, Stringer 1989). In my opinion, the Omo 1, Border Cave 1, and Klasies River Mouth (KRM 16425) cranial specimens exhibit a morphological pattern that justifies their classification as early modern Africans (Smith 1985, Smith et al. 1989a). They retain larger brow ridges, particularly in the supero-inferior dimension (thickness), than is typical of more recent southern Africans; but these structures are reduced in size compared to more archaic Africans like Florisbad or Ngaloba. Zygomatic bones from Border Cave (1) and KRM (16651) are basically modern in form, with horizontally oriented lateral zygomatico-alveolar crests. The KRM specimen interstingly exhibits a rather columnar frontal process, a feature described as characteristic of archaic humans (Smith 1976, Smith & Ranyard 1980). The mandibles from this group are more variable. Border Cave 2 and 5 could be lost in a modern sample. KRM 41815 is quite robust, but conforms to a modern pattern – including the possession of a distinct mental eminence. KRM 21776 and 13400 exhibit rather vertical symphyses, with only slight development (21776), or absence (13400), of a mental eminence and are thus morphologically somewhat more primitive than the Border Cave mandibles. Interestingly, new research by Deacon (1989, Deacon & Geleijnse 1988) suggests that KRM 41815 is not chronologically older than the other KRM mandibles, as Singer & Wymer (1982) originally claimed, but broadly contemporary with them.

Thus, the total morphological pattern of the Border Cave, KRM, and Omo 1 hominids is commensurate with their classification as modern humans. Furthermore, the somewhat primitive aspects of certain features in this sample (particularly in the KRM fossils), like brow ridge development and mandibular symphyseal morphology in some specimens, are to be expected in early representatives of modern humans and help to establish their phylogenetic connection to earlier, more archaic Africans. The presence of these more archaic folk at KRM is suggested by yet unpublished maxillae, which are stratigraphically older (Deacon 1989) than the KRM specimens discussed above.

The problem with the hominid remains from these sites does not really involve their morphology but their context and date. Despite tendencies to argue that these hominids date in excess of about 100 kyr, there is no unequivocal evidence that this is the case (Smith 1985, Smith & Paquette 1989, Smith et al. 1989a Wolpoff 1989). At Klasies River Mouth, Uranium disequilibrium dating places the boundary between the MSA 1 (Middle Stone Age) and MSA 2 at 98 kyr to 110 kyr ago (Deacon 1989, Deacon & Geleijnse 1988). The hominids (except the new maxillae and some other fragments) are from MSA 2 and thus are younger than the dated boundary. The Howiesonspoort levels above the MSA 2 deposits are dated at 95 kyr by Butzer (1978) on the basis of sedimentology, maximally 50 kyr by Shackleton (in Wymer & Singer 1982) on the basis of oxygen isotope analysis, and between 18 kyr and >38 kyr on admittedly old ^{14}C determinations (Sampson 1974). While it would seem reasonable to accept an age of between about 40 kyr, or perhaps 80 kyr (Deacon 1989), and 100 kyr for the KRM MSA 2 sample, it is not presently possible to be certain of their exact age within this range.

The dating of Omo Kibish 1 has similar problems. The Thorium/Uranium date of 130 kyr for Omo 1 is generally considered unreliable, because it was derived from molluscs. A ^{14}C date of > 37 kyr has been obtained for a level in Member 3, considerably higher up in the stratigraphic sequence (also on mollusc material), and has also been used to support a relatively high antiquity for Omo 1. However, Howell (1978:216) suggests that the meager mammalian fauna associated with Omo 1 '... is frankly unhelpful and unconvincing of any very remote antiquity.' Thus, an age of greater than 37 kyr may be reasonable for Omo 1, but a secure radiometric age estimate for this specimen is not available at present.

The Border Cave specimens have problems with both context and date. These have been outlined in detail by Klein (1983), Smith (1985), and Stringer (1989). Even if we assume the Border Cave 1, 2 and 5 specimens are associated with MSA levels at the site (which is supported by Beaumont's [1980] chemical analyses), it is not clear exactly how old these deposits are. Radiocarbon analyses from above the deposits presumed to have contained these hominids yield values that suggest an age of > 49 kyr (Beaumont et al. 1978). Butzer (pers. comm.) has recently argued that the Border Cave hominids are between 65 kyr and 95 kyr old, on the basis of correlation with oxygen isotope stages, but Parkington (1990) has noted that correlation between an inland site like Border Cave and isotope changes in sea water involves making some very questionable assumptions.

My own opinion is that an age estimate of > 50 kyr for the KRM, Border Cave and Omo 1 hominids is not unreasonable on the basis of our present knowledge, but a dogmatic claim for modern humans in Africa at 100 kyr or earlier has no solid basis in the available paleoanthropological record. If an age approaching 100 kyr for the emergence of modern human anatomical form in Africa were to be firmly established, it would certainly support the Stringer & Andrews prediction regarding the earliest appearance of modern humans. However, if the earliest modern Africans only have an antiquity of 40 to 60 kyr, and if the 92 kyr thermoluminescence date for the early modern hominids from Qafzeh (Valladas et al. 1988) or the 80-100 kyr electron spin resonance data for the early modern hominids from Skhūl (Stringer et al. 1989) are correct, it might well be that modern human anatomical form has a greater antiquity in southwest Asia than southern Africa. If so, much of the genetic evidence cited in support of a recent common origin for all modern humans and the view that modern and archaic humans represent separate biological species would have to be substantially re-evaluated, since both are based on an African rather than Asian origin for modern people.

On the basis of our knowledge at present, it does seem likely that the appearance of modern human anatomical form in Europe and other 'peripheries' is a more recent phenomenon. The earliest modern humans from Europe almost certainly postdate 40 kyr; and there is no convincing evidence that modern humans appear earlier than this in either mainland East Asia or Australasia (Habgood 1989a, this volume), although chronological uncertainty makes dogmatic statements regarding the timing of the earliest appearance of modern humans impossible for either region (Smith 1985). If modern human anatomical form appears earlier in Africa and/or southwest Asia than Europe, eastern Asia and Australasia, it would suggest that some form of gene flow into these 'peripheral' regions was involved in the emergence of modern Europeans, east Asians and Australians. However, such a situation is by no means excluded by any regional continuity model and certainly does not demonstrate that a biological speciation event was involved in the emergence of modern anatomical form or that Neanderthals and Asian archaic hominids were irrelevant to the origins of modern Europeans and Asians.

2.2 *Regional continuity: Africa*

On the other hand, if Stringer and Andrews' other two predictions are true, namely that no transitional fossils can be found outside Africa and that no non-African regional clade features have much antiquity, rejection of the Recent African speciation/replacement model would be very difficult. Stringer (1988:566) has recently listed the specimens he considers representative of the African transition between archaic and early modern humans. They are Florisbad, Ngaloba, Omo Kibish 2, and the Jebel Irhoud sample. Stringer also implies that Zuttiyeh from Israel may fit into this group, but recent analysis of this specimen casts considerable doubt on

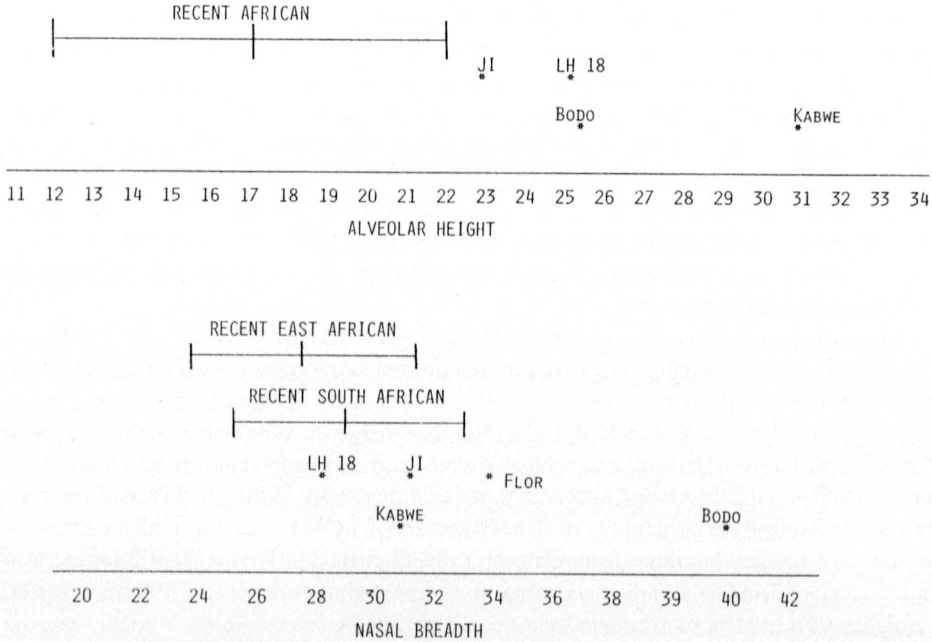

Figure 1. Alveolar heights and nasal breadths of African fossil specimens compared to recent African ranges for the same measurements. Data for the fossil specimens were collected by the author, and the data for recent Africans are from Howells (1973). All measurements are in mm. Key to fossil specimens: LH 18 = Ngaloba, JI = Jebel Irhoud 1, FLOR = Florisbad and KABWE = Kabwe 1.

its 'transitional' nature (Simmons et al., in press). I would agree that the African specimens listed by Stringer are an excellent transitional group between more archaic forms like Kabwe and the early modern Africans discussed previously (see Smith 1985, Smith et al. 1989a). I have some reservations about the Jebel Irhoud specimens being included in this group, since North Africa has been culturally and geographically more closely linked to the circum-Mediterranean world than to southern Africa since at least the end of the Acheulean. It would seem to me that, since the speciation event leading to modern humans presumably was a sub-Saharan African phenomenon, transitional forms should be restricted to that portion of Africa only. But having registered my concern on this point, I have included Jebel Irhoud 1 in the comparisons that follow.

If one compares this African Transitional Group with specimens like Kabwe, which everyone seems to agree represents the ancestral African condition, the African Transitional Group clearly exhibits changes in the direction of the early modern African condition. These include aspects of facial reduction (particularly in height), more horizontally oriented lateral zygomatico-alveolar crests, reduced brow ridges and development of canine fossae. I have chosen to specifically evaluate three features which reflect these changes and allow comparisons of pattern between these hominids and potential transitional forms outside of Africa. The features examined are nasal breadth, alveolar height, and brow ridge form.

Nasal breadth range (Fig. 1) shows a general trend toward reduction in the African Transitional Group compared to the earlier Kabwe and Bodo specimens but remains above the mean for recent South and East Africans. Alveolar height (Fig. 1) indicates an even clearer intermediate position for the African Transitional Group. Finally, metric values for both antero-posterior projection and supero-inferior thickness of brow ridges are reduced in the African Transitional Group compared to earlier hominids, yet are larger than in recent Africans (Smith, pers. observ.). Despite the uncertainty associated with the small available sample sizes, these data, along with other features (Bräuer 1984, Day & Stringer 1982, Smith 1985), suggest that designation of this group as a transitional sample in Africa is quite defendable.

2.3 *Regional continuity: Europe and elsewhere*

Using the same line of reasoning and the same characteristics, I would argue that other, non-African fossil groups make equally viable transitional samples. For example, the Vindija hominids from Yugoslavia are an excellent transitional sample between earlier south-central European Neanderthals and early modern humans from the same region (Smith 1984, Smith et al. 1989b, Wolpoff et al. 1981). The Vindija sample is admittedly fragmentary; but for most pertinent morphological comparisons, Vindija sample sizes are equal to or larger than the African Transitional Group. Nasal breadths and alveolar heights for the two available Vindija maxillae, for instance, fall closer to the modern central European mean than to the European Neanderthal mean (Fig. 2). Vindija adult brow ridges are also reduced compared to the earlier Krapina Neanderthals in both thickness and projection (Smith & Ranyard 1980, Smith et al. 1989b). Finally, the Vindija mandibles exhibit a less archaic symphyseal morphology than the Krapina mandibles (Smith et al. 1985, Wolpoff et al. 1981) and are, in some ways, only slightly more archaic than the Klasies River Mouth mandibles discussed previously. While it is true that we do not have a complete picture of total cranial morphology in the Vindija hominids, what is observable exhibits a clearly transitional pattern for central Europe.

It has been argued that these transitional features in the Vindija sample are parallelisms and therefore irrelevant to the origin of modern Europeans. Howell (1984:xx), for example, finds them 'equivocal' because the Vindija total morphological pattern is Neanderthal and not anatomically modern. While this interpretation could be correct, it seems to me that Neanderthals exhibiting some features approaching the modern human condition would be exactly what one would expect of a 'transitional' form in Europe. It is possible that the changes seen in the Vindija sample reflect the effects of some form of gene flow from modern human populations further east, like the Qafzeh people for instance. But this would certainly not constitute replacement, since the Vindija hominids are obviously much more similar to other Neanderthals than to the Qafzeh sample.

Stringer & Andrews (1988:1266) argue that 'genuine' features indicating regional continuity in Europe are essentially non-existent, '... although there are undoubtedly shared primitive characters, such as relatively large brow ridges and teeth compared with modern Europeans and homoplasies found elsewhere, such as protruding occipital regions.' The evidence for continuity in the brow ridges of South-Central Europeans is, however, not simply one of size but also shape. Smith & Ranyard (1980) have shown that the temporal sequence from Krapina to Vindija to early modern Europeans (Mladeč, Velika Pécina, Brno 2, etc.) is characterized by a pattern of increasingly greater diminution of the brow ridge at mid-orbit (a region lying slightly lateral to the true midpoint distance over each orbit) relative to move lateral or medial regions. This ultimately results in separation of the brow ridge into a superciliary arch and supraorbital trigone in early modern Europeans.

This pattern can be demonstrated graphically (Fig. 3) by use of a simple index of mid-orbital

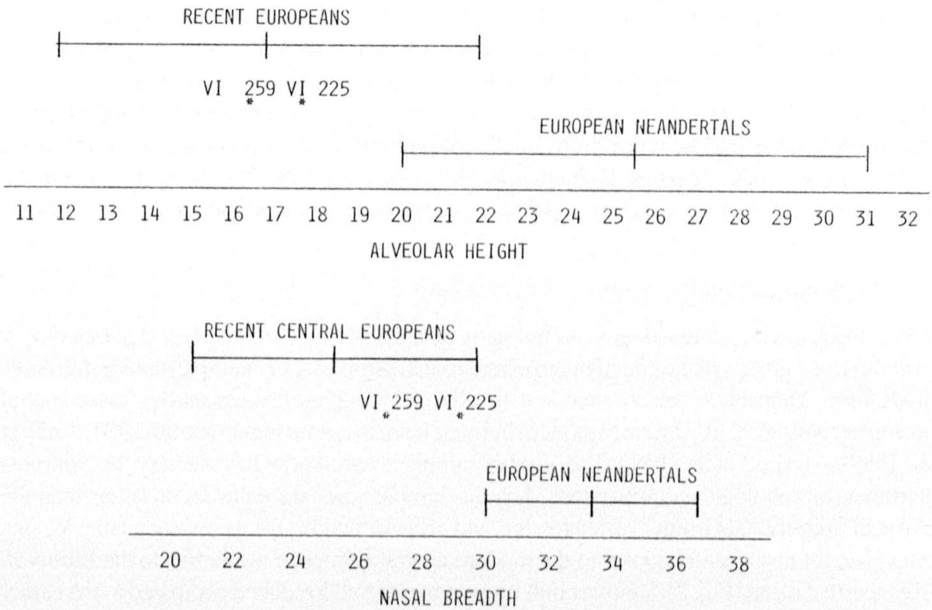

Figure 2. Alveolar heights and nasal breadths for the Vindija maxillae compared to other European Neanderthals and recent Europeans. All measurements are in mm and were made by the author.

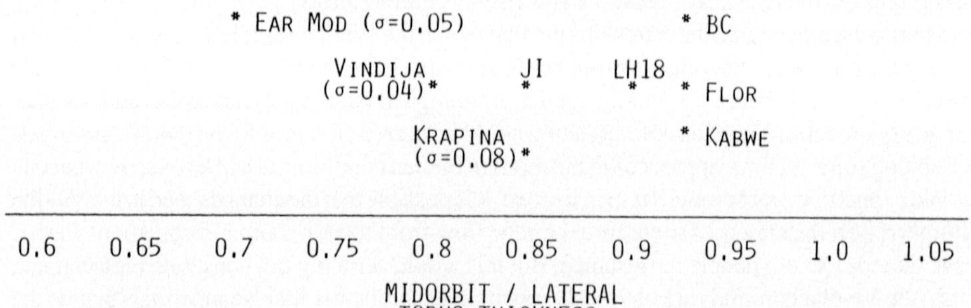

Figure 3. Indices of midorbital to lateral thickness of the brow ridges. The Index is calculated by dividing the midorbital thickness by the lateral thickness for individual specimens. All measurements and calculations were made by the author. Key to fossil specimens: LH 18 = Ngaloba, JI = Jebel Irhoud 1, FLOR = Florisbad, KABWE = Kabwe 1, BC = Border Cave 1 and EAR MOD = early modern central Europeans.

to lateral brow ridge measurements (measurements are defined in Smith & Ranyard 1980). For this index, the smaller the value, the greater the mid-orbit reduction relative to the lateral part of the brow ridge. The pattern for brow ridge thickness in Central Europe is clear (projection follows the same pattern): Vindija is intermediate in degree of midorbit reduction between Krapina and early modern Europeans. The pattern in African hominids is different. Except for

Jebel Irhoud 1, reduction in the African specimens occurs rather evenly across the torus. The pattern of relative mid-orbit reduction is not evident in the southern African sample, even in Border Cave 1. Thus I would argue that the pattern of brow ridge reduction, not simply the occurrence of reduction per se, is a potential regional characteristic for Europe. It certainly is not evident in the African hominids. Frayer (this volume) documents other features also suggesting morphological continuity in Europe.

The high frequency of occipital bunning in European Neanderthal and early modern Central Europeans has also been cited by me as evidence of European continuity (Smith 1984). Stringer & Andrews (1988) note the presence of this feature in modern North African groups and suggest that it represents a homoplasy. In my opinion, the presence of bunning in North Africa, even in the Mousterian-associated Jebel Irhoud hominids, is most likely to be the result of genetic influence from Europe, where bunning occurs earlier (e.g. in the Riss-aged Biache and other European specimens) and more frequently in archaic and early modern samples. Certainly bunning is not characteristic of southern African populations at this time, and I think the spread of this feature into North Africa is an example of the fact that patterns of human gene flow are complex and multidirectional during the Late Pleistocene. I would also explain the similarity of brow ridge pattern between Jebel Irhoud 1 and European Neanderthals (Fig. 3) as suggesting European influence in North Africa during the Late Pleistocene.

Similar indications of continuity have been documented for other peripheral areas, namely mainland East Asia and Australasia. While Habgood (1989a, this volume) has argued that some of the indications of regional continuity in both regions presented by Wolpoff and his co-workers (Thorne & Wolpoff 1981, Wolpoff 1985, Wolpoff et al. 1984) are not demonstrable clade features for these regions, he has isolated a smaller, but convincing, set of characteristics which do suggest some degree of regional continuity in each of these regions.

It is interesting to note that, despite the phenetic evidence of transition from archaic to modern humans in Africa, no conclusive African clade features have been isolated that traverse this transition (Habgood 1989b, Rightmire 1986, Smith et al. 1989a). As Rightmire (1986:218) puts it, it is difficult to '...identify any "clade features" that may link archaic specimens from Broken Hill, Elandsfontein, and Lake Ndutu with later populations'. In a detailed assessment of the pertinent southern African fossils, Habgood (1989b) was unable to identify unique African regional clade features, with the possible exception of a wide interparietal groove in Florisbad. The pattern of brow ridge change (Fig. 3) may be a possible African clade feature, but this has yet to be established conclusively. Thus, the Stringer and Andrews prediction that modern regional characteristics should be of greater antiquity in Africa than in the 'peripheries' does not appear to be supported in the paleoanthropological record.

3 CONCLUDING STATEMENT

Based on this evidence, I assert that Stringer and Andrews' first two theoretical predictions regarding the human fossil record (Table 1) are not supported by existing paleontological data. To refute a speciation/replacement model of modern human origins, it is only necessary to show that there is some degree of regional continuity across the archaic/modern boundary in some geographic regions other than the proposed homeland of moderns. Obviously, I believe this has been done; there are reasonable indications of transitional groups and continuity across the archaic/modern human transition outside Africa. Thus, the paleoanthropological data do not support the Recent African speciation/replacement model.

Finally, there are 'Out of Africa' models which do not require that cladogenesis or complete replacement be involved in modern Eurasian origins. Bräuer's Afro-European Sapiens Model (1984, 1989, this volume), for example, clearly portrays the origin of modern Europeans as resulting from a movement of modern peoples into Europe. However, the presence of archaic features in the earliest modern Europeans is explained as an assimilation of Neanderthals into the invading modern population (Bräuer 1980). Bräuer describes his model as replacement with hybridization, and such a perspective is more compatible with the argument presented in this paper than is a speciation/replacement model. However, Bräuer's model still emphasizes replacement of archaic by modern populations, with 'hybridization' playing a rather minor role – particularly in Europe. I would argue that the data from Central Europe and other 'peripheries' would suggest that local continuity was much more significant, at least in certain regions, than Bräuer's model would allow (Smith 1984, 1985, Smith et al. 1989a, b). At the present time, however, it is not possible to conclusively establish the relative importance of replacement and continuity in 'peripheral' areas such as Europe.

For Europe, and probably Asia and Australasia as well, I suspect that non-indigenous influences, probably ultimately of African origin, were of considerable importance in the emergence of modern human anatomical form. However, the indications of continuity strongly suggest that this influence did not take the form of invading groups and certainly not invading groups of a new and different biological species. Rather, if the emergence of modern humans is a speciation event at all, it is best considered an example of what Van Valen (1986:412) has termed geographically extended phyletic speciation. Thus, Stephen Jay Gould is perhaps correct to state that modern humans are 'an entity, not a tendency' (Gould 1987:19); but if this is so, Neanderthals and other non-African archaic humans are a part of the entity.

ACKNOWLEDGEMENTS

My participation in the 12th International Congress of Anthropological and Ethnological Sciences in Zagreb was facilitated by the scholarly exchange program of the U.S. Academy of Sciences and the Jugoslavenska Akademija Znanosti i Umjetnosti. I am grateful to both academies for their support. Aspects of the work reported here were also supported by grants from the Alexander von Humboldt Foundation, the University of Tennessee and the US National Science Foundation (BSN-8606674). The ideas presented here benefitted greatly from discussions with many of my colleagues. In particular I thank G. Bräuer, M.H. Wolpoff, P. Habgood, A.B. Falsetti, M. Green, S. Donnelly, M. Liston, S. Myster and J. Kidder.

REFERENCES

Beaumont, P.B. 1980. On the age of Border Cave hominids 1-5. *Palaeotol. Afr.* 23: 21-33.

Beaumont, P.B., H. de Villiers & J.C. Vogel 1978. Modern man in sub-Saharan Africa prior to 49 000 years BP: A review and evaluation with particular reference to Border Cave. *S. Afr. J. Sci.* 74: 409-419.

Binford, L.R. 1985. Human ancestors: Changing views of their behavior. *J. Anthropol. Archaeol.* 4: 292-327.

Binford, L.R. 1989. Isolating the transition to cultural adaptions: An Organizational Approach. In E. Trinkaus (ed), *Biocultural Emergence of Humans in the Later Pleistocene*. Cambridge: Cambridge University Press.

Bräuer, G. 1980. Die morphologischen Affinitäten des jungpleistozänen Stirnbeines aus dem Elbmündungsgebiet bei Hahnöfersand. *Z. Morphol. Anthropol.* 71: 1-42.

Bräuer, G. 1984. A craniological approach to the origin of anatomically modern *Homo sapiens* in Africa and implications for the appearance of modern Europeans. In F.H. Smith & F. Spencer (eds), *The Origins of Modern Humans.* New York: A.R. Liss, pp. 327-410.

Bräuer, G. 1989. The evolution of modern humans: A comparison of the African and non-African evidence. In P. Mellars & C.B. Stringer (eds), *The Human Revolution: Behavioral and Biological Perspectives in the Origins of Modern Humans.* Edinburgh: Edinburgh University Press, pp. 123-154.

Butzer, K. 1978. Sediment stratigraphy of the Middle Stone Age sequence at Klasies River Mouth, Tzitzikama coast, South Africa. *S. Afr. Archaeol. Bull.* 33: 141-151.

Cann, R.L., M. Stoneking & A.C. Wilson 1987. Mitochondrial DNA and human evolution. *Nature* 325: 31-36.

Coon, C.S. 1963. *The Origin of Races.* New York: A.A. Knopf.

Day, M. & C.B. Stringer 1982. A reconsideration of the Omo Kibish remains and the *erectus-sapiens* transition. *Pret. I. Cong. Int. Palaeont. Hum.,* Nice, pp. 814-846.

Deacon, H.J. 1989. Late Pleistocene palaeoecology and archaeology in the Southern Cape. In P. Mellars & C.B. Stringer (eds), *The Human Revolution: Behavioral and Biological Perspectives on the Origins of Modern Humans.* Edinburgh: Edinburgh University Press, pp. 547-564.

Deacon, H.J. & V.B. Geleijnse 1988. The stratigraphy and sedimentology of the main site sequence, Klasies River, South Africa. *S. Afr. Arch. Bull.* 43: 5-14.

Delson, E. 1988. One source not many. *Nature* 332: 206.

Excoffier, L. & A. Langaney 1989. Origin and differentiation of human mitochondrial DNA. *Am. J. Hum. Genet.* 44: 73-85.

Excoffier, L., B. Pellegrini, A. Sanchez-Mazas, C. Simon & A. Langaney 1987. Genetics and the history of sub-Saharan Africa. *Yrbk. Phys. Anthropol.* 30: 151-194.

Gould, S.J. 1987. Brushes all the way down. *Nat. His.* 96: 12-19.

Gould, S.J. 1988. A novel notion of Neanderthal. *Nat. His.* 97: 16-21.

Habgood, P. 1989a. The evolution of modern humans: Evidence from Australasia seen in a global perspective. In P. Mellars & C.B. Stringer (eds), *The Human Revolution: Behavioral and Biological Perspectives on the Origins of Modern Humans.* Edinburgh: Edinburgh University Press, pp. 245-273.

Habgood, P. 1989b. An examination of regional features on Middle and early Late Pleistocene sub-Saharan African hominids. *S. Afr. J. Sci.* 44: 17-22.

Howell, F.C. 1978. Hominidae. In V.J. Maglio & H.B.S. Cooke (eds), *Evolution of African Mammals.* Cambridge, MA: Harvard University Press, pp. 154-248.

Howell, F.C. 1984. Introduction. In F.H. Smith & F. Spencer (eds), *The Origins of Modern Humans.* New York: A.R. Liss, pp. xiii-xxii.

Howells, W.W. 1973. Cranial Variation in Man: A Study by Multivariate Analysis of Patterns of Differences Among Recent Human Population. Cambridge, MA: *Peabody Museum Papers,* Vol. 67.

Klein, R.G. 1983. The stone age prehistory of southern Africa. *Ann. Rev. Anthropol.* 12: 25-48.

Lewin, R. 1988. Modern human origins under close scrutiny. *Science* 239: 1240-1241.

Nei, M. 1985. Human evolution at the molecular level. In T. Ohta & K. Akoi (eds), *Population Genetics and Molecular Evolution.* Berlin: Springer Verlag, pp. 41-64.

Parkington, J. 1990. A critique of the consensus view on the age of Howieson's Poort assemblages in South Africa. In P. Mellars (ed.), *The Emergence of Modern Humans.* Edinburgh: Edinburgh University Press, pp. 34-55.

Rightmire, G.P. 1984. *Homo sapiens* in sub-Saharan Africa. In F.H. Smith & F. Spencer (eds), *The Origins of Modern Humans.* New York: A.R. Liss, pp. 295-325.

Rightmire, G.P. 1986. Africa and the origins of modern humans. In R. Singer & J.K. Lundy (eds), *Variation, Culture and Evolution in African Populations: Papers in Honor of Dr. Hertha de Villiers.* Johannesburg: Witwatersrand University Press, pp. 255-264.

Sampson, C.G. 1974. *The Stone Age Prehistory of Southern Africa.* New York: Academic Press.

Simmons, T., A.B. Falsetti & F.H. Smith in press. Evolutionary patterns in the frontal bones of late Pleistocene hominids from western Asia. *J. Hum. Evol.*

Singer, J. & J. Wymer 1982. *The Middle Stone Age at Klasies River Mouth in South Africa.* Chicago: University of Chicago Press.

Smith, F.H. 1976. The Neanderthal Remains from Krapina: A Descriptive and Comparative Study. University of Tennessee, Department of Anthropology. *Report of Investigations* 15.

Smith, F.H. 1984. Fossil hominids from the Upper Pleistocene of Central Europe and the origin of modern Europeans. In F.H. Smith & F. Spencer (eds), *The Origin of Modern Humans.* New York: A.R. Liss, pp. 137-209.

Smith, F.H. 1985. Continuity and change in the origin of modern *Homo sapiens. Z. Morphol. Anthropol.* 75: 197-222.

Smith, F.H., D.C. Boyd & M. Malez 1985. Additional Upper Pleistocene human remains from Vindija Cave, Croatia, Yugoslavia. *Am. J. Phys. Anthropol.* 68: 375-383.

Smith, F.H. & A.B. Falsetti & S.M. Donnelly 1989a. Modern human origins. *Yrbk. Phys. Anthropol.* 32: 35-68.

Smith, F.H & S.P. Paquette 1989. The adaptive basis of Neanderthal facial form with some thoughts on the nature of modern humans origins. In E. Trinkaus (ed.), *Biocultural Emergence of Humans in the Later Pleistocene.* Cambridge: Cambridge University Press.

Smith, F.H. & G.C. Ranyard 1980. Evolution of the supraorbital region in Upper Pleistocene hominids from south – central Europe. *Am. J. Phys. Anthropol.* 53: 589-610.

Smith, F.H., J.F. Simek & M.S. Harrill 1989b. Geographic variation in supraorbital tours reduction during the later Pleistocene. In P. Mellars & C.B. Stringer (eds), *The Human Revolution: Behavioral and Biological Perspectives on the Origins of Modern Humans.* Edinburgh: Edinburgh University Press, pp. 172-193.

Spencer, F. 1984. The Neanderthals and their evolutionary significance: A brief historical survey. In F.H. Smith & F. Spencer (eds), *The Origins of Modern Humans.* New York: A.R. Liss, pp. 1-49.

Spencer, F. & F.H. Smith 1981. The significance of Aleš Hrdlička's 'Neanderthal phase of man': A historical and current assessment. *Am. J. Phys. Anthropol.* 56: 435-459.

Spuhler, J.N. 1988. Evolution of mitchondrial DNA in monkeys, apes and humans. *Yrbk. Phys. Anthropol.* 31: 15-48.

Stringer, C.B. 1988. The dates of Eden. *Nature* 331: 565-566.

Stringer, C.B. 1989. Documenting the origin of modern humans. In E. Trinkkaus (ed.), *Biocultural Emergence of Humans in the Later Pleistocene.* Cambridge: Cambridge University Press, pp. 67-96.

Stringer, C.B. & P. Andrews 1988. Genetic and fossil evidence for the origin of modern humans. *Science* 239: 1263-1268.

Stringer, C.B., R. Grün, H.P. Schwarcz & P. Goldberg 1989. ESR dates for the hominid burial sites of ES Skhūl in Israel. *Nature* 338: 756-758.

Tattersall, E., E. Delson & J. Van Couvering (eds) 1988. *Encyclopedia of Human Evolution and Prehistory.* New York: Garland.

Thorne, A.G. & M.H. Wolpoff 1981. Regional continuity in Australasian Pleistocene hominid evolution. *Am. J. Phys. Anthropol.* 55: 337-349.

Trinkaus, E. 1982. A history of *Homo erectus* and *Homo sapiens* paleontology in America. In F. Spencer (ed.), *A History of American Physical Anthropology 1930-1980.* New York: Academic Press, pp. 261-280.

Valladas, H., J.L. Reyss, J.L. Joron, G. Valladas, O. Bar-Yosef & B. Vandermeersch 1988. Thermoluminescence dating of Mousterian 'Proto-Cro-Magnon' remains from Israel and the origin of modern man. *Nature* 331: 614-616.

Van Valen, L.M. 1986. Speciation and our own species. *Nature* 322: 412.

Wainscoat, J. 1987. Out of the garden of Eden. *Nature* 325: 13.

Wiley, E.O. 1981. *Phylogenetics: The Theory and Practice of Phylogenetic Systematics.* New York: John Wiley and Sons.

Wolpoff, M.H. 1985. Human evolution at the peripheries: The pattern at the eastern edge. In P.V. Tobias (ed.), *Hominid Evolution: Past, Present and Future.* New York: A.R. Liss, pp. 355-365.

Wolpoff, M.H. 1986. Describing anatomically modern *Homo sapiens:* A distinction without a definable difference. *Anthropos (Brno)* 23: 41-53.

Wolpoff, M.H. 1989. The place of Neanderthals in human evolution. In E. Trinkaus (ed.), *Biocultural Emergence of Humans in the Later Pleistocene.* Cambridge: Cambridge University Press, pp. 97-141.

Wolpoff, M.H., F. Smith, M. Malez, J. Radovčić & D. Rukavina 1981. Upper Pleistocene human remains from Vindija Cave, Croatia, Yugoslavia. *Am. J. Phys. Anthropol.* 54: 499-545.

Wolpoff, M.H., X. Wu & A.G. Thorne 1984. Modern *Homo sapiens* origins: A general theory of hominid evolution involving the fossil evidence from East Asia. In F.H. Smith & F. Spencer (eds), *The Origins of Modern Humans.* New York: A.R. Liss, pp. 411-483.

Defining modern humans: A multivariate approach

James H. Kidder & Richard L. Jantz
Department of Anthropology, University of Tennessee, Knoxville, USA

Fred H. Smith
Department of Anthropology, Northern Illinois University, De Kalb, USA

1 INTRODUCTION

Establishing a taxonomic definition of anatomically modern *Homo sapiens* is integral to the debate concerning the origin of modern humans. The development of such a definition has been the subject of considerable controversy in recent years (see Day & Stringer 1982, Wolpoff 1986, Brown 1981). Despite this, few systematic attempts to unequivocally define *Homo sapiens* have been undertaken to date. The importance that such a definition holds for the understanding of Late Pleistocene hominid evolution is clearly expressed by Day & Stringer, who state that 'indeed, if we cannot provide a valid definition of the only living human species there is little hope of our recognizing extinct taxa' (1982: 831). Day and Stringer, as well as others (e.g. Stringer & Andrews 1988, Tattersall 1986, Gould 1987) believe that a precise definition of modern *Homo sapiens*, applicable to this taxon throughout its temporal and geographic distribution, should be possible. This belief, in part, derives from the perspective that all modern humans shared a relatively recent common ancestry and represent a species separate from archaic Eurasian hominids.

Day & Stringer (1982) propose just such a definition for anatomically modern *Homo sapiens* based on cranial morphometrics. Their definition comprises five metric characteristics and one observable feature (see also Stringer & Andrews 1988). These are as follows:

1) a short, high cranium with a ratio of basion-bregma height to length of 0.70 or a vertex radius to glabello-occipital length ratio of 0.64;

2) a parietal angle of less than 138°;

3) a bregma-asterion chord/biasterionic breadth ratio of more than 1.19;

4) a frontal angle of less than 133°;

5) a discontinuous supra-orbital torus;

6) an occipital angle of greater than 114° (Day & Stringer 1982).

Day and Stringer state that while all characteristics need not be present on all modern human crania, those containing 3 or more of these 6 features should be classified anatomically modern *Homo sapiens*.

Wolpoff (1986), applied these criteria to two Australasian samples: One from Lake Alexandrina and one presumably from Tasmania. From his analysis, Wolpoff found that as many as 34% of the frontal bone angles (Tasmanian females) and 43% of the occipital bone angles (Lake Alexandrina females) possess measurements that are outside the range of modern *Homo sapiens*, according to the criteria set forth by Day and Stringer. Wolpoff (1986:51) goes on to state that '...any morphological definition (of modern *Homo sapiens*) is unlikely to be valid' and suggests that the reason no single definition will apply adequately to all cases relates to the role of local continuity or multiregional evolution in modern human origins.

While Wolpoff's results demonstrate the limitations of the Day and Stringer definition, they also point out those of a strictly univariate approach to this problem. Given that a transition of any kind can, according to Wolpoff (1989), only apply to a population, some of Wolpoff's results are understandable. Even in a recent modern skeletal sample, not all individuals are necessarily going to fall within the modern range of variation in all of a series of univariate analyses. Wolpoff's analysis, however, shows a large percentage of the Lake Alexandrina and Tasmanian individuals to be outside the modern human range, as defined by Day and Stringer's criteria. Despite these results all these individuals are, by virtue of being documented representatives of recent populations, anatomically modern *Homo sapiens*. Howells (1973:21) describes the Lake Alexandrina sample as a '...well defined local population, and the linguistic and archaeological evidence indicates long-time stability (2000 years?) in the area given'. The Tasmanian sample, while possibly not as homogeneous, nonetheless is probably from the same general time period (Howells 1973) and is unquestionably drawn from a modern human sample.

Given these conflicting results, we maintain that it is potentially instructive to approach the question of a definition of anatomically modern *Homo sapiens* using a multivariate analytical technique. This methodology has the advantage of permitting the analysis of a total metric pattern rather than several cranial measurements individually. While this strategy in itself is not new (Stringer 1974, 1978, Howells 1973, Rightmire 1976, Bräuer 1984, Habgood 1988), its use as an approach to the definition of anatomically modern humans has been limited. Additionally, very little has been done in attempting to separate size and shape elements in such studies. Indeed, Day & Stringer (1982:831) state that 'some of the metrical differences recognized can be shown to lie behind multivariate discrimination...although this is sometimes masked by amalgamation of size and shape differences'. If the known (or at least suspected) size and particularly shape ranges of recent *Homo sapiens* are documented, then one should be able to use such a range to establish the shape-related metric boundaries of modern *Homo sapiens*. The question then becomes not whether a definition of modern *Homo sapiens* is possible, but rather, at what time do hominids conforming to such a definition appear.

In this study, we will show that, using a series of analyses based on a multivariate technique, when a sample of fossil hominids is compared to a known anatomically modern (i.e. recent human) sample, it is certainly possible to determine an unequivocally modern human metric pattern. We also show that a shape related metric pattern commensurate with that of recent modern humans appears late, near the end of the Pleistocene, in western and central Europe.

2 METHODS AND MATERIALS

The primary data set employed in this study consists of 49 archaic and early modern hominid fossils. These fossils are grouped in approximate temporal order and presented in Table 1. To facilitate separation of size and shape elements, these data are also logged. Numerous studies have attempted to relate the phylogenetic and chronological relationships of these hominids (Stringer 1974, 1978, Stringer et al. 1984, Smith 1984, 1985, Wolpoff 1980). For this analysis, primary emphasis will be placed on the European and Near Eastern modern samples traditionally considered to represent 'modern humans'. It is widely considered that the European and

Near Eastern Neanderthal fossils fall well outside the modern range of human cranial variation. They are included here both as comparative samples and also to further examine this assumption.

Measurements for the fossil material are taken directly from the fossil crania or from previously published sources (see Table 1). Seventeen measurements, incorporating both cranial and facial dimensions and corresponding to the system developed by Howells (1973) are used. These measurements and their codes are: Glabello-occipital length (GOL), nasio-occipital length (NOL), maximum cranial breadth (XCB), maximum frontal breadth (XFB), bisterphanic breadth (STB), biasterionic breadth (ASB), mastoid length (MDH), mastoid width (MDB), bifrontal breadth (FMB), nasio-frontal subtense (NAS), frontal chord (FRC), nasion-bregma subtense (FRS), parietal chord (PAC), bregma-lambda subtense (PAS), frontomalare radius (FMR), occipital chord (OCC), and occipital subtense (OCS). It is necessary that the measurements used conform to the Howells' definitions so they are comparable to the reference sample. These measurements are chosen primarily because they are obtainable with the greatest degree of regularity in the fossil crania. All measurements taken by the authors are assumed to be accurate within three millimeters. Some data taken from the literature are considered to be of limited accuracy and are contained in double parentheses in Table 1. None of these latter data are used in the statistical analyses reported here.

In order to accurately determine the 'modernity' of a given fossil, a reference sample of known anatomically modern *Homo sapiens* is required. This is important because any definition constructed around fossil data is arbitrary and, thus, potentially circular. Anatomically modern *Homo sapiens* is represented in our study by a broad sample of 595 crania from six populations: South Dakota Arikara (N = 66; Howells 1973), Peruvian Amerindians (N = 110; Howells 1973), American Whites (N = 100; Moore-Jansen 1989), American Blacks (N = 100; Moore-Jansen 1989), South west Pacific Tolai (N = 110; Howells 1973), and Austrians (N = 109; Howells 1973). This sample does not include some of the extreme (in the morphometric sense) populations of modern *Homo sapiens*, but it should provide a reliable base line against which to judge European and Near Eastern hominids.

We use Darroch and Mosimann's (1985) definition of shape to construct shape variables. Each variable for each individual (row in the data matrix) is divided by the geometric mean for all variables. This is equivalent to taking the log of each variable and subtracting the arithmetic mean of the logged variables from each logged variable. The result is a dimensionless log shape variable. The log transformed variables, prior to being converted to shape variables, contain both size and shape information.

The multivariate technique used for this study is the Mahalanobis distance or D^2. This method has been widely employed to calculate relationships betweeen hominids (Ambergen & Schaafsma 1984, Howells 1973, Fatti 1986, Rightmire 1976, Stringer 1974, 1978, Van Vark 1970).

Darroch & Mosimann (1985) demonstrate how Mahalanobis squared distances between two sample means can be decoupled into squared shape distance and a squared residual difference of log size. We have adapted their procedure to our analyses by calculating a log size and shape squared distance of a fossil from the centroid of the modern human sample. The log covariance matrix is total for the modern human sample. We then calculate a log size squared distance from the size variable, and log shape squared distances are obtained by subtraction.

A typicality probability is obtained by comparing the squared Mahalanobis distance of a fossil to the modern reference sample for the appropriate degrees of freedom (Campbell 1984). Where p = the number of variables, degrees of freedom are p for size and shape, p-1 for shape,

Table 1. Fossil hominid data. All measurements are listed in mm. Data taken from the following sources: (1) Stringer et al. (1979), (2) Howell (1960), (3) Suzuki & Takai (1970), (4) McCown & Keith (1937-39), (5) Trinkaus (1983), (6) Taken personally (F.H.S.), (7) Wolpoff (unpubl.), (8) Vandermeersch (1981), (9) Matiegka (1934).

Fossil (source)	GOL	NOL	XCB	XFB	STB	ASB	MDH	MDB	FMB	NAS	FRC	FRS	PAC	PAS	FMR	OCC	OCS
Petralona (1)	209.0	200.0	150.0	120.0	113.0	((119.5))			127.0	22.0	111.0	21.0	105.0	19.0	81.0		
Steinheim (2)	185.0	177.0	132.5	118.0		107.0			117.0		99.0		97.5				
Amud (3)	215.0		154.0	124.0		134.0					120.0		112.0			105.0	
Tabun 1 (4)	183.0	178.0	141.0	121.5		120.0					96.0	19.0	105.0			90.0	29.0
Shanidar 5 (5)	207.2	205.5	154.0	128.0	126.5	118.2	27.5	16.1	115.0		111.3	17.7	118.0	25.4	82.5	101.3	33.4
Saccopastore (6)	181.5	182.0	142.0	116.0		117.0			118.0		105.5		112.0				
Neanderthal (6)	201.0	194.0	147.0	120.0	113.0				111.0	19.0	116.0	22.0	103.0	17.0			
La Quina H5 (6)	201.0	199.0	139.0	104.0	105.0	113.0	26.0	13.0	100.0	27.0	107.0	20.0	104.0	12.0	84.0		
Spy 1 (6)	202.0	200.0	145.0	121.0	112.0	116.0	24.0	9.0	114.0		101.0	16.0	115.0	22.0	90.0		
Spy 2 (6)	200.0		153.0	127.0	120.0	123.0	26.0	8.0	114.0				104.0	17.0	84.0		
Monte Circeo (8)	204.0		155.0	127.0		124.0			120.0		117.0		109.0			88.0	
La Ferrassie (6)	208.0	204.0	159.0	131.0	125.0	124.0	38.0	12.0	111.0	28.0	117.0	22.0	110.0	18.0	82.0	102.0	32.0
La Chapelle (6)	211.0	207.0	157.0	125.0	122.0	121.0	30.0	11.0	114.0	23.0	108.0	21.0	112.0	18.0	83.0	94.0	38.0
Jebel Irhoud (7)	198.0	194.6	156.0	120.0	115.6	120.1	40.9		127.0		113.0		117.4				
Qafzeh 6 (8)	196.0		144.5	125.0		124.0			124.0		114.0						
Qafzeh 9 (8)	198.0		140.0	117.0		111.0			111.5		115.0		129.0			101.0	
Skhūl 4 (4)	206.0	193.0	148.0	121.0		132.0					118.0	28.0	122.0			86.0	26.0
Skhūl 5 (6)	193.0	184.0	146.0	112.0	111.0	116.0	28.0	15.0	112.0	17.0	109.0	25.0	109.0	27.0	80.0	92.0	30.0
Mladeč 1 (6)	203.0	196.0	145.0	131.0	124.0	112.0	30.0	16.0	102.0	12.0	114.0	30.0	115.0	19.0	84.0	112.0	37.0
Mladeč 2 (6)	185.0	174.0	139.0	117.0	107.0	116.0	32.0	12.0	107.0	13.0	102.0	29.0	108.0	27.0	90.0		
Mladeč 5 (6)	208.0	204.0	153.0	134.0	112.0	122.0	30.0	22.0	114.0	19.0	116.0	33.0	117.0	22.0	95.0		
Cro-Magnon 1 (6)	203.0	198.0	151.0	135.0	123.0	107.0	34.0	18.0	109.0	17.0	124.0	31.0	120.0	24.0	90.0	98.0	35.0
Cro-Magnon 2 (6)	192.0	190.0	140.0	127.0	121.0		33.0	18.0	101.0	17.0	114.0	28.0	122.0	24.0			
Cro-Magnon 3 (6)	204.0	199.0	154.0	133.0	120.0	119.0			102.0	16.0	127.0	33.0	112.0	19.0			
Cro-Magnon 8 (6)	191.0	186.0	155.0	126.0	119.0	124.0	34.0	22.0	107.0	20.0	125.0	26.0	125.0	34.0			
Dolní Věstonice 3 (6)	187.0	183.0	132.0	113.0	103.0	105.0	21.0	15.0	91.0	16.0	106.0	29.0	118.0	23.0	87.0	91.0	26.0
Předmostí I (9)	189.0		139.0	120.0		102.0					107.0		125.0			94.0	
Předmostí III (9)	201.5		145.0	128.0		110.0			116.3		120.0		120.0			100.0	
Předmostí IV (9)	191.5		144.0	122.0		109.0			106.3		114.0		120.0			101.0	
Předmostí IX (9)	196.0		145.0	128.0		105.0			114.9		115.0		120.0			95.0	

Předmostí X (9)	185.5		144.0	123.0		106.3			112.2		112.0		107.0			96.0	
Brno 1 (6)	190.0	185.0	143.0	119.0	110.0				96.5	12.0	117.0	27.0	109.0	23.0			
Brno 2 (6)	205.0	202.0	134.0	127.0	127.0	115.0	26.0	16.0	96.0	25.0	118.0	27.0	128.0	25.0	96.0		
Paderborn (6)	196.0	191.0	154.0	126.0	123.0	132.0	34.0	14.0	103.0	19.0	113.0	29.0	127.0	28.0	78.0	95.0	31.0
Abri Pataud (6)	182.0	178.0	138.0	118.0	114.0	108.0	29.0	11.0	99.0	18.0	111.0	28.0	108.0	20.0	76.0	98.0	33.0
Engis (6)	190.0		133.0	116.0	114.0	111.0	24.0	11.0			113.0	28.0	122.0	28.0		99.0	32.0
Oberkassel 1 (6)	195.0	190.0	140.0	113.0	112.0	118.0	35.0	17.0	107.0	21.0	114.0	26.0	120.0	22.0	81.0	102.0	28.0
Oberkassel 2 (6)	181.0	178.0	136.0	111.0	109.0	105.0	33.0	7.0	95.0	15.0	104.0	25.0	123.0	26.0	76.0	94.0	24.0
Laugerie Basse (6)	170.0		145.0	116.0	115.0		27.0	11.0	88.0				113.0	29.0	73.0	99.0	23.0
Kaufertsberg (6)	184.0	179.0	142.0	121.0	119.0	116.0	38.0	12.0	94.0	20.0	121.0	29.0	116.0	25.0	74.0	95.0	24.0
Ofnet 4 (6)	176.0	174.0	143.0	122.0	117.0	111.0	30.0	12.0	93.0	13.0	105.0	30.0	114.0	26.0	71.0		
Ofnet 6 (6)	183.0	176.0	140.0	114.0	114.0	106.0	39.0	13.0			104.0	27.0	120.0	30.0		94.0	28.0
Ofnet 8 (6)	182.0	176.0	146.0	119.0	114.0	122.0	33.0	11.0	92.0		106.0	30.0	107.0	20.0	74.0	102.0	36.0
Ofnet 13 (6)	186.0	183.0	130.0	105.0	107.0	110.0	29.0	13.0	92.0		110.0	25.0	108.0	19.0	79.0	95.0	33.0
Ofnet 15 (6)	179.0	174.0	140.0	118.0	111.0	122.0	29.0	9.0	93.0		117.0	28.0	115.0	25.0	72.0		
Ofnet 18 (6)	182.0	179.0	139.0	118.0	115.0		33.0	12.0	96.0		105.0	25.0	117.0	26.0	72.0	100.0	30.0
Ofnet 24 (6)	184.0	182.0	140.0	129.0	120.0	114.0	34.0	13.0	104.0		110.0	27.0	111.0	18.0	76.0	103.0	34.0
Ofnet 25 (6)	178.0	175.0	136.0	113.0	111.0	115.0	33.0	11.0	93.0		106.0	26.0	113.0	21.0	73.0	96.0	26.0
Ofnet 29 (6)	177.0	175.0	133.0	107.0	102.0	103.0	28.0	10.0	84.0		104.0	24.0	107.0	20.0	69.0		
Kelsterbach (6)	185.0	181.0	140.0	117.0	117.0	105.0			97.0		113.0	24.0	115.0	24.0		92.0	22.0
Modern means	177.1	174.8	136.5	114.6	110.1	107.1	27.5	11.9	96.7	16.9	107.8	24.6	111.2	24.2	75.3	94.2	27.7

and 1 for size. In the analyses which follow, a fossil is considered typical of modern *Homo sapiens* if it falls within the 95% concentration contour of the modern sample and atypical if it falls beyond. All computations are done using a basic program written for this purpose.

Due to the fragmentary nature of the fossil record, the data are divided into subsets of variables so that fossils with certain variables in common are analysed as a group. A minimum of five variables is used per analysis. This is beneficial in that many fossils are included in a large number of analyses. Certain fossils (i.e. Qafzeh 9 and Laugerie Basse) possess peculiar measurement sets, but are deemed important to the analysis and are, therefore, included in some analyses. Eight analyses are done for both cranial size/shape and shape. It was felt that this number of analyses would be sufficient to delineate patterns present in the fossil record.

3 RESULTS AND DISCUSSION

The results of the individual analyses are presented in Table 2. For each analysis, specimens included are listed along with their size/shape D^2s, shape D^2s, size D^2s, and whether the fossil falls within (W) or outside (O) the recent human range for the size/shape and shape analyses. Each analysis also lists the variables used per analysis and the 95% chi-square confidence level

Table 2. Distance results. For all samples: O = outside the modern human range, W = within the modern human range.

Fossil	Log size/shape D^2	Log shape D^2	Log size D^2	Log fossil size	Size/shape typicality	Shape typicality
Analysis No. 1						
All variables						
95% level	27.587	26.296				
La Ferrassie	51.659	45.082	6.577	4.257	O	O
La Chapelle	90.996	87.939	3.030	4.222	O	O
Skhūl V	68.497	66.898	1.599	4.202	O	O
Mladeč 1	51.425	47.996	3.459	4.227	O	O
Cro-Magnon 1	37.132	26.935	10.197	4.283	O	O
Dolní Věstonice 3	36.117	36.116	0.001	4.150	O	O
Paderborn	27.655	20.012	7.643	4.265	O	W
Abri Pataud	8.337	8.007	0.330	4.173	W	W
Oberkassel 1	19.666	15.364	4.302	4.236	W	W
Oberkassel 2	33.751	33.100	0.651	4.115	O	O
Kaufertsberg	23.554	21.640	1.914	4.207	W	W
Ofnet 8	42.107	42.024	0.083	4.161	O	O
Ofnet 13	19.866	19.866	0.000	4.148	W	W
Ofnet 24	44.603	44.054	0.549	4.180	O	O
Ofnet 25	14.302	14.302	0.000	4.148	W	W

Mean logged fossil size for modern sample 4.149

Table 2 (Cont.).

Fossil	Log size/shape D^2	Log shape D^2	Log size D^2	Log fossil size	Size/shape typicality	Shape typicality
Analysis No. 2						
Variables: GOL NOL XCB XFB ASB FMB FRC PAC OCC						
95% Level	16.919	15.507				
Shanidar 1	40.408	29.899	10.509	4.910	O	O
La Ferrassie	37.450	26.371	11.079	4.913	O	O
La Chapelle	68.559	60.833	7.726	4.894	O	O
Skhūl V	43.822	42.335	1.487	4.840	O	O
Mladeč 1	30.504	23.861	6.643	4.887	O	O
Cro-Magnon 1	25.517	14.820	10.697	4.911	O	W
Dolní Věstonice 3	10.119	10.116	0.003	4.796	W	W
Paderborn	24.391	16.009	8.382	4.898	O	O
Abri Pataud	3.521	3.274	0.247	4.815	W	W
Oberkassel 1	11.843	8.073	3.770	4.865	W	W
Oberkassel 2	6.475	6.456	0.019	4.793	W	W
Kaufertsberg	12.709	11.291	1.418	4.839	W	W
Ofnet 8	20.257	19.639	0.618	4.825	O	O
Ofnet 13	10.962	10.924	0.038	4.791	W	W
Ofnet 24	14.232	11.957	2.275	4.850	W	W
Ofnet 25	4.785	4.770	0.015	4.802	W	W
Kelsterbach	3.975	3.698	0.277	4.816	W	W
Mean logged fossil size for modern sample 4.798						
Analysis No. 3						
Variables: GOL NOL XCB XFB ASB FRC PAC						
95% level	14.067	12.592				
Steinheim	46.454	46.054	0.400	4.843	O	O
Tabun 1	34.686	34.584	0.102	4.878	O	O
Shanidar 1	32.702	24.290	8.412	4.973	O	O
Jebel Irhoud	20.900	15.086	5.814	4.955	O	O
Saccopastore	11.446	10.918	0.528	4.893	W	W
La Quina H5	41.079	40.698	0.381	4.889	O	O
Spy 1	42.539	39.725	2.814	4.928	O	O
La Ferrassie	33.241	23.179	10.062	4.983	O	O
La Chapelle	53.334	45.841	7.493	4.967	O	O
Skhūl IV	52.653	43.598	9.055	4.977	O	O
Skhūl V	32.837	31.894	0.943	4.902	O	O
Mladeč 1	26.666	21.612	5.054	4.949	O	O
Mladeč 2	44.580	44.575	0.005	4.869	O	O
Mladeč 5	26.834	16.248	10.586	4.986	O	O
Cro-Magnon 1	24.762	16.507	8.255	4.972	O	O
Cro-Magnon 3	21.518	13.263	8.255	4.972	O	O
Cro-Magnon 8	18.836	9.944	8.892	4.976	O	W
Dolní Věstonice 3	4.077	3.992	0.085	4.877	W	W
Brno 2	22.305	15.395	6.910	4.963	O	O
Paderborn	21.792	14.739	7.053	4.964	O	O
Abri Pataud	3.193	3.054	0.139	4.880	W	W

Table 2 (Cont.).

Fossil	Log size/shape D²	Log shape D²	Log size D²	Log fossil size	Size/shape typicality	Shape typicality
Oberkassel 1	8.798	5.164	2.634	4.962	W	W
Oberkassel 2	6.185	6.172	0.013	4.862	W	W
Kaufertsberg	10.937	8.644	2.293	4.922	W	W
Ofnet 4	3.582	3.323	0.259	4.885	W	W
Ofnet 6	13.242	13.122	0.120	4.879	W	O
Ofnet 8	17.187	16.490	0.697	4.897	O	O
Ofnet 13	7.977	7.969	0.008	4.863	W	W
Ofnet 15	16.900	15.488	1.412	4.910	O	O
Ofnet 24	11.632	10.220	1.412	4.910	W	W
Ofnet 25	4.169	4.113	0.056	4.875	W	W
Ofnet 29	3.488	2.855	0.633	4.837	W	W
Kelsterbach	3.641	3.113	0.528	4.893	W	W
Mean logged fossil size for modern sample 4.866						
Analysis No. 4						
Variables: GOL XCB XFB ASB FRC PAC OCC						
95% Level	14.067	12.592				
Amud 1	45.202	33.629	11.573	4.898	O	O
Tabun 1	40.722	40.719	0.003	4.780	O	O
Shanidar 1	30.139	23.334	6.805	4.870	O	O
Monte Circeo	42.471	38.185	4.286	4.851	O	O
La Ferrassie	35.700	26.669	9.031	4.884	O	O
La Chapelle	64.304	59.659	4.645	4.854	O	O
Qafzeh 9	6.556	2.836	3.720	4.846	W	W
Skhūl IV	43.737	38.063	5.674	4.862	O	O
Skhūl V	24.226	23.721	0.505	4.803	O	O
Mladeč 1	25.681	19.023	6.658	4.869	O	O
Cro-Magnon 1	24.114	17.010	7.104	4.872	O	O
Dolní Věstonice 3	4.973	4.972	0.001	4.777	W	W
Předmostí I	9.613	9.025	0.588	4.805	W	W
Předmostí III	13.179	7.903	5.276	4.859	W	W
Předmostí IV	5.336	2.535	2.801	4.837	W	W
Předmostí IX	14.281	11.666	2.615	4.835	O	W
Předmostí X	7.582	7.077	0.505	4.803	W	W
Paderborn	24.263	15.061	9.202	4.885	O	O
Abri Pataud	2.002	1.795	0.207	4.794	W	W
Engis	4.113	2.824	1.289	4.818	W	W
Oberkassel 1	8.048	5.247	2.801	4.837	W	W
Oberkassel 2	6.220	6.182	0.038	4.771	W	W
Kaufertsberg	7.704	5.357	2.347	4.832	W	W
Ofnet 6	5.614	5.516	0.098	4.789	W	W
Ofnet 8	11.545	10.191	1.354	4.819	W	W
Ofnet 13	9.383	9.319	0.064	4.769	W	W
Ofnet 24	12.221	10.208	2.013	4.828	W	W
Ofnet 25	4.127	4.029	0.098	4.789	W	W
Kelsterbach	3.139	2.877	0.262	4.796	W	W
Mean logged fossil size for modern sample 4.778						

Table 2 (Cont.).

Fossil	Log size/shape D^2	Log shape D^2	Log size D^2	Log fossil size	Size/shape typicality	Shape typicality
Analysis No. 5						
Variables: GOL NOL ASB FRC PAC OCC						
95% level	12.592	11.070				
Tabun 1	38.873	38.818	0.055	4.816	O	O
Shanidar 1	26.755	26.067	0.688	4.858	O	O
La Ferrassie	29.816	23.179	6.637	4.927	O	O
La Chapelle	55.100	51.124	3.976	4.904	O	O
Skhūl IV	65.606	60.209	5.397	4.917	O	O
Skhūl V	29.303	28.696	0.607	4.856	O	O
Mladeč 1	19.281	13.766	5.515	4.918	O	O
Cro-Magnon 1	14.750	10.036	4.714	4.911	O	W
Dolní Věstonice 3	4.800	4.660	0.140	4.840	W	W
Paderborn	23.688	17.562	6.126	4.923	O	O
Abri Pataud	2.869	2.710	0.159	4.841	W	W
Oberkassel 1	6.773	2.488	4.285	4.907	W	W
Oberkassel 2	3.300	3.141	0.159	4.841	W	W
Kaufertsberg	10.424	8.707	1.717	4.877	W	W
Ofnet 6	10.204	10.130	0.074	4.836	W	W
Ofnet 8	16.727	16.080	0.647	4.857	O	O
Ofnet 13	4.538	4.288	0.250	4.845	W	W
Ofnet 24	3.015	1.897	1.118	4.867	W	W
Ofnet 25	3.395	3.236	0.159	4.841	W	W
Kelsterbach	3.111	2.885	0.226	4.844	W	W
Mean logged fossil size of modern sample 4.825						
Analysis No. 6						
Variables: GOL NOL XCB XFB FRC PAC						
95% level	12.592	11.070				
Petralona	52.947	49.628	3.319	4.967	O	O
Steinheim	46.349	45.682	0.487	4.872	O	O
Tabun 1	29.926	29.906	0.020	4.893	O	O
Shanidar 1	32.480	24.322	8.158	5.006	O	O
Jebel Irhoud	20.854	15.927	4.927	4.982	O	O
Saccopastore	8.659	8.463	0.196	4.915	W	W
Neanderthal	29.327	26.986	2.341	4.956	O	O
La Quina H5	41.073	40.877	0.196	4.915	O	O
Spy 1	42.499	40.076	2.423	4.957	O	O
La Ferrassie	33.238	24.463	8.775	5.010	O	O
La Chapelle	53.228	46.663	6.565	4.995	O	O
Skhūl IV	44.887	38.844	6.043	4.991	O	O
Skhūl V	32.748	32.161	0.578	4.927	O	O
Mladeč 1	25.303	19.770	5.533	4.987	O	O
Mladeč 2	42.989	42.928	0.061	4.889	O	O
Mladeč 5	26.795	16.886	9.909	5.017	O	O
Cro-Magnon 1	16.167	5.405	10.762	5.022	O	W
Cro-Magnon 2	9.026	5.206	3.820	4.972	W	W
Cro-Magnon 3	21.317	13.460	7.857	5.004	O	O

Table 2 (Cont.).

Fossil	Log size/shape D^2	Log shape D^2	Log size D^2	Log fossil size	Size/shape typicality	Shape typicality
Cro-Magnon 8	17.137	9.574	7.563	5.002	O	W
Dolní Věstonice 3	3.843	3.670	0.173	4.914	W	W
Brno 1	7.707	6.239	1.468	4.944	W	W
Brno 2	21.837	14.717	7.120	4.999	O	O
Paderborn	8.596	1.046	7.550	5.002	W	W
Abri Pataud	3.071	2.919	0.152	4.913	W	W
Oberkassel 1	6.987	4.959	2.028	4.952	W	W
Oberkassel 2	5.890	5.889	0.001	4.897	W	W
Kaufertsberg	9.962	8.009	1.953	4.951	W	W
Ofnet 4	3.434	3.261	0.173	4.914	W	W
Ofnet 6	12.591	12.395	0.196	4.915	W	O
Ofnet 8	12.696	12.564	0.132	4.912	O	O
Ofnet 13	7.152	7.103	0.049	4.890	W	W
Ofnet 15	9.028	8.408	0.620	4.928	W	W
Ofnet 18	2.846	2.545	0.301	4.919	W	W
Ofnet 24	11.079	9.971	1.108	4.938	W	W
Ofnet 25	0.519	0.515	0.004	4.896	W	W
Ofnet 29	3.142	2.578	0.564	4.870	W	W
Kelsterbach	1.897	1.099	0.798	4.932	W	W
Mean logged fossil size for modern sample 4.898						
Analysis No. 7						
Variables: GOL XCB XFB STB FMB PAC PAS						
95% level	14.067	12.592				
Petralona	65.486	64.370	1.116	4.616	O	O
Shanidar 1	29.410	21.537	7.873	4.689	O	O
La Quina H5	63.418	57.094	6.324	4.467	O	O
Neanderthal	29.111	29.106	0.005	4.569	O	O
Spy 1	19.414	18.586	1.558	4.624	O	O
Spy 2	31.263	30.872	0.391	4.598	O	O
La Ferrassie	30.505	28.532	1.937	4.630	O	O
La Chapelle	38.344	36.602	1.742	4.627	O	O
Skhūl V	41.283	38.283	1.680	4.626	O	O
Mladeč 1	20.293	19.276	1.017	4.614	O	O
Mladeč 2	14.911	14.520	0.391	4.598	O	O
Mladeč 5	33.175	29.399	3.776	4.653	O	O
Cro-Magnon 1	17.751	11.998	5.753	4.672	O	W
Cro-Magnon 2	8.570	6.357	2.213	4.634	W	W
Cro-Magnon 3	21.502	20.917	0.585	4.540	O	O
Cro-Magnon 8	19.873	9.547	10.326	4.706	O	W
Dolní Věstonice	9.218	9.072	0.146	4.556	W	W
Brno 1	6.897	6.813	0.084	4.584	W	W
Brno 2	28.413	25.001	3.412	4.649	O	O
Paderborn	7.386	0.000	7.739	4.688	W	W
Abri Pataud	2.623	2.527	0.096	4.559	W	W
Oberkassel 1	14.716	13.218	1.498	4.623	O	O
Oberkassel 2	6.866	6.829	0.037	4.580	W	W

Table 2 (Cont.).

Fossil	Log size/shape D^2	Log shape D^2	Log size D^2	Log fossil size	Size/shape typicality	Shape typicality
Laugerie Basse	10.113	9.722	0.391	4.598	W	W
Kaufertsberg	2.847	2.057	0.790	4.609	W	W
Ofnet 4	3.355	2.764	0.591	4.604	W	W
Ofnet 8	6.768	6.622	0.146	4.556	W	W
Ofnet 13	14.892	12.689	2.203	4.510	O	O
Ofnet 15	2.559	2.428	0.131	4.587	W	W
Ofnet 18	1.011	0.344	0.667	4.606	W	W
Ofnet 24	22.488	22.467	0.021	4.578	O	O
Ofnet 25	4.309	3.980	0.329	4.548	W	W
Ofnet 29	13.175	9.776	3.399	4.495	W	W
Mean logged fossil size for modern sample 4.572						
Analysis No. 8						
Variables: GOL XCB XFB ASB FRC						
95% level	11.070	9.488				
Steinheim	14.761	14.719	0.042	4.830	O	O
Amud 1	30.884	16.110	14.774	4.983	O	O
Tabun 1	27.417	27.005	0.412	4.862	O	O
Shanidar 1	25.219	17.773	7.446	4.941	O	O
Saccopastore	5.972	5.413	0.559	4.866	W	W
La Quina H5	25.989	25.756	0.233	4.856	O	O
Spy 1	32.347	30.435	1.912	4.890	O	O
Monte Circeo	17.132	7.006	10.126	4.958	O	W
La Ferrassie	23.558	10.748	12.810	4.973	O	O
La Chapelle	41.044	33.286	7.758	4.943	O	O
Qafzeh 6	11.230	5.652	5.578	4.927	O	W
Qafzeh 9	5.870	4.030	1.840	4.889	W	W
Skhūl IV	22.376	12.749	9.627	4.955	O	O
Skhūl V	14.824	13.452	1.372	4.882	O	O
Mladeč 1	16.908	11.698	5.210	4.924	O	O
Mladeč 2	10.241	9.925	0.316	4.859	W	O
Mladeč 5	21.261	9.922	11.339	4.965	O	O
Cro-Magnon 1	21.466	13.260	8.206	4.946	O	O
Cro-Magnon 3	17.246	7.120	10.126	4.958	O	W
Cro-Magnon 8	13.683	3.891	9.792	4.956	O	W
Dolní Věstonice 3	3.704	3.683	0.023	4.832	W	W
Předmostí I	9.418	9.276	0.142	4.852	W	W
Předmostí III	11.749	6.418	5.331	4.925	O	W
Předmostí IV	4.685	2.699	1.986	4.891	W	W
Předmostí IX	13.439	10.637	2.802	4.901	O	O
Předmostí X	4.572	3.493	1.079	4.877	W	W
Brno 2	20.497	16.209	4.228	4.916	O	O
Paderborn	20.982	13.371	7.611	4.942	O	O
Abri Pataud	0.693	0.377	0.316	4.859	W	W
Engis	3.945	3.386	0.559	4.866	W	W
Oberkassel 1	8.020	6.108	1.912	4.890	W	W
Oberkassel 2	5.417	4.839	0.578	4.809	W	W
Kaufertsberg	7.201	4.219	2.982	4.903	W	W

Table 2 (Cont.).

Fossil	Log size/shape D^2	Log shape D^2	Log size D^2	Log fossil size	Size/shape typicality	Shape typicality
Ofnet 4	3.232	2.885	0.347	4.860	W	W
Ofnet 6	5.262	5.261	0.001	4.839	W	W
Ofnet 8	10.477	10.476	0.001	4.839	W	O
Ofnet 13	5.931	4.899	0.042	4.830	W	W
Ofnet 15	11.720	9.682	2.138	4.893	O	O
Ofnet 24	9.967	7.981	1.986	4.891	W	W
Ofnet 25	4.003	3.955	0.048	4.846	W	W
Ofnet 29	2.740	1.943	0.797	4.804	W	W
Kelsterbach	2.802	2.423	0.379	4.861	W	W
Mean logged fossil size for modern sample 4.838						

based on the recent modern human range for the analysis. Tables 3 and 4 present these results in cross-tabulation form. The expression 'within the modern range of variation' will refer to fossils that fall within the range of the recent anatomically modern data set. It should be pointed out that comparisons between fossil D^2s within individual analyses should be regarded carefully as this analytical technique probably does not accurately reflect morphological distances between these fossils, but rather, how far each is from the recent human data set.

3.1 *Early archaic Homo sapiens*

This is a restricted sample consisting only of Steinheim, Petralona, and Saccopastore 1, each of which is generally considered to predate 'typical' European Neanderthals (Stringer et al. 1984). For all of the analyses in which either Steinheim or Petralona are included, both are outside the range of modern humans. These results are not surprising given the markedly archaic nature of these hominids and generally confirm the assessment of these fossils in other comparable studies (e.g. Stringer et al. 1979, Habgood 1988). Additionally, the Steinheim cranium suffers from considerable postmortem distortion. It is not clear what effect this distortion has on the specimen's placement in our analyses but one might expect, given the nature of this distortion, that it would tend to make the specimen fall closer to the modern range than would be the case without it.

The hominid from this group of greatest interest in our analyses is Saccopastore 1. This fossil is used in Analyses 3 and 6. For both analyses, it falls within the range of the modern sample. Because of the estimated date of this fossil, this must be considered an anomaly. Condemi (1988:42) has recently concluded that both Saccopastore 1 and 2 are Neanderthals by virtue of their possession of certain Neanderthal characteristics but that 'they are also distinguished from classic Neanderthals by the presence of a great number of archaic features, as well as of a large quantity of incompletely developed Neanderthal traits'. This is in agreement with Stringer & Trinkaus (1981), who believe the Saccopastore face to exhibit Neanderthal characteristics. Stringer et al. (1984), however, point out that Saccopastore 1 as a whole, does not exhibit a typical Neanderthal pattern, being more rounded in the occipital region. Additionally, the skull is reported to be small, with a cranial capacity of 1250 cc (Condemi 1988). These last two factors probably explain the modern D^2s derived for this specimen.

Table 3. Size/shape analysis. For all samples: O = outside the modern human range, W = within the modern human range.

Fossil	Analysis 1	Analysis 2	Analysis 3	Analysis 4	Analysis 5	Analysis 6	Analysis 7	Analysis 8
Petralona	-	-	-	-	-	O	O	-
Steinheim	-	-	O	-	-	O	-	O
Amud 1	-	-	-	O	-	-	-	O
Tabun 1	-	-	O	O	O	O	-	O
Shanidar 1	-	O	O	O	O	O	O	O
Jebel Irhoud	-	-	O	-	-	O	-	-
Saccopastore	-	-	W	-	-	W	-	W
La Quina H5	-	-	O	-	-	O	O	O
Neanderthal	-	-	-	-	-	O	O	-
Spy 1	-	-	O	-	-	O	O	O
Spy 2	-	-	-	-	-	-	O	-
Monte Circeo	-	-	-	O	-	-	-	O
La Ferrassie	O	O	O	O	O	O	O	O
La Chapelle	O	O	O	O	O	O	O	O
Qafzeh 6	-	-	-	-	-	-	-	O
Qafzeh 9	-	-	-	W	-	-	-	W
Skhūl IV	-	-	O	O	O	O	-	O
Skhūl V	O	O	O	O	O	O	O	O
Mladeč 1	O	O	O	O	O	O	O	O
Mladeč 2	-	-	O	-	-	O	O	W
Mladeč 5	-	-	O	-	-	O	O	O
Cro-Magnon 1	O	O	O	O	O	O	O	O
Cro-Magnon 2	-	-	-	-	-	W	W	-
Cro-Magnon 3	-	-	O	-	-	O	O	O
Cro-Magnon 8	-	-	O	-	-	O	O	O
Dolní Věstonice 3	O	W	W	W	W	W	W	W
Předmostí I	-	-	-	W	-	-	-	W
Předmostí III	-	-	-	W	-	-	-	O
Předmostí IV	-	-	-	W	-	-	-	W
Předmostí IX	-	-	-	O	-	-	-	O
Předmostí X	-	-	-	W	-	-	-	W
Brno 1	-	-	-	-	-	W	W	-
Brno 2	-	-	O	-	-	O	O	O
Paderborn	O	O	O	O	O	W	W	O
Abri Pataud	W	W	W	W	W	W	W	W
Engis	-	-	-	W	-	-	-	W
Oberkassel 1	W	W	W	W	W	W	O	W
Oberkassel 2	O	W	W	W	W	W	W	W
Laugerie Basse	-	-	-	-	-	-	W	-
Kaufertsberg	W	W	W	W	W	W	W	W
Ofnet 4	-	-	W	-	-	W	W	W
Ofnet 6	-	-	W	W	W	W	-	W
Ofnet 8	O	O	O	W	O	O	W	W
Ofnet 13	W	W	W	W	W	W	O	W
Ofnet 15	-	-	O	-	-	W	W	O
Ofnet 18	-	-	-	-	-	W	W	-
Ofnet 24	O	W	W	W	W	W	O	W
Ofnet 25	W	W	W	W	W	W	W	W
Ofnet 29	-	-	W	-	-	W	W	W
Kelsterbach	-	W	W	W	W	W	-	W

Table 4. Shape analysis. For all samples: O = outside the modern human range, W = within the modern human range.

Fossil	Analysis 1	Analysis 2	Analysis 3	Analysis 4	Analysis 5	Analysis 6	Analysis 7	Analysis 8
Petralona	-	-	-	-	-	O	O	-
Steinheim	-	-	O	-	-	O	-	O
Amud 1	-	-	-	O	-	-	-	O
Tabun 1	-	-	O	O	O	O	-	O
Shanidar 1	-	O	O	O	O	O	O	O
Jebel Irhoud	-	-	O	-	-	O	-	-
Saccopastore	-	-	W	-	-	W	-	W
La Quina H5	-	-	O	-	-	O	O	O
Neanderthal	-	-	-	-	-	O	O	-
Spy 1	-	-	O	-	-	O	O	O
Spy 2	-	-	-	-	-	-	O	-
Monte Circeo	-	-	-	O	-	-	-	W
La Ferrassie	O	O	O	O	O	O	O	O
La Chapelle	O	O	O	O	O	O	O	O
Qafzeh 6	-	-	-	-	-	-	-	W
Qafzeh 9	-	-	-	W	-	-	-	W
Skhūl IV	-	-	O	O	O	O	-	O
Skhūl V	O	O	O	O	O	O	O	O
Mladeč 1	O	O	O	O	O	O	O	O
Mladeč 2	-	-	O	-	-	O	O	O
Mladeč 5	-	-	O	-	-	O	O	O
Cro-Magnon 1	O	W	O	O	W	W	W	O
Cro-Magnon 2	-	-	-	-	-	W	W	-
Cro-Magnon 3	-	-	O	-	-	O	O	W
Cro-Magnon 8	-	-	W	-	-	W	W	W
Dolní Věstonice 3	O	W	W	W	W	W	W	W
Předmostí I	-	-	-	W	-	-	-	W
Předmostí III	-	-	-	W	-	-	-	W
Předmostí IV	-	-	-	W	-	-	-	W
Předmostí IX	-	-	-	W	-	-	-	O
Předmostí X	-	-	-	W	-	-	-	W
Brno 1	-	-	-	-	-	W	W	-
Brno 2	-	-	O	-	-	O	O	O
Paderborn	W	O	O	O	-	W	W	O
Abri Pataud	W	W	W	W	W	W	W	W
Engis	-	-	-	W	-	-	-	W
Oberkassel 1	W	W	W	W	W	W	O	W
Oberkassel 2	O	W	W	W	W	W	W	W
Laugerie Basse	-	-	-	-	-	-	W	-
Kaufertsberg	W	W	W	W	W	W	W	W
Ofnet 4	-	-	W	-	-	W	W	W
Ofnet 6	-	-	O	W	W	O	-	W
Ofnet 8	O	O	O	W	O	O	W	O
Ofnet 13	W	W	W	W	W	W	O	W
Ofnet 15	-	-	O	-	-	W	W	O
Ofnet 18	-	-	-	-	-	W	W	-
Ofnet 24	O	W	W	W	W	W	O	W
Ofnet 25	W	W	W	W	W	W	W	W
Ofnet 29	-	-	W	-	-	W	W	W
Kelsterbach	-	W	W	W	W	W	-	W

3.2 *European and Near Eastern Neanderthals*

For the European Neanderthals, a clear pattern emerges: No fossil in this group shows a pattern of analyses within the recent modern range. In only one instance does a Neanderthal cranium fall within the recent modern range for either the size/shape or the shape analyses. Monte Circeo lies well within the modern range for shape Analysis 8. This result is confusing given that, morphologically, this fossil has been described as similar to other Neanderthals (Sergi 1974, Stringer & Trinkaus 1981, Stringer et al. 1984). However, Monte Circeo is well outside the range of modern humans for both the size/shape and the shape analyses in Analysis 4.

Among the Neanderthal group, the distances from the recent modern range for La Chapelle are routinely greater than those for the other European and Near Eastern Neanderthals, a finding which provides further confirmation that La Chapelle hardly represents a 'typical' or average Neanderthal. On the other extreme, La Ferrassie 1 exhibits an overall distance profile closer to the modern human range than other European Neanderthals (except Spy 2 in Analysis 7). Additionally, the shape distances for the Shanidar 1 specimen are, overall, smaller than those for most of the European Neanderthals. Tabun 1 tends to exhibit shape distances that are considerably greater than those of Shanidar 1 and are closer to those found for European Neanderthals. Amud 1 exhibits somewhat a mixed pattern in the two analyses (4 and 8) in which it is included, although its shape distances from the modern human range are always less than Tabun 1's. The results for the Near Eastern Neanderthals are particularly interesting given that multivariate assessment of the frontal bone reveals Shanidar 1 to be more divergent from early modern Near Easterns than either Tabun 1 or Amud 1 is (Simmons et al., in press).

The north African fossil Jebel Irhoud, although not strictly a Neanderthal, is also included in this group. It is used in 2 analyses and in neither instance falls within the modern range. While the phylogenetic significance of this fossil is currently unclear, Bräuer has described the supraorbital region as having 'its greatest affinities with the torus morphology of the Neanderthal(oids) of the Near East and Europe' (1984:359). In addition, the vault is low and exhibits occipital bunning similar to that of European Neanderthals. There is, however, more parietal bossing and a higher biparietal breadth than that typically found in the European Neanderthals.

3.3 *Early European and Near Eastern non-archaic samples*

The non-archaic or 'modern' *Homo sapiens* sample has been loosely partitioned into an early group and a late group. The early modern group consists of fossils that have been dated earlier than ca. 26-28 Ky BP and include samples from the sites of Qafzeh, Skhūl, Mladeč and Cro-Magnon.

Qafzeh 6 and 9 are dated to between 90 and 100 kya by both thermoluminescence (TL) and electron spin resonance (ESR) (Valladas et al. 1988, Schwarcz et al. 1989) and are used in Analyses 4 and 8. Qafzeh 6 is found to be outside the modern range for Analysis 8 (the only analysis in which it is included) but Qafzeh 9 is within the modern range for both Analyses 4 and 8. Qafzeh 6, while having a small face still retains relatively well-developed brow-ridges and, in comparison to Qafzeh 9 has a somewhat flattened, more prognathic face. This fossil also has a large mastoid process and a noticeable canine fossa. Qafzeh 9 has a facial morphology similar to recent humans and considerably reduced brow-ridges. It also exhibits a frontal morphology that is generally more like recent humans than are those of the Qafzeh 3 and 6 hominids (Simmons et al. in press). The more modern position of Qafzeh 9 in both Analysis 4 and Analysis 8 may be due to its subadult developmental status.

The two Skhūl specimens also recently dated between ca. 80-100 Ky BP by ESR (Stringer et al. 1989) exhibit a non-modern pattern. For Skhūl IV, which is used for Analyses 3-6, both size/shape and shape D^2s are well outside the recent modern range of variation. Of the eight analyses for Skhūl V, all D^2s are outside the modern range. Although generally modern in appearance, this skull is robust and possesses distinct brow ridges. Its placement outside the modern range is therefore, as the shape D^2s show, not primarily a function of size.

The recent ESR and TL dates place both the Qafzeh and Skhūl hominids in close temporal approximation to each other and raise the possibility that they represent the same or similar regional populations. If this is the case, the analyses reported in this paper portray the combined Skhūl/Qafzeh sample as a highly variable one, straddling the boundary between archaic humans and the modern human range. Except for Qafzeh 9, the Skhūl/Qafzeh shape D^2s are consistently more divergent from the modern human range than the earliest modern Europeans.

The earliest post-Neanderthal European fossils, those from Mladeč, exhibit distances that are, almost without exception, outside the modern range. In only one analysis for the specimens of Mladeč 1, 2 and 5 is a distance found to be within the modern range (Analysis 8 for Mladeč 2). In most analyses, Mladeč 2 seems unusually divergent from the modern human range for such a gracile specimen. This may indicate problems with its reconstruction. When Mladeč 1 and 5 are viewed from a metric standpoint, the reasons for their non-modern D^20s become apparent. Five of the seventeen measurements for Mladeč 1 are between 12% and 18% greater than the modern means (GOL, NOL, XFB, STB and OCC) and one (OCS) is 33% greater. Although this individual is probably a female and the splanchnocranium is mostly modern in appearance, it is overall a long, wide skull. The same is also true of Mladeč 5, which exhibits five dimensions (GOL, NOL, XCB, XFB and ASB) between 12% (XCB) and 18% (GOL) higher than the recent means for these measurements.

Although these crania are not Neanderthals by any definition, these specific measurements are close to those of the Neanderthal means and the resulting distance statistics are not primarily size-weighted. Thus, their position outside the modern range is largely a function of the shape of the Mladeč hominids and suggests closer relationships to archaic humans than is the case for more recent samples.

The fossil sample from Cro-Magnon deviates from the Mladeč pattern in that two of the eighteen D^2s for the size/shape analyses are within the modern range (Analyses 6 and 7 for Cro-Magnon 2), while in the shape analyses, eleven of eighteen D^2s are within the modern range (Analyses 2, 5, 6 and 7 for Cro-Magnon 1, 6-7 for Cro-Magnon 2, 8 for Cro-Magnon 3, and Analyses 3, 6, 7 and 8 for the anomalous Cro-Magnon 8). It is in this group that the differences between the size/shape D^2 and the shape D^2 are most pronounced. When size compensation is employed, the pattern for the Cro-Magnon fossils becomes distinctly modern. This is not altogether unexpected, considering that the Cro-Magnon hominids are probably several thousand years more recent than those from Mladeč (Smith 1984, Wolpoff 1989). The face of Cro-Magnon 1 has been described as being flatter in appearance than Neanderthals but still pulled forward from the vault (Stringer et al. 1984). Conversely, Cro-Magnon 2 has measurements considerably reduced over Cro-Magnon 1, 3 and 8 and is even more modern in overall appearance. This fossil is similar to presumably later female crania such as Abri Pataud and Předmostí IV (Stringer et al. 1984). Only Cro-Magnon 3, which is used for Analyses 3, 6, 7 and 8 and falls outside the modern range for all analyses. This fossil possesses three measurements (GOL, NOL, and ASB) that are between 11.08 and 15.20% larger than the recent means.

3.4 *Later European post-Neanderthal samples*

This group continues the trend toward gracilization with fossils exhibiting a general decrease in size compared to the Mladeč/Cro-Magnon samples. In fact throughout this entire analysis, the pattern for this late group is clearly within the range of modern human variation. Of twenty-five fossils in this group, only four do not contain a majority of analyses within the modern range. It is also in this group of fossils that distinctions between the size/shape set of analyses and shape analyses significantly diminish, as these fossils begin to assume sizes closer to those of the recent modern sample. Results for specific individuals and sites samples in this group are discussed in detail in the following paragraphs.

Of eight analyses calculated for the 26 Ky old Dolní Věstonice 3 cranium, seven are found to be within the modern range of variation. For Analysis 1, in which all variables are used, this fossil is outside the modern range. This analysis contains three measurements (FMR, MDH and MDB) that are between 13.47% and 31.11% greater in Dolní Věstonice than the modern means. Also displaying a modern pattern is the Předmostí material, also approximately 26 Ky old, in which for Analyses 4 and 8, Předmostí I, III, IV and X fall within the range of modern variation for the size/shape D^2. Only Předmostí IX is outside the modern range. When size is compensated for, this cranium also falls within the modern range for Analysis 4. Only in Analysis 8 does Předmostí IX fall outside the modern range even when size is adjusted for. Thus despite arguments that both Dolní Věstonice 3 and the Předmostí sample exhibit morphological features which are considered to be Neanderthal-reminiscent (Smith 1984), their metric shape pattern is clearly modern human.

The results for the two crania from Brno are a bit more equivocal in nature. While the Brno 1 cranium is found to have distances within the modern range in both analyses in which it is used (6 and 7), Brno 2 has no distances within the modern range in any of the four analyses for either the size/shape analyses or the shape analyses that include it. Since Brno 2 is thought to be 26.32 ± 0.24 Ky BP and has been characterized as both robust and Neanderthal-like in morphology (Smith 1984), this result is not unexpected. The results for Brno 1 is commensurate with the suggestion that it dates toward the end of the Pleistocene (Smith 1984).

Also from this general time-period is the Paderborn cranium dated to 27.4 ± 0.6 Ky BP by radiocarbon and around 26 Ky BP by amino acid racemization (Henke & Protsch 1978). Although modern in appearance, this fossil falls outside the modern range of variation in five of eight analyses even when adjusted for size. A primary reason for these results may be the width of this cranium. The biasterionic breadth is 132 mm. This measurement matches that of Skhūl IV and, within this set of fossils is only exceeded by that of Amud 1 (134 mm). This variable is employed in four of the five analyses for which non-modern D^2s are recorded. Additionally, two of the three analyses for which modern distances are found do not contain this variable.

Abri Pataud, which is considered to be Proto-Magdalenian in age, or approximately 22 Ky BP (Stringer et al. 1984), is used in eight analyses and has distances within the modern range in all eight. Although this skull may be an adolescent (maxillary M3's only partially erupted), it maintains affinities to other modern humans.

Both Oberkassel fossils, also thought to be Magdalenian, are used in all eight analyses and both show a clearly modern pattern, having modern D^2 for all but one analyses each (Analysis 7 for Oberkassel 1 and Analysis 1 for Oberkassel 2). These are clearly modern humans with little to distinguish them from the recent sample.

The Engis adult cranium also exhibits a modern pattern. Although used for only Analyses 4 and 8, the D^2 for both are within the modern range. This skull is attributed to either Aurignacian or Upper Perigordian levels and is described as being 'entirely modern in morphology' (Stringer et al. 1984:112). The D^2 analysis for this fossil is in agreement with this interpretation.

Two fossils, Laugerie Basse and Kaufertsberg are associated with Magdalenian and late Magdalenian or Proto-Mesolithic assemblages respectively. Laugerie Basse is used in Analysis 7 for which a D^2 in the modern range is derived. Kaufertsberg exhibits a similar pattern as that for Abri Pataud in that for all eight analyses modern distances are present.

Slightly later in time (possibly early Mesolithic), the fossils from Ofnet appear to continue the trend of increasing modernity for the European later modern sample. Of the nine fossils from this site, seven possess a majority of analyses within the modern range. In the cases of Ofnet 4, 18 and 29, all analyses contained D^2s in the modern range. Only in Ofnet 8 and Ofnet 15 is a modern pattern not evident. Only half (two of four) of the size/shape and shape analyses for Ofnet 15 have modern distances and for Ofnet 8 only three of eight for the size/shape set and two of eight for the shape set of analyses are in the modern range.

Although the overall pattern of the D^2s for the Ofnet fossils is modern, the results for Ofnet 8 are puzzling. Two possible explanations exist for these aberrant D^2s. As stated earlier, even in a recent modern sample it is not uncommon to find single individuals who do not fall within the recent modern range of variation. When using a 95% confidence level, it is expected that 5% of those individuals who have metric patterns that should fall within the range of recent modern humans do not.

The additional possibility exists that, due to the necessity of reconstructing most of the crania for this site, inaccuracies in the reconstruction of Ofnet 8 may result in the D^2s being outside the modern range of variation. The size D^2s for this fossil are not large enough to attribute the discrepancies to variation in size.

The last fossil used in the European modern group, Kelsterbach, has been dated by R. Protsch at 15.8±0.41 Ky BP by radiocarbon, although a collagen date of 31.2±0.6 Ky BP and an amino acid racemization date of 32 Ky BP is associated with the cranium (Protsch & Semmel 1978). This fossil is modern in appearance and is quite gracile (Smith 1984). Not only does it fall within the modern range for all analyses in which it is used, but it accounts for many of the lowest D^2s relative to the recent human sample recorded in presumably fossil specimens. Of all of the hominids analysed, Kelsterbach appears to have the most 'modern' morphology. The D^2s would tend to support the more recent date for this fossil.

4 CONCLUSION

These results suggest that providing an unequivocal definition of modern *Homo sapiens* is an exceedingly difficult undertaking. Even in Europe and the Near East, where the distinction between early 'modern' humans and their Neanderthal forerunners is considered by some to be obvious (Stringer et al. 1984, Stringer & Andrews 1988), a definition based on morphological patterns does not totally coincide with the metric patterns isolated in this study. In Europe, hominid samples that exhibit both size/shape and shape metric patterns within the range of recent human variation first appear between 26 and 28 Ky BP. These hominids would also qualify as modern humans on morphological grounds (Smith 1984, Stringer et al. 1984). However, for both Europe and Near Eastern samples dating earlier than 26 to 28 Ky BP and yet also considered to exhibit a modern morphological pattern (Smith 1984, 1985, Stringer et al. 1984, Trinkaus 1984), the situation is much more complex.

The data from Europe support a dual-staged shift in cranial form. The earliest 'modern'

Europeans, the Mladeč sample, exhibit both a number of Neanderthal-reminiscent morphological features (Frayer 1984, 1986, Smith 1984, Wolpoff 1980) and a metric pattern that falls out of the recent human range of variation in both the size/shape and shape D^2 analyses reported here. Thus, we would argue that the Mladeč sample does not qualify unequivocally as a modern human sample. There does, however, appear to be a gradual shift toward a modern metric pattern with the Cro-Magnon sample. This shift, though, is mosaic and becomes apparent only when size is controlled. The Cro-Magnon hominids exhibit shape patterns that approach the recent human range to a greater extent than the Mladeč hominids do. Nonetheless, the shape analyses for the Cro-Magnon sample do not totally place it within the modern range. This occurs, as mentioned above, only with the post-26 Ky European sample.

In the Near East, the Qafzeh 6 and Skhūl IV and V D^2s are generally even further removed from the modern human range than are those of the early post-Neanderthal Europeans. Qafzeh 9 is an exception, falling within the recent range in both analyses in which it is included. Still, the Skhūl/Qafzeh hominids as a sample do not conform to the modern human metric pattern as defined in this study.

Thus, it appears that demonstrably modern humans (humans falling in both metric and morphological ranges of recent human variation) do not occur in either Europe or the Near East without passing through stages (i.e. Mladeč and Skhūl/Qafzeh respectively) which fall out of the recent human metric pattern and which, on the basis of other studies (Frayer 1984, 1986, Simmons et al., in press, Smith 1984, 1985, Wolpoff 1980, 1989), exhibit distinct similarities to archaic *Homo sapiens*. It should be noted, however, that the analyses reported here are not designed to test questions of regional continuity between archaic and modern humans.

Since craniometrics seem to have a reasonably strong genetic basis, patterns derived from metric analyses are no less indicative of biological relationships than is morphology (Nakata et al. 1974). Thus, an unequivocal definition of modern *Homo sapiens*, or simply *Homo sapiens* as per Stringer & Andews (1988), should encompass both total morphological and metric patterns. We conclude that formulating such a definition is not yet possible even for European and Near Eastern samples. We believe this reflects archaic human contribution to the earliest 'modern' samples from both of these regions. Our results, then, basically support Wolpoff's (1986) claims regarding the precise definition of modern humans and reaffirm our thoughts that archaic and modern *Homo sapiens* cannot be clearly differentiated into distinctly separate taxonomic units when the earliest 'modern' samples are included.

REFERENCES

Ambergen, A. & W. Schaafsma 1984. Interval estimates for posterior probabilities, applications to Border Cave. In G. van Vark & W. Howells (eds), *Multivariate Statistical Methods in Physical Anthropology*. Dordrecht: Riedel Publishing Company, pp. 115-134.

Bräuer, G. 1984. A craniological approach to the origin of anatomically modern *Homo sapiens* in Africa and implications for the appearance of modern Europeans. In F. Smith & F. Spencer (eds), *The Origin of Modern Humans: A World Survey of the Fossil Evidence*. New York: Alan R. Liss Inc., pp. 327-410.

Brown, P. 1981. Sex determination of Australian Aborigine crania from the Murray Valley: A reassessment of the Larnach and Freedman technique. *Archaeology in Oceania* 16: 53-63.

Campbell, N. 1984. Some aspects of allocation and discrimination. In G. van Vark & W. Howells (eds), *Multivariate Statistical Methods in Physical Anthropology*. Dordrecht: Riedel Publishing Company, pp. 177-192.

Condemi, S. 1988. A review and analysis of the Riss-Wurm Saccopastore skulls, can they provide evidence in regard to the origin of Near Eastern Neandertals? *L'Homme de Neandertal, v.3 L'anatomie*, Liege, pp. 39-48.

Day, M. & C. Stringer 1982. A reconsideration of the Omo Kibish remains and the *erectus-sapiens* transition. In M. de Lumley (ed.), L'*Homo erectus* et la Place de l'Homme de Tautavel Parmi les Hominides Fossiles. *1er Cong. Inst.* Palaeontol. *Hum. Nice* 2, pp. 814-846.

Darroch, J. & J. Mosimann 1985. Canonical and principal components of shape. *Biometrika* 72(2): 241-252.

Fatti, L. 1986. Discriminant analysis in prehistoric physical anthropology. In R. Singer & J. Lundy (eds), *Variation, Culture and Evolution in African Populations.* Witwatersrand: Witwatersrand University Press, pp. 27-34.

Frayer, D. 1984. Biological and cultural change in the European late Pleistocene and early Holocene. In F. Smith & F. Spencer (eds), *The Origins of Modern Humans: A World Survey of the Fossil Evidence*. New York: Alan R. Liss Inc., pp. 211-250.

Frayer, D. 1986. Cranial variation at Mladeč and the relationship between Mousterian and Upper Palaeolithic hominids. *Anthropos (Brno)* 23: 243-256.

Habgood, P. 1988. A morphometric investigation into the origin(s) of anatomically modern humans. Ph.D thesis, University of Sydney.

Henke, W. & R. Protsch 1978. Die Paderborner Calvaria: Ein diluvialer *Homo sapiens*. *Anthropol. Anz.* 36: 85-108.

Howell, F. 1960. European and northwest African middle Pleistocene hominids. *Cur. Anthropol.* 1: 195-232.

Howells, W. 1973. *Cranial Variation in Man.* Cambridge: Harvard University Press.

Mahalanobis, P. 1936. On the generalized distance in statistics. *Proceedings of the National Institute of Science, India* 12: 49-55.

Matiegka, J. 1934. *Homo Předmostensis: Fosiliní Človek z Předmostí na Moravě. 1. Lebky.* Prague: Cěske Akademie Věd a Umění

Nakata, M., P. Yu, B. Davis & W. Nance 1974. Genetic determinants of cranio-facial morphology: A twin study. *Ann. Hum. Genet. Lond.* 37: 431-443.

Protsch, R. & A. Semmel 1978. Zur Chronologie des Kelsterbach-Hominiden. *Eiszeitalter Gegenw.* 28: 200-210.

Rao, C. 1952. *Advanced statistical methods in biometric research.* New York: John Wiley.

Rightmire, G. 1969. On the computation of Mahalanobis' Generalized Distance (D^2). *Amer. J. of Phys. Anthropol.* 30: 157-160.

Rightmire, G. 1976. Relationships of Middle and Upper Pleistocene hominids from sub-Saharan Africa. *Nature* 260: 238-240.

Schwarcz, H., R. Grün, B. Vandermeersch, O. Bar-Yosef, H. Valladas & E. Tchernov 1988. ESR dates for the hominid burial site of Qafzeh in Israel. *J. Hum. Evol.* 17: 733-737.

Sergi, S. 1974. *Il Cranio Neanderthaliano de Monte Circeo.* Rome: Accademia Nazionale die Lincei.

Simmons, T., A. Falsetti & F. Smith (in press). Evolutionary patterns in the frontal bones of late Pleistocene hominids from western Asia. *J. Hum. Evol.*

Smith, F. 1984. Fossil hominids from the Upper Pleistocene of central Europe and the origin of modern Europeans. In F. Smith & F. Spencer (eds), *The Origin of Modern Humans: A World Survey of the Fossil Evidence*. New York: Alan R. Liss Inc., pp. 137-209.

Smith, F. 1985. Continuity and change in the origin of modern *Homo sapiens*. *Z. Morphol. Anthropol.* 75: 197-222.

Stringer, C. 1974. Population relationships of later Pleistocene hominids: A multivariate study of available crania. *J. Archaeol. Sci.* 1: 319-342.

Stringer, C. 1978. Some problems in Middle and Upper Pleistocene hominid relationships. In E. Chivers & K. Joysey (eds), *Recent Advances in Primatology, Vol. 3. Evolution*. London: Academic Press, pp. 395-418.

Stringer, C. 1988. The Dates of Eden. *Nature* 331: 565-566.

Stringer, C. & P. Andrews 1988. Genetic and fossil evidence for the origin of modern humans. *Science* 239: 1263-1268.

Stringer, C., R. Grün, H. Schwarcz & P. Goldberg 1989. ESR dates for the hominid burial site of Es Skhūl in Israel. *Nature* 338: 756-758.

Stringer, C., F. Howell & J. Melentis 1979. The significance of the fossil hominid skull from Petralona, Greece. *J. Archaeol. Sci.* 6: 235-253.

Stringer, C., J.-J. Hublin & B. Vandermeersch 1984. The origin of anatomicaly modern humans in western

Europe. In F. Smith & F. Spencer (eds), *The Origins of Modern Humans: A World Survey of the Fossil Evidence*. New York: Alan R. Liss, Inc., pp. 51-135.

Stringer, C. & E. Trinkaus 1981. The Shanidar Neanderthal crania. In C. Stringer (ed.), *Aspects of Human Evolution*. London: Taylor and Francis, pp. 129-165.

Suzuki, H. & F. Takai 1970. *The Amud Man and His Cave Site*. Tokyo: Academic Press of Japan.

Tattersall, I. 1986. Species recognition in human paleontology. *J. Hum. Evol.* 15: 165-175.

Trinkaus, E. 1983. *The Shanidar Neandertals*. New York: Academic Press.

Trinkaus, E. 1984. Western Asia. In F. Smith & F. Spencer (eds), *The Origins of Modern Humans: A World Survey of the Fossil Evidence*. New York: Alan R. Liss, Inc., pp. 251-294.

Valladas, H., J. Reyss, J. Joron, G. Valladas, O. Bar-Yosef & B. Vandermeersch 1988. Thermoluminescence dating of Mousterian 'Proto-Cro-Magnon' remains from Israel and the origin of modern man. *Nature* 331: 614-616.

Van Vark, G. 1970. Some statistical procedures for the investigation of prehistoric human skeletal material. Thesis, Groningen University.

Vandermeersch, B. 1981. *Les Hommes Fossiles de Qafzeh (Israel)*. Paris: Editions du C.N.R.S.

Wolpoff, 1980. *Palaeoanthropology*. New York: Knopf.

Wolpoff, M. 1986. Describing anatomically modern *Homo sapiens*: A distinction without a definable difference. *Anthropos (Brno)* 23: 41-53.

Wolpoff, M. 1989. The place of the Neandertals in human evolution. In E. Trinkaus (ed.), *Biocultural Emergence of Modern Human in the Later Pleistocene*. Cambridge: Cambridge University Press, pp. 97-141.

The persistence of Neanderthal features in post-Neanderthal Europeans

David W. Frayer
Department of Anthropology, University of Kansas, Lawrence, USA

1 INTRODUCTION

With the recent developments regarding the application of mitochondrial DNA for dating phylogenetic events (Cann et al. 1987), a new surge of proclamations about the relationship between Middle and Upper Paleolithic groups has swept across the scientific and popular literature (Fisher 1988, Gould 1987, Lewin 1987, 1988, Stringer & Andrews 1988a, Tierney et al. 1988). It is interesting how rapidly and confidently the mitochondrial DNA data have been accepted as the final proof for eliminating the Neanderthals from any contribution to subsequent European populations. At the same time, it is puzzling how a collection of scattered, fragmentary and poorly-dated African fossils have been substituted as the ancestors of us all. In many ways, this reflects the tendency in paleoanthropology to prefer sensational catastrophism over mundane gradualism and the willingness of some to accept without criticism (or even suspicion) the results of molecular biology. Thus, if the results of the mitochondrial DNA analysis are correct, the replacement of Neanderthals in Europe was only a minor incident compared to the Old World-wide event in which all archaic human populations were totally replaced by African migrants.

Interpretations of mitochondrial DNA variation among living people and inferences from it about the fossil record are controversial (Darlu & Tassy 1987, Spuhler 1987, Wolpoff 1989, Wolpoff et al. 1988), but, if there is a single advantage to these developments, the mitochondrial DNA results so polarize the issues that they provide some clear-cut, testable predictions about the fossil record. For example, accepted at face-value, the low level of variation in mitochondrial DNA suggests that modern humans have a recent, single geographic origin, presumed to be southern Africa or the Levant (Stringer & Andrews 1988a). This makes statements about hybridization between Mousterian and Upper Paleolithic groups (Bräuer 1981) no longer tenable, since the standard interpretation of the mitochondrial DNA model allows no gene flow between archaic *Homo* and modern *Homo* populations. Similarly, specific predictions about regional and chronological continuity in the fossil record are no longer possible, even though there is a wealth of information to support local continuity (Hrdlička 1927, Weidenreich 1947, 1951, Thorne & Wolpoff 1981, Wolpoff et al. 1984). Unborn historians of paleoanthropology will one day marvel at the willingness of some human paleontologists in the late 1980's to eliminate the rich, pre-modern human fossil record from any kind of contribution to the course of human evolution.

Stringer & Andrews (1988a: Table 1) have outlined some of the basic predictions of the two models. In condensed form these relate to continuity or discontinuity in morphological patterning over time and to the widespread or regionally restricted occurrence of transitional fossils. Specifically for Europe, Stringer and Andrews argue that 'there is an absence of ... transitional fossils' (1988a:1266), a statement which ignores numerous, extensive publications on what is considered by some to be good evidence for a transitional sequence in central Europe (Jelínek 1969, Smith 1982, 1984, Wolpoff 1980, Wolpoff et al. 1981). Regardless of this oversight, it is perhaps difficult to ascertain what are true transitional specimens, since age, sex and individual variation (along with problems in chronological placement and the incomplete nature of some of the fossil remains) can prejudice judgments about what is and is not a transitional fossil.

Patterns of change in morphological features, however, offer a more reliable test of the separate predictions of the gradual or replacement models. With gradual change, morphological states considered unique to Middle Paleolithic samples of a region would be expected to occur in the samples that immediately succeed them; and, due to selective and stochastic factors, certain geographic areas should be characterized by unique regional features. If replacement accounts for the change, the samples succeeding Middle Paleolithic groups should not show features considered as uniquely characterizing the earlier period. Indeed, Stringer & Andrews (1988a:1266) argue with regard to the European sequence that 'there is little or no continuity of genuine regional features, for the most distinctive and well-established characteristics of Neanderthals are poorly represented or absent in contemporaneous or immediately succeeding *Homo sapiens* fossils.'

Stringer and Andrews do not provide a list of the features which are 'absent' or 'poorly represented' in the post-Neanderthal Europeans, but apparently these do not include details of browridge development, tooth size patterning or 'protruding occipitals', since they consider such traits as symplesiomorphies or homoplasies. This is somewhat inconsistent and definitely confusing since it is precisely these features which have been used to characterize the anatomically modern state of the African (or Levant) replacing populations (e.g. Border Cave, Klasies River and Qafzeh). Although I think it is unjustified to eliminate these from consideration, I have chosen a different set of features to test for patterns of continuity or discontinuity between the European Middle and Upper Paleolithic.

2 ANATOMICAL FEATURES

The features reviewed in this paper do not derive from my own assessment of what might be unique morphologies of European Middle Paleolithic hominids. Rather, I have selected three traits which proponents of the replacement model have suggested to be uniquely derived (autapomorphic) or common traits of European Neanderthals. These come from a compilation by Stringer et al. (1984:55), who include the supra-iniac fossa, horizontal-oval (H-O) mandibular foramen, and dorsal orientation of the scapular axillary border in their list of about 20 Neanderthal features. The same authors have separately considered some (or all) of these features as characterizing European Neanderthals (Hublin 1978a, 1980, Stringer 1982, 1984, Vandermeersch 1985) and used them for distinguishing Neanderthals from Upper Paleolithic, Mesolithic or more recent Europeans.

The data for each sample come primarily from my own analysis of the relevant skeletal material. For much of the Neanderthal sample (which includes only western and central European specimens, dating mainly to Würm I and II contexts), I have relied upon key publications which have provided individual data for the specific traits (Hublin 1978a, Smith 1978, Trinkaus

1977, 1983), but these have been supplemented by my own observations on many of the European Neanderthal remains. For the post-Mousterian groups, data derive from my personal examination of the material or, in some cases where the specimens no longer exist or were unavailable for study, from the photographs or descriptions in the literature. I have also utilized casts for determination of the trait on some specimens and verified some of the data with Smith (pers. comm.) and Trinkaus (pers. comm.). The Upper Paleolithic sample is divided into an Early and Late phase following the rationale outlined in Frayer (1978). The only difference concerns the placement of Azilian material. Earlier, I included this material in the Mesolithic, but now consider the Azilian as terminal Late Upper Paleolithic.

For the supra-iniac fossa and the mandibular foramen type, I provide comparative data for recent Europeans based on the Medieval Hungarians from Halimba and Zalavár, which are dated between the 9th-12th centuries. I do not contend that these 900 year old Hungarians are typical of a pan-European pattern for the traits nor that they are direct lineal descendants of the any of the earlier populations. More work documenting the variation of these traits in recent Europeans is necessary, but for now the reported frequencies provide a bench mark for comparison with more ancient Europeans. Unfortunately, due to time constraints, I was unable to collect data for the frequency of the axillary border types in these Hungarian samples. Consequently, modern European data for this trait come from von Eickstedt (1925:225).

2.1 *The supra-iniac fossa*

Although this trait was recognized as a special Neanderthal feature at the turn of the century by Gorjanović-Kramberger (1902) and Klaatsch (1902), in the past 10 years the development of a fossa above inion has received much attention. The renewed interest is due to the work of Hublin (1978a, 1978b) and Santa Luca (1978), who almost simultaneously presented evidence to show that the supra-iniac fossa was primarily limited to European and Near Eastern Neanderthals. This feature, which consists of a depression located on the occipital plane above the nuchal torus, occurs in both sexes and in subadult and adult Neanderthals. According to the literature (Hublin 1978a), the supra-iniac fossa is extremely rare in recent Europeans. Questions remain about the cause for the development of the trait, but it does not seem to be related to a specific muscle insertion. Hublin (1978a:171) considers the supra-iniac fossa as a 'homologue' to the external occipital protuberance (which is absent in Neanderthals) and suggests the fossa is produced by the contraction of the space between the superior and supreme nuchal lines. On the other hand, Heim (1982) argues that the presence of the fossa relates to the retarded osteogenic development of the midsagittal region above inion. He illustrates five contemporary human infants with porosity in the supra-iniac region, and I have observed a similar condition in Upper Paleolithic and Mesolithic children. This pitting occurs in some cases when the external occipital protuberance is developed. In the post-Neanderthal adults, a supra-iniac fossa occurs with the external occipital protuberance, which leads me to conclude that the supra-iniac fossa and external occipital protuberance are not homologous structures.

Whatever the biological cause of this feature, it does seem to be primarily limited to Neanderthals. Indeed, the supra-iniac fossa is considered a 'probable autapomorphy' of Neanderthals by Stringer et al. (1984:55). However, it is important to note that the supra-iniac fossa is not unique to Neanderthals, since a shallow supra-iniac fossa occurs on Qafzeh 6 (Hublin 1978a: 154, Vandermeersch 1981:106) and on Skhūl IX (McCown & Keith 1939:280). In addition, the the supra-iniac fossa appears sporadically on occipitals outside of Europe (e.g. Sinanthropus III, Weidenreich 1951), while about half of the European pre-Würm occipitals (e.g. Bilzingsleben

A1, Vértesszöllös 2 and Petralona) completely lack the feature. Among the Neanderthals the supra-iniac fossa is polymorphic, showing a variable expression from a deep, sometimes doubled fossa as in La Chapelle-aux-Saints to a large, extensive fossa in Salzgitter-Lebenstedt or to the single depression as in Gibraltar (Hublin 1978a). Despite variation in form, virtually all European Neanderthals possess a supra-iniac fossa, so that this feature would appear to be a reasonable candidate for a regional morphological marker. The only possible exception is a single Krapina occipital fragment (Krapina 11, Occipital 5). This specimen preserves most of the left occipital and nuchal plane, but is broken lateral to inion. However, from what remains of the bone near the midline and above the occipital torus, there is no evidence for the development of a supra-iniac fossa. According to Radovčić et al. (1988), this occipital derives from the same level as Krapina 1 (Cranium A) and is therefore relatively late at the site. Because of questions concerning its taxonomic affiliation and uncertainties related to its fragmentary state, I have not included it in the Neanderthal tabulations.

In the Upper Paleolithic and Mesolithic samples, there are no specimens possessing a supra-iniac fossa which exactly match the Neanderthal range of variation. In contrast, Upper Paleolithic and Mesolithic supra-iniac fossae are nearly always associated with an external occipital protuberance; and, although an individual fossa may be as deep as in the Neanderthals, it is generally not as extensive in its vertical and transverse dimensions. Another difference relates to shape, in that the Neanderthal fossa is often triangular with the apex superiorly directed, while in the Upper Paleolithic/Mesolithic specimens the apex of the triangle points inferiorly. Thus, while present in a similar position in post-Neanderthal Europeans, the fossa is not identical and seems to be more clearly an artifact of the very strong supreme nuchal lines and generally prominent external inion.

Besides size differences, the fossa occurs in much lower frequencies in the Upper Paleolithic and Mesolithic. As shown in Table 1, frequencies for presence of the fossa in adults range between 25% and about 19% for Upper Paleolithic and Mesolithic samples, contrasting with the 100% rate for the Mousterian specimens. However, given the low occurrence of any development of a supra-iniac fossa in recent Europeans (6% in Medieval Hungarians and 2% as reported by Hublin 1978:44, for modern French), the relatively high incidence in the Upper Paleolithic and Mesolithic samples is unexpected for models requiring complete replacement. Certainly, the supra-iniac fossa is neither absent nor even poorly represented in the Upper Paleolithic or Mesolithic. Yet, given differences in morphology and the substantially lower incidence in post-Mousterian samples, the phylogenetic implications of these observations alone are subject to debate.

Table 1. Occurrence (in percent) of the supra-iniac fossa for European Mousterian to recent groups. [1]Not including Krapina occipital 5 (11), [2]From Hublin (1978a).

	Present	Absent	(n)
Neanderthals[1]	100.0	0.0	24
Early Upper Paleolithic	25.0	75.0	20
Late Upper Paleolithic	23.7	76.3	38
Mesolithic	19.3	80.7	161
Medieval Hungarians	5.9	94.1	237
Modern French[2]	2.0	98.0	100

2.2 *The shape of the mandibular foramen*

In his studies of the Krapina hominids, Smith (1976, 1978) noted the high frequency of the horizontal-oval (H-O) mandibular foramen in Neanderthals. The morphological state consists of a broad bridge across the most superior aspect of the mandibular foramen, contrasting with the normal condition where the lingula is isolated at the anterior border and a V-shaped sulcus separates the anterior and posterior borders of the foramen. First reported by Gorjanović-Kramberger (1906), the H-O mandibular foramen rarely occurs in recent humans. Smith (1978) argues the trait corresponds to the attachment of the sphenomandibular ligament and is either a discrete trait (without functional significance) or an indicator of a broadened ligamental band related to increased masticatory stress. The former explanation is better supported by his data, and Smith (1978:530) proposes that the H-O trait may serve as a 'genetic marker' for Neanderthals. Stringer et al. (1984:55) apparently concur with this assessment, since they include this feature in their table of proposed Neanderthal autapomorphies with the description: 'H-O pattern of mandibular foramen common'.

Following Smith's definitions, the mandibular foramen shape (H-O or Normal) was scored for Upper Paleolithic, Mesolithic and Medieval groups and frequencies for European Neanderthals were recalculated. In each of these samples, only late adolescents and adults were included and presence/absence frequencies were determined by the number of mandibles examined rather than the number of foramina studied. In some cases (limited to the Mesolithic and Medieval samples) the trait does not occur bilaterally. Despite the absence of the trait on one of the sides, an H-O mandibular fossa was scored as present when at least one side exhibited the anomalous morphology.

Unlike the supra-iniac fossa, the morphological state of the Neanderthal mandibular foramen is exactly matched in Upper Paleolithic and Mesolithic hominids as well as the recent sample from Hungary. Table 2 reviews frequencies of H-O and Normal mandibular foramen types for five European samples. It is obvious the H-O character state is common in European Neanderthals (52.6%) and uncommon in the Medieval Hungarians (1.4%). Yet, if one drew phylogenetic implications from the frequency of this trait in the five samples, the break in incidence occurs not between the Neanderthal and Early Upper Paleolithic samples, but between the Early and Late Upper Paleolithic. In fact, the Neanderthal and Early Upper Paleolithic samples have very similar frequencies, both of which are substantially higher than the incidence of the H-O trait for the Late Upper Paleolithic, Mesolithic and Medieval samples. Given the fact that the H-O mandibular foramen is a rarely found variant in Amerind, African and Australian *Homo sapiens* samples and occurs infrequently in non-European fossil mandibles, as Smith (1978:530) argues, it may represent a characteristic regionally restricted to Europe and, possibly, the Near East. If so, the high incidence in the Neanderthal and Early

Table 2. Occurrence (in percent) of mandibular fossa types for European Mousterian to recent groups.

	H-O	Normal	(n)
Neanderthals	52.6	47.4	19
Early Upper Paleolithic	44.4	55.5	9
Late Upper Paleolithic	5.3	94.7	38
Mesolithic	1.9	98.1	161
Medieval Hungarians	1.4	98.6	208

Upper Paleolithic samples argues for continuity between these groups. If complete replacement of Neanderthals occurred, these Early Upper Paleolithic specimens come to possess the same unique feature which characterizes the group they displaced. At the same time, a substantial discontinuity occurs between the Early Upper Paleolithic and the Late Upper Paleolithic samples for which there is no question of an ancestral/descendant relationship. Those who hold the view that autapomorphic features signal populational (or species) discontinuity are required to interpret these data for the H-O mandibular foramen as indicating a break within the Upper Paleolithic, not between the Mousterian and Upper Paleolithic.

2.3 *The axillary scapular border*

The distinctive morphology of the scapular axillary border has long been recognized as a feature characteristic of Neanderthals. Following the classification of von Eickstedt (1925), three basic axillary border patterns have been defined. These are dorsal, bisulcate and ventral – with dorsal the common pattern in Neanderthals and ventral the most frequent variant in samples of recent *Homo sapiens*. As Smith (1976) and Trinkaus (1977, 1983) have argued, due to the probable interaction of heredity and environment, the phylogenetic significance of this characteristic is uncertain. Yet Stringer et al. (1984:55) include a dorsal axillary sulcus in their list of proposed Neanderthal autapomorphies and common characteristics. Despite questions concerning the hereditary nature of the feature, this allocation is reasonable considering the high incidence of the dorsal type border in European Neanderthals.

In general, axillary borders can be placed rather confidently in one of these three types, whether they derive from a Neanderthal or a Mesolithic hominid. However, on some specimens

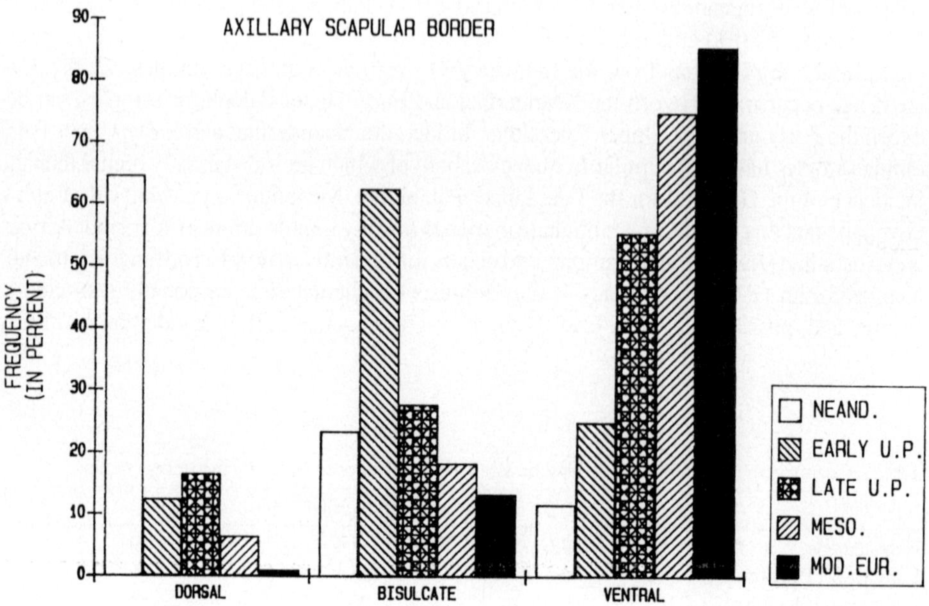

Figure 1. Frequencies in percent of axillary border types for Neanderthal, Early Upper Paleolithic, Late Upper Paleolithic, Mesolithic and Modern European samples.

the pattern can change from dorsal to bisulcate (or bisulcate to ventral) below the proximal third of the border. In addition, most subadults of whatever prehistoric period possess a ventral type of axillary border. Thus, to minimize scoring problems, the frequencies reported only include adults and primarily describe the morphological appearance of the proximal third of the axillary border. As with the mandibular foramen type, exact morphological matches for each of these character states can be found in all samples.

Differences between frequencies of the three types are obvious; Neanderthals are dominated by the dorsal (and bisulcate) pattern, while recent Europeans are characterized by the ventral pattern (Fig. 1). It is important to recognize that Neanderthal frequencies (n = 17) are primarily represented by 11 scapulas from Krapina. Without them, the European Neanderthal incidence would be 100% dorsal, represented by La Ferrassie 1 & 2, Neanderthal 1, Spy 2, St. Césaire and Vindija 209. Frequencies of the three border types in Early and Late Upper Paleolithic and Mesolithic samples show a trend for reduction in the number of dorsal and bisulcate types through time, although all three are distributed through these populations.

Depending on one's bias, the data presented in Figure 1 can be read in different ways. For example, from frequencies of the dorsal type, there is marked decrease in incidence between the Neanderthal and Early Upper Paleolithic samples followed by slight decreases from the Upper Paleolithic samples to the modern Europeans. These observations on variation of the dorsal pattern suggest that the Neanderthals represent a unique pattern. Yet, variation in the bisulcate and ventral patterns lead to an opposite conclusion. For the bisulcate frequencies, the Early Upper Paleolithic has the highest incidence and is nearly as distinct from chronologically later populations as the Neanderthals are in their high frequency of the dorsal pattern. From the incidence of the ventral type, there is a regular increase through time from the low occurrence in Neanderthals to the high occurrence in recent Europeans. The Early and Late Upper Paleolithic frequencies are clearly intermediate, while the Mesolithic frequency closely approaches the modern condition.

If 'the bisulcate pattern and full dorsal sulcus strongly imply a larger, more powerful *M. teres minor*' (Trinkaus 1977:233), the combined frequencies of both patterns in Neanderthals and Early Upper Paleolithic hominids indicate that these samples share a level of muscular robusticity which differs from chronologically later Europeans. For this feature, it is not sufficient to simply conclude that Neanderthals differ from the modern condition. The Early (and to a lesser extent the Late) Upper Paleolithic individuals also differ and occupy an intermediate position.

A final observation on axillary border patterns concerns variation in the Mesolithic. Although Figure 1 shows frequencies of the ventral sulcus in the Mesolithic at 75%, this incidence is not typical of all Mesolithic samples. Figure 2 provides frequencies for 4 site complexes: Skateholm/Henriksholm-Bøgebakken from Scandinavia; Téviec/Hoëdic from France; Vlasac from Yugoslavia; and Muge from Portugal. It is obvious that the frequencies at Muge deviate from the pattern found in the three other Mesolithic samples, with a high incidence of the dorsal (26.7%) and the bisulcate (46.7%) patterns. These frequencies at Muge are also substantially different from frequencies reported by von Eickstedt (1925:225) for modern groups with sample sizes larger than 10 individuals. From these relatively high incidences of dorsal and bisulcate types at Muge, one would predict these individuals show greater overall robusticity; but compared to the other three samples, the specimens from Muge are the most gracile in cranial and postcranial features. Frequencies for axillary border types in the Mesolithic indicate that inter-population variation occurs which may relate to specific posturing habits. However, long-term term shifts in the frequency of axillary border types demonstrate a decrease in the more dorsal teres minor muscle attachment.

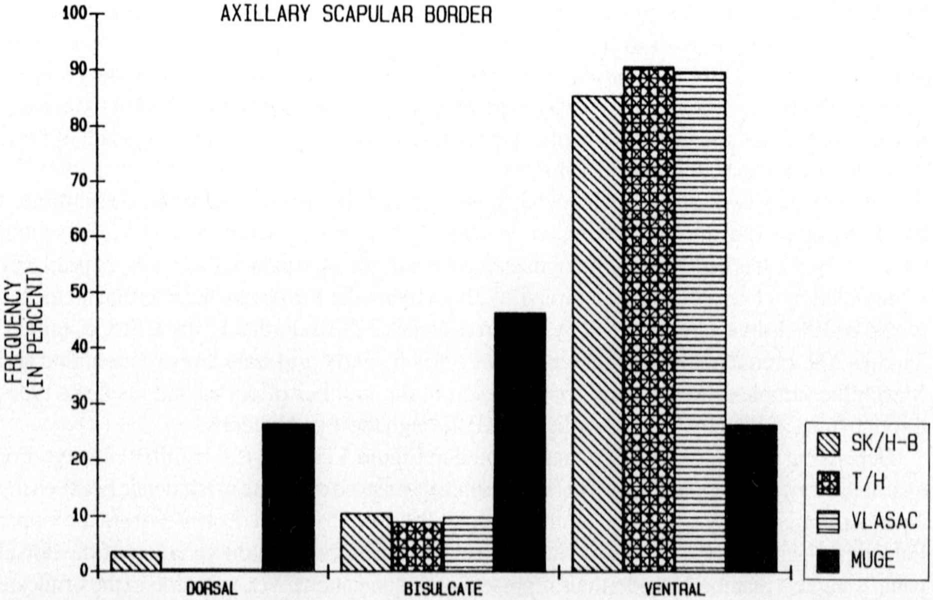

Figure 2. Frequencies in percent of axillary border types in four Mesolithic sites or site complexes. (SK-H/B = Skateholm-Henriksholm/Bøgebakken from Sweden and Denmark; T/H = Téviec & Hoëdic from France; Vlasac from Yugoslavia; Muge from Portugal).

3 CONCLUSIONS

At the IXth ICAES meetings in Chicago, in a paper entitled 'Neanderthal Man: Facts and Figures' W.W. Howells noted that everyone agrees that Neanderthals existed and that their morphology differed from modern Europeans, but that the 'agreement stops here' (1975:389). Two ICAES meetings and 15 years have passed and Howells' evaluation remains accurate. There is still an ongoing dispute about the relationship (or lack of one) between Neanderthals and subsequent European *Homo sapiens*, and no prospect that the controversy will be resolved in the near future. Part of this stems from preconceived bias and attitudes about the evolutionary process and, no less important, from the insufficient fossil record. But, one thing is clear. The uniqueness of Neanderthals cannot be judged without some consideration of variation occurring in the humans which directly succeeded them, whether these humans evolved from Neanderthals or replaced them.

In the features reviewed here, Upper Paleolithic hominids possess, albeit in lower frequencies, some of the same morphological characteristics which have been used to define Neanderthals as unique. These data, then, support arguments for a frequency shift explanation for the morphological differences between Neanderthal and later Europeans, rather than models requiring complete replacement. Adherents of the mitochondrial DNA model must either eliminate these features from the list of unique (or common) Neanderthal characteristics, or explain why two successive hominid populations of Europe (which are required to be unrelated genetically) share the particular features. Furthermore, judged by the common occurrence of these so-called Neanderthal autapomorphies, statements which describe Upper Paleolithic hominids as 'instantly recognizable as anatomically modern' (Howells 1981:74) or as 'distinctively anatomically modern' (Howell 1984:xxi) need to be modified to account for morphological overlap. Finally, Stringer & Andrews (1988b:774) ask for the 'presentation of further data to test' the gradual and replacement models. Variation over time in these three traits provides no evidence to support the 'Out of Africa' or other replacement hypotheses. Rather, the persistence of these traits suggests a degree of continuity which is only feasible with a model accepting some measure of genetic contribution of Neanderthals to subsequent *Homo sapiens* populations.

ACKNOWLEDGEMENTS

Much of this work could not have been accomplished without the help of European scientists who allowed me access to their Upper Paleolithic and Mesolithic collections. Their cooperation is greatly appreciated. For permission to examine the Hungarian Medieval samples I thank Prof. Tóth Tibor at the Természettudományi Muzeum, Budapest. Finally, R. Caspari and M.H. Wolpoff (Michigan) provided important comments on the manuscript. The research was supported by NSF Grant BNS 8419057 and a General Research Grant from the University of Kansas.

REFERENCES

Bräuer, G. 1981. New evidence on the transitional period between Neanderthal and modern man. *J. Hum. Evol.* 10: 467-474.

Cann, R.L., M. Stoneking & A.C. Wilson 1987. Mitochondrial DNA and human evolution. *Nature 325: 31-36.*

Darlu, P. & P. Tassy 1987. Roots (a comment on the evolution of human mitochondrial DNA and the origins of modern humans). *Hum. Evol.* 2: 407-412.

Eickstedt, E. von 1925. Variationen am Axillarrand der Scapula. *Anthropol.Anz.*2:217-228.

Fisher, A. 1988. The more things change... *Mosaic* 19(1): 22-33.

Frayer, D.W. 1978. *Evolution of the Dentition in Upper Palaeolithic and Mesolithic Europe.* Lawrence: University of Kansas Publications in Anthropology.

Gorjanović-Kramberger, D. 1902. Der paläolithische Mensch und seine Zeitgenossen aus dem Diluvium von Krapina in Kroatien, III. *Mitt. Anthropol. Gesell. Wien* 32: 189-216.

Gorjanović-Kramberger, D. 1906. *Der Diluviale Mensch von Krapina in Kroatien: Ein Beitrag zur Paläoanthropologie.* Wiesbaden: Kreidel.

Gould, S.J. 1987. Bushes all the way down. *Nat. Hist.* 87(6): 12-19.

Heim, J.-L. 1982. *Les Enfants Néanderthaliens de la Ferrassie.* Paris: Masson.

Howell, F.C. 1984. Introduction. In F.H. Smith & F. Spencer (eds), *The Origins of Modern Humans.* New York: Liss, pp. xiii-xxii.

Howells, W.W. 1975. Neanderthal man: Facts and figures. In R.H. Tuttle (ed.), *Paleoanthropology: Morphology and Paleoecology.* The Hague: Mouton, pp. 389-408.

Howells, W.W. 1981. Current theories on the origin of *Homo sapiens sapiens.* In D. Ferembach (ed.), *Les Processus de l'Hominisation.* Paris: CNRS, pp. 73-78.

Hrdlička, A. 1927. The Neanderthal phase of man. *J. Anthropol. Inst.* 57: 249-274.

Hublin, J.-J. 1978a. Le Torus Occipital Transverse et les Structures Associées: Évolution dans le Genre Homo. Univ. Paris: Thèse de Troisième Cycle.

Hublin, J.-J. 1978b. Quelques caracterès apomorphes du crane néandertalien et leur interprétation phylogénique. *C.R. Acad. Sc. Paris (Sèrie D)* 287: 923-924.

Hublin, J.-J. 1980. La Chaise-Suard, Engis 2 et la Quina H 18. Développement de la morphologie occipitale externe chez l'enfant prénéandertalien et néandertalien. *C.R. Acad. Sci. Paris (Sèrie D)* 291: 669-672.

Jelínek, J. 1969. Neanderthal man and *Homo sapiens* in central and eastern Europe. *Curr. Anthropol.* 10: 475-503.

Klaatsch, H. 1902. Occipitalia und Temporalia der Schädel von Spy verglichen mit denen von Krapina. *Zeitschr. Ethnol.* 34: 392-409.

Lévêque, F. & B. Vandermeersch 1980. Découverte de restes humains dans un niveau castelperronien à Saint-Césaire (Charente-Maritime). *C.R. Acad. Sc. Paris (Sèrie D)* 291: 187-189.

Lewin, R. 1987. Africa: Cradle of modern humans. *Science* 237: 1292-1295.

Lewin, R. 1988. Modern humans under close scrutiny. *Science* 239: 1240-1241.

McCown, T.D. & A. Keith 1939. *The Stone Age of Mount Carmel II*. Oxford: Clarendon.

Radovčić, J., F.H. Smith, E. Trinkaus & M.H. Wolpoff 1988. *The Krapina Hominids: An Illustrated Catalog of the Skeletal Collection*. Zagreb: Mladost.

Santa Luca, A.P. 1978. A re-examination of presumed Neanderthal-like fossils. *J. Hum. Evol.* 7: 619-636.

Smith, F.H. 1976. *The Neandertal Remains from Krapina*. Knoxville: Department of Anthropology, University of Tennessee.

Smith, F.H. 1978. Evolutionary significance of the mandibular foramen area in Neandertals. *Amer. J. Phys. Anthropol.* 48: 523-532.

Smith, F.H. 1982. Upper Pleistocene hominid evolution in South-Central Europe: A review of the evidence and an analysis of trends. *Curr. Anthropol.* 23: 667-703.

Smith, F.H. 1984. Fossil hominids from the Upper Pleistocene of Central Europe and the origin of modern Europeans. In F.H. Smith & F. Spencer (eds), *The Origins of Modern Humans*. New York: Liss, pp. 211-250.

Spuhler, J.N. 1987. The evidence of mitochondrial and nuclear DNA on the origins of *Homo erectus* and *Homo sapiens. Paper presented at the 86th Annual Meeting of the American Anthropological Association, Chicago.*

Stringer, C.B. 1982. Towards a solution to the Neanderthal problem. *J. Hum. Evol.* 11: 431-438.

Stringer, C.B. 1984. Fate of the Neanderthal. *Nat. Hist.* 93(12): 6-12.

Stringer, C.B. & P. Andrews 1988a. Genetic and fossil evidence for the origin of modern humans. *Science* 239: 1263-1268.

Stringer, C.B. & P. Andrews 1988b. Response to 'Modern human origins'. *Science* 241: 773-774.

Stringer, C.B., J.-J. Hublin & B. Vandermeersch 1984. The origin of anatomically modern humans in Western Europe. In F.H. Smith & F. Spencer (eds), *The Origins of Modern Humans*. New York: Liss, pp. 51-135.

Thorne, A.G. & M.H. Wolpoff 1981. Regional continuity in Australasian Pleistocene hominid evolution. *Amer. J. Phys. Anthropol.* 55: 337-349.

Tierney, J., L. Wright & K. Springen 1988. The search for Adam and Eve. *Newsweek* (January 11): 46-52.

Trinkaus, E. 1977. A functional interpretation of the axillary border of the Neanderthal scapula. *J. Hum. Evol.* 6: 231-234.

Trinkaus, E. 1983. Neanderthal postcrania and the adaptive shift to modern humans. In E. Trinkaus (ed.), The Mousterian Legacy. Oxford: *B.A.R. Series* No. 164, pp. 165-200.

Vandermeersch, B. 1981. *Les Hommes Fossiles de Qafzeh (Israël)*. Paris: CNRS.

Vandermeersch, B. 1985. The origin of the Neandertals. In E. Delson (ed.), *Ancestors: The Hard Evidence*. New York: Liss, pp. 306-309.

Weidenreich, F. 1947. Facts and speculations concerning the origin of *Homo sapiens. Amer. Anthropol.* 49: 187-203.

Weidenreich, F. 1951. Morphology of Solo man. *Anthropol. Pap. Amer. Mus. Nat. Hist.* 43(3): 205-290.

Wolpoff, M.H. 1980. *Palaeoanthropology*. New York: Knopf.

Wolpoff, M.H. 1989. Multiregional evolution: The fossil alternative to Eden. In P. Mellars & C.B. Stringer (eds), *The Human Revolution: Behavioural and Biological Perspectives on the Origins of Modern Humans*. Edinburgh: Edinburgh University Press, pp. 62-108.

Wolpoff, M.H., F.H. Smith, M. Malez, J. Radovčić & D. Rukavina 1981. Upper Pleistocene hominid remains from Vindija Cave, Croatia, Yugoslavia. *Amer. J. Phys. Anthropol.* 54: 499-545.

Wolpoff, M.H., J.N. Spuhler, F.H. Smith, J. Radovčić, G. Pope, D.W. Frayer, R.B. Eckhardt & G. Clark 1988. Modern human origins. *Science* 241: 772-773.

Wolpoff, M.H., X. Wu & A.G. Thorne 1984. Modern *Homo sapiens* origins: A general theory of hominid evolution involving the fossil evidence from East Asia. In F.H. Smith & F. Spencer (eds), *The Origins of Modern Humans*. New York: Liss, pp. 411-484.

Homo sapiens sapiens remains from the island of Crete

Fiorenzo Facchini & Gianni Giusberti
Istituto di Antropologia dell' Universita', Bologna, Italy

1 INTRODUCTION

The skeletal remains discussed in this report come from the island of Crete. They were found in the district of Kania in 1893 by Vittorio Simonelli (1897), geologist and paleontologist at Bologna University, and given to us by the late Vittorio Vialli, Head of the Department of Geology and Paleontology at the same University, to conduct an anthropological study. The human remains consist of several fragments of cranium and postcranium cemented by a very hard calcareous breccia, which could be related to the formation of a littoral bench. The ossiferous breccia is made up of a sandy sediment with calcareous skeletons and the shells of many marine species from the Upper, Middle and Lower Superior Littoral, which are typical of sandbanks.[1] Fragments of bones and horns of Cervidae were also found. Among the mineralogical components of the breccia are some angular clasts of eocenic limestone, as well as some intruding volcanic rock clasts (granite) and red earth. All of these are cemented by a calcareous breccia. The bones appear to be heavily mineralized but rather fragile. Erosion, likely of marine origin, has damaged both the skull and the long bones. The erosion surface of the natural endocranial mould shows a Lithophagus hole (Fig. 1).

At the 'Centre des Faibles Radioactivité de Gif-Sur-Ivette in Paris' Yuji Yokoyama chronometrically dated a sample of breccia in direct contact with the human bones by using the Protactinium/Uranium method. This method produced an age of 51 000 ± 12 000 years (Facchini et al. 1989). Additionally pollen analysis has been carried out on 3 samples of ossiferous matrix, one taken from a point in close contact with the left parietal exocranial surface (sample A) and others taken at distances of approximately 5 to 12 cm from sample A (respectively samples B and C). Samples A and B indicate slightly different climatic conditions: Drier for A and more humid for B. Sample C, probably stratigraphically below and thus more ancient than the other samples denotes a cooler and more humid climatic situation. Analogous climatic fluctuations are pointed out in the Near East about 50 000 years ago (Leroi-Gourhan 1988). Chronologically speaking, the correlations with other sequences available for the island of Crete (Bottema 1980) seem to suggest a pre-Holocene deposit, consistent with the chronometric age estimate.[2]

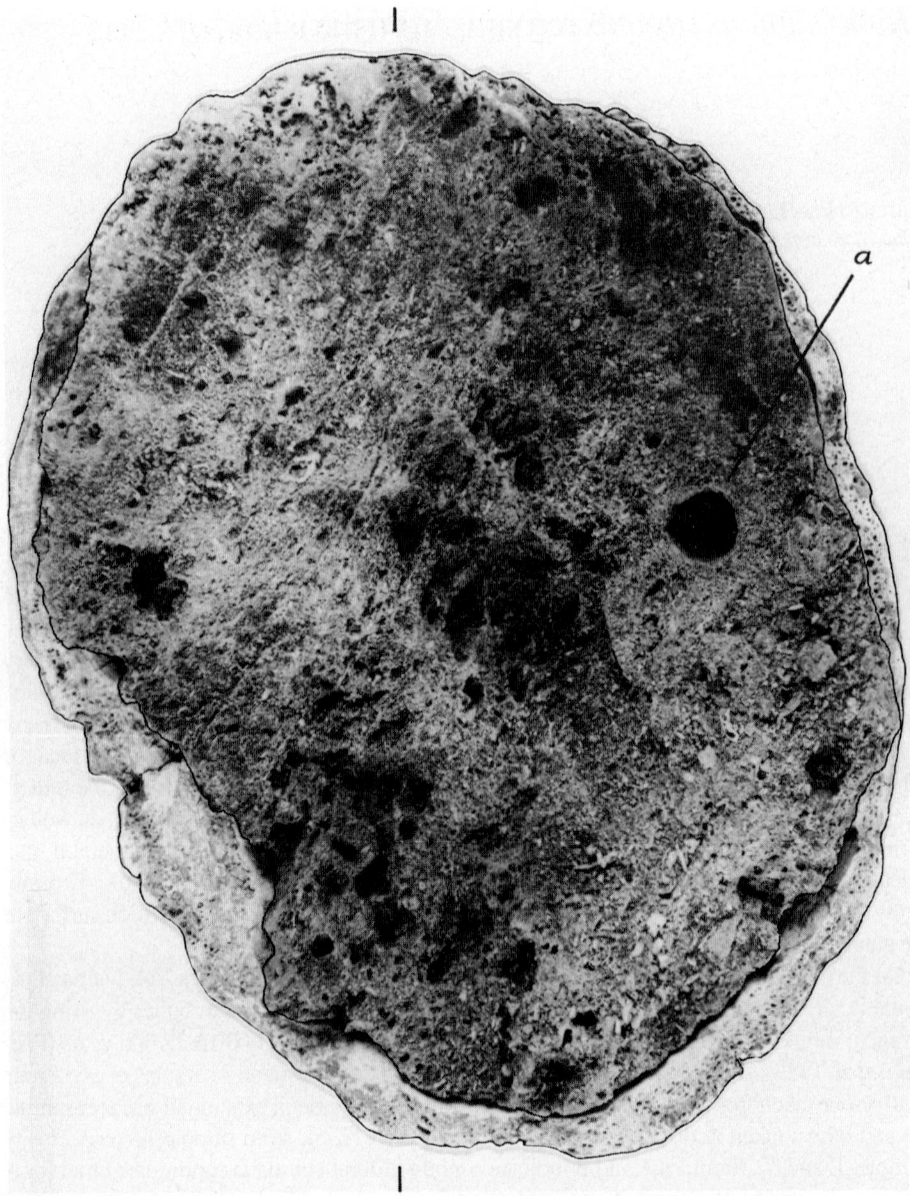

Figure 1. The Crete skullcap fragment with the natural endocranial mould. The Lithophagus hole is indicated by (a).

2 THE HUMAN REMAINS

The skeletal parts at our disposal for study are the following:

a) Skull: A fragment of skullcap consisting of a broad portion of the frontal as well as an

Figure 2a. The Crete skullcap in exocranial view. Anterior region is oriented superiorly.

incomplete parietal bone, incomplete natural exocranial and endocranial moulds, a portion of right upper orbit edge, the right medial orbit margin and part of the frontal process of the right maxilla (Fig. 2).

b) Postcranial Skeleton: Fragment of left scapula, incomplete right and left humeri, proximal and distal fragments of left femur, distal fragments of right and left tibiae, and fragments of fibula (Fig. 3).

Figure 2b. The Crete skullcap in endocranial view. Anterior region is oriented superiorly.

2.1 *Description of the skullcap*

The skullcap is 6 mm thick at the frontal sinus and has a 7 mm maximum value on the left parietal bone near the tuberosity (Fig. 4). The sutures are totally obliterated except for some traces of coronal suture on the exocranial surface. Examination of the sagittal profile of the specimen, integrated with the study of endocranial moulds (to fill in gaps in the preserved bone surface), reveals weakly developed superciliary ridges, a receding and low forehead, and a rather low and flat cranial vault (Fig. 5). The curvature index of the frontal bone is rather high, thus confirming its flattened character. The curvature of frontal bone is also reflected by a frontal angle of 136° (FRA of Howells 1973).

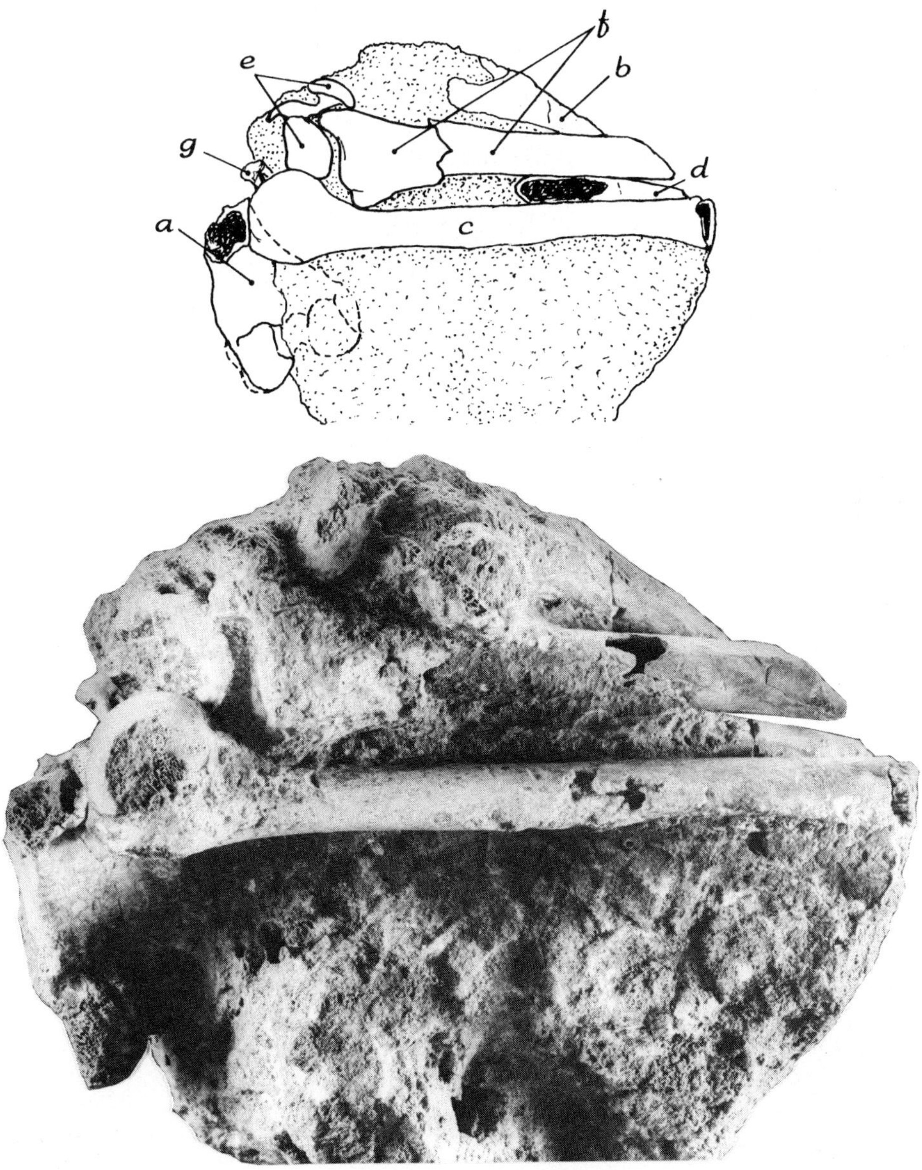

Figure 3. A block of the ossiferous breccia from Crete. Postcranial skeletal remains observable in this piece: distal (a) and proximal (b) left femoral fragments; incomplete [mesio-proximal region] left humerus (c); midshaft of the right humerus (d); left scapula [articular region] (e); right tibia [lower midshaft] (f); the right second lower molar (g) is shown.

The endocranial imprints of the middle meningeal vessels are represented by the anterior branch, from which branches off the middle (with two divisions), and by the posterior branch (Fig. 6). A lower, secondary division extends its branches to the middle and posterior branches. On the whole the anastomoses between the middle and the posterior branches form a very

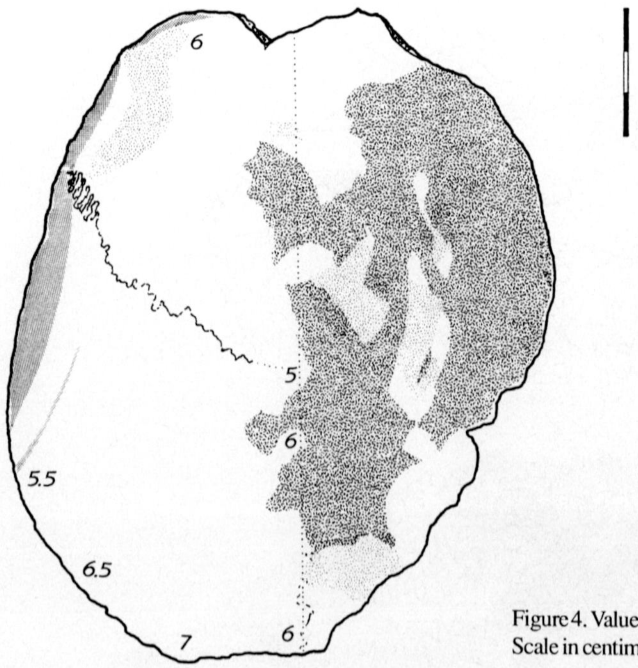

Figure 4. Values of Crete skullcap thickness (mm). Scale in centimeters.

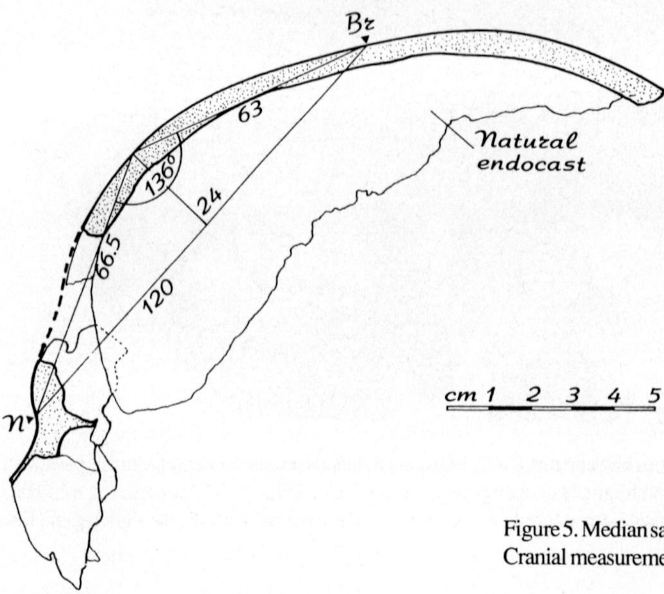

Figure 5. Median sagittal section of the Crete skull. Cranial measurements are in mm.

simple network in the left hemisphere. No anastomoses were found between the anterior and middle branches.

The portion of facial skeleton, that is possible to reconstruct can be connected to the frontal bone by using the natural endocranial mould. The face reveals a modern upper orbital margin,

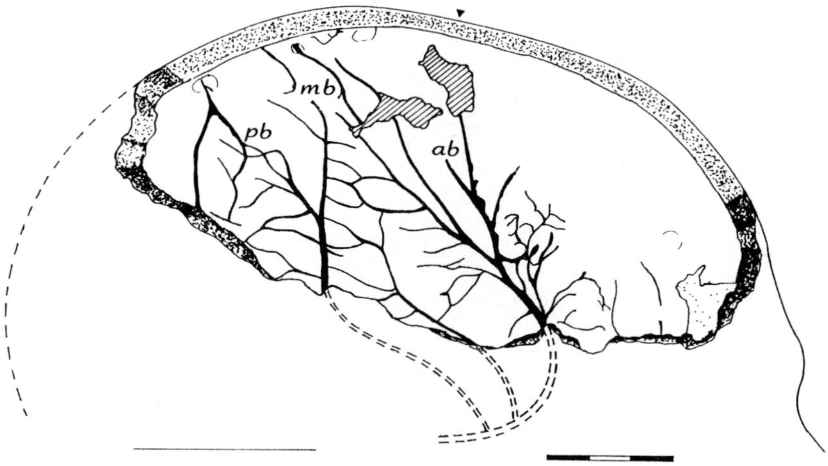

Figure 6. Middle meningeal vessels of the Crete skull are shown in left endocranial view (ab = anterior branch; mb = middle branch; pb = posterior branch).

a certain prominence of the nasal bones, and a medium-high orbit (Figs. 7 and 8). The right frontal sinus is broad and stretches to the upper part of the orbit.

The right second lower molar, with worn cusps but dentine not yet exposed, is also preserved (Fig. 3). It exhibits a carious lesion in the external neck, extending into the dental pulp cavity. The tooth is rather small, judging from the crown dimensions (buccolingual diameter: 9.3 mm; mesiodistal diameter: 9.7 mm, which yield a surface area of 90.2 mm^2). These values, combined with a rather small skull, is comensurate with a female sex estimate, for which small dental sizes are often characteristic (cf. Chamla 1980, Hillson 1986).

2.2 *The postcranial skeleton*

As far as the postcranial skeleton is concerned, it probably represents the same individual as the skull. The peculiar positioning of the bones of the limbs is worthy of note: The bones of the right forearm are *grosso modo* flexed next to the humerus on the same side, while the left humerus, parallel to the former, is anatomically positioned near the left scapula. Concerning the lower limbs, the fragments of both tibiae are positioned parallel to each other and flexed near the left femur. This might suggest that the remains belonged to a burial and that the buried individual was laid with its limbs flexed (Fig. 3).

Though study of the postcranial elements cannot be complete because they have not yet been entirely freed from the matrix, it is possible to make some remarks. The left humerus, almost complete, lacks its distal extremity and is still partially encased in the breccia. Its length was minimally 27-28 cm. It has been possible to take the following measurements on the right humerus at midshaft: Circumference 54 mm, transverse diameter 17.2 mm, anteroposterior diameter 18 mm. The index is 95. The left femur, represented only by the proximal extremity and lacking its head, shows a slight subtrocanteric flattening with an 80.1 platimeric index. The right tibia, represented by the distal half, is slightly rounded with a 70.6 cnemic index (euricne-

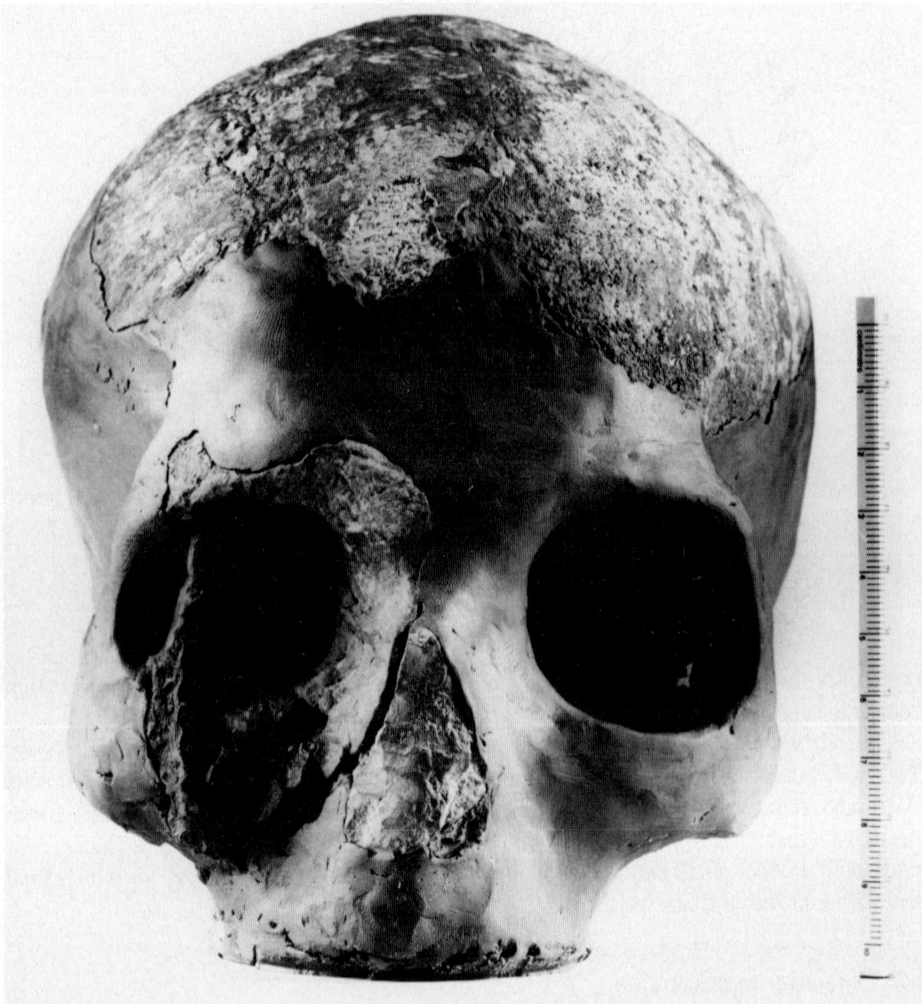

Figure 7. The Crete skull in *norma facialis* (reconstruction). Scale in centimeters.

my). The transverse diameter is comparatively larger than the anteroposterior measurement and the cross-section has an amygdaloid shape. The distal fragment of the left tibia is almost completely encased in the breccia. On the whole, these remains may be ascribed to an adult, probably female individual. The characteristics observed lead us to classify this individual as *Homo sapiens sapiens*.

3 COMPARISONS

In order to further evaluate the Crete specimen, a comparison with other *Homo sapiens sapiens* and Neanderthal remains is in order. With regard to the thickness, of the skullcap, the values

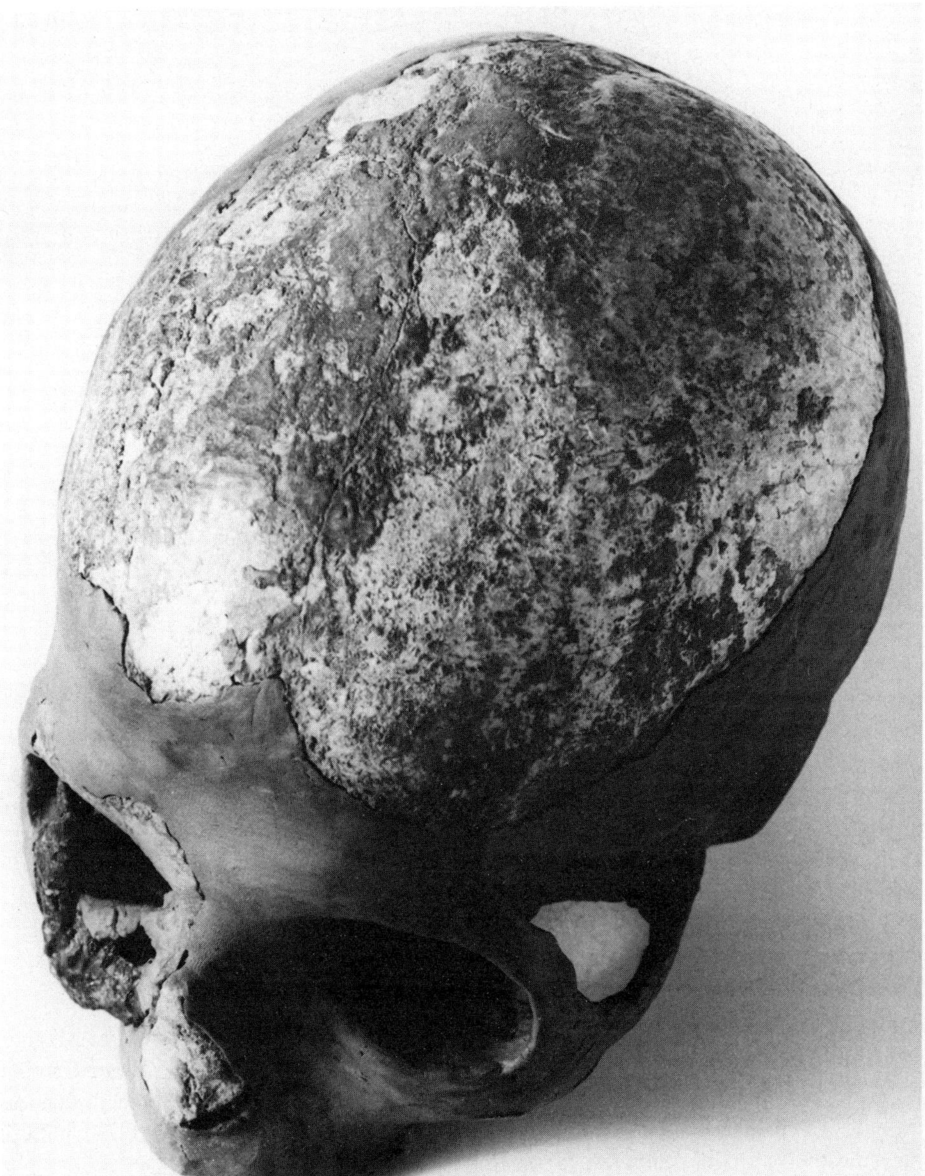

Figure 8. The reconstruction of the Crete skull in antero-superior view.

observed on the Crete specimen are slightly below the Neanderthal values and fall within the *Homo sapiens sapiens* range, albeit slightly above the values of recent humans. The frontal bone curvature index for the Crete specimen (89.5), though it cannot be determined with certainty, falls near the indices of Combe Capelle (89.1), Tabun (89.8) as well as Circeo (89.3)

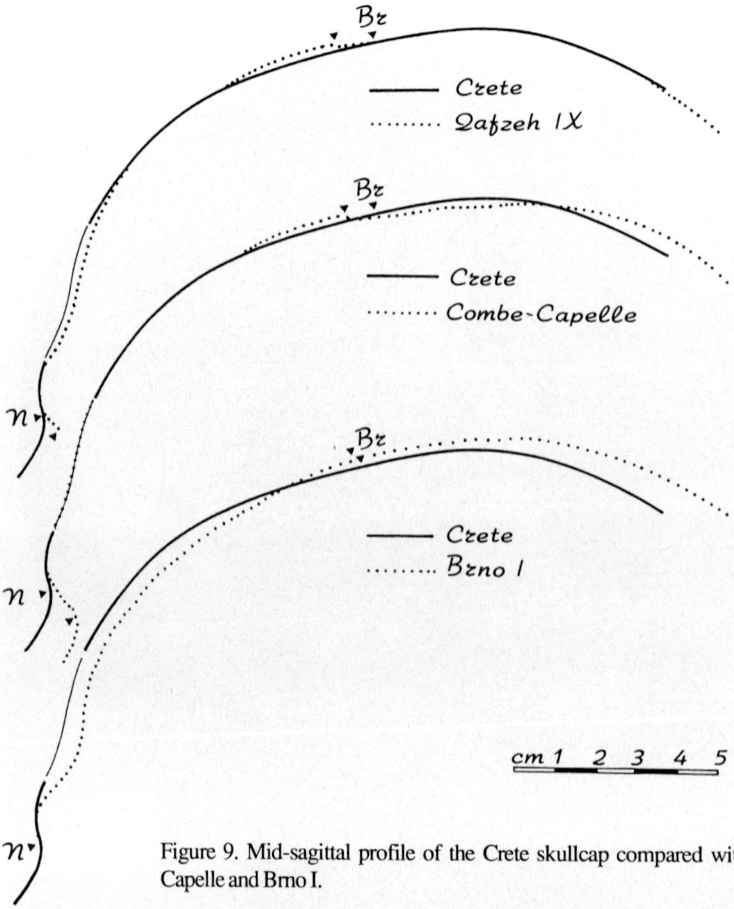

Figure 9. Mid-sagittal profile of the Crete skullcap compared with Qafzeh IX, Combe Capelle and Brno I.

and thus lies in the Neanderthal range of variation. Lower values are observed in Cro-Magnon 1 (85.0), Cro-Magnon 2 (87.1), Cro-Magnon 3 (85.1), Qafzeh 6 (85.7) and Arene Candide (82.6). It must be pointed out that the degree of prominence of the glabellar region influences the index value. In fact, if the prominence is pronounced, it may reduce the index even if this is not indicative of a higher curvature for the frontal bone (Tab. 1; Fig. 9).

More informative is the frontal angle (FRA of Howells 1973). The frontal angle of the Crete skullcap (136°) is intermediate between the Qafzeh 6 (132°) and Neanderthal ones (Neanderthal: 139°, Shanidar 1: 144°; Shanidar 5: 147°). Furthermore, it falls in an intermediate position between Omo I (135°) and Florisbad (137.9°). It is interesting to note that, according to Day & Stringer (1982), values higher than 136.5° are to be considered typical of *Homo erectus*, while values below 133° are observed in modern *Homo sapiens*.

A principal components analysis based on four frontal variables (frontal arc, frontal chord, min. and max. frontal breadths) reveals a greater dispersion of Upper Paleolithic specimens compared to Neanderthals. The Crete frontal lies in the range of variation of the Upper Paleolithic specimens and is differentiated from Neanderthals, Předmostí and Qafzeh (Fig. 10).

Table 1. Comparative craniometrical data (measurements in mm). [a] Sergi (1974), [b] This study, [c] Coon (1962), [d] Howell FC (1951), [e] Mean value in Neanderthal sample (9 obs.) by Trinkaus (1983), p.130, [f] Wolpoff (1980), [g] de Lumley (1973), p.303, [h] Mean value in Skhūl-Qafzeh sample (4 obs.) by Trinkaus (1983), p.130, [i] Vallois & Vandermeersch (1972), [j] Trinkaus (1983), [k] Matiegka (1934), [m] Vallois & Billy (1965), [n] Sergi et al. (1974), [p] Smith (1976), [q] Angel (1944), * Measure defined by Martin & Saller (1957), ** Measure defined by Howells (1973).

	Frontal arc (M 26)*	Frontal chord (M 29)* FRC**	Frontal curvature index (M 29/ M 26)	Min frontal breadth (M 9)*	Max frontal breadth (M 10)* XFB**	Index (M 9/ M 10)*	Frontal angle (M 32.5)* FRA**	Parietal thickness (tuber)
Circeo I[a] ♂	131	117	89.3	113	127	88.9	-	7[b]
Neanderthal ♂	133[c]	116[c]	87.2	107[d]	122[d]	87.7	139.0°(9)[e]	10.5[f]
La Chapelle ♂	121[c]	107[c]	88.4	109[c]	122[d]	89.3	-	6[g]
Qafzeh VI ♂	133	114	85.7	109.5	125	87.6	132.2°(4)[h]	8[i]
Shanidar 1[j] ♂	119	111.3	93.5	-	128	-	144°	8
Shanidar 5[j] ♂	129	118.0	91.5	-	128	-	147°	9
Předmost IV, X[k] ♀	131.3(2)	113.(2)	87.2(2)	100.5(2)	122.5(2)	82.0(2)	-	5.5
Předmost I, III, IX[k] ♂	129.5(3)	114(3)	86.7(3)	102.3(3)	125.3(3)	82.0(3)	-	5.7
Combe Capelle[c]	138	123	89.1	107	114	93.8	-	-
Cro-Magnon 1[m] ♂	147	125	85.0	102.5	126	81.3	-	10
Cro-Magnon 2[m] ♀	132	115	87.1	97.5	120	81.2	-	6.5
Cro-Magnon 3[m] ♂	148	126	85.1	96.5	123	78.4	-	6
Arene Candide[n] ♂	144	119	82.6	(97)	116	(83.6)	-	6
Mladeč I[k] ♂	133	-	-	94?	126?	74.6?	-	-
Mladeč V[k] ♂	140	-	-	105	127?	82.7?	-	-
Brno I[k] ♂	135	-	-	90	-	-	-	-
Brno II[k] ♂	130	-	-	101[p]	125[p]	82.6	-	-
Neolithic and Early Helladic 3000-2000 BC[q] ♂ ♀	129.1(14)	113.7(13)	-	96.1(20)	115.8(12)	-	-	-
Crete ♀	134	120	89.5	90?	117?	76.9?	136°	7

Some skullcap features are reminiscent of archaic characteristics. Specifically the branching patterns of the middle meningeal vessels which exhibit primitive features in the Crete specimen. The lack of anastomoses between the anterior and the middle branches, as well as the relative simplicity of connecting branches between the middle and the posterior one, aligns the Crete find with evolved *Homo erectus* and archaic *H. sapiens* (e.g. Broken Hill, Omo 2, Mechta el Arbi, Qafzeh) more than to the modern *H. sapiens* forms, which exhibit more meningeal pathways and anastomoses (cf. Saban 1984).

Concerning the postcranial skeleton (see Table 2), the platymeric index of the left femur (81.1) places the Crete specimen nearer to the Neanderthals (Shanidar 1: 81.0; Shanidar 4: 84.9; Shanidar 6: 80.7; Amud 1: 83.0 & 79.7; Spy: 80.3; Skhūl-Qafzeh: 83.5, N = 5; European Neanderthals: 80.4, N = 10) (Fig. 11) than the Cro-Magnon type from Grimaldi Barma Grande (76.9), Předmostí (70.9; N = 18) and an Upper Paleolithic group (74.3; N = 18), where a more developed flattening of the subtrochanteric diaphysis is evident (Fig. 11). Concerning the midshaft index of the tibia, its 70.6 value places the Crete sample closer to Shanidar 2 and 6

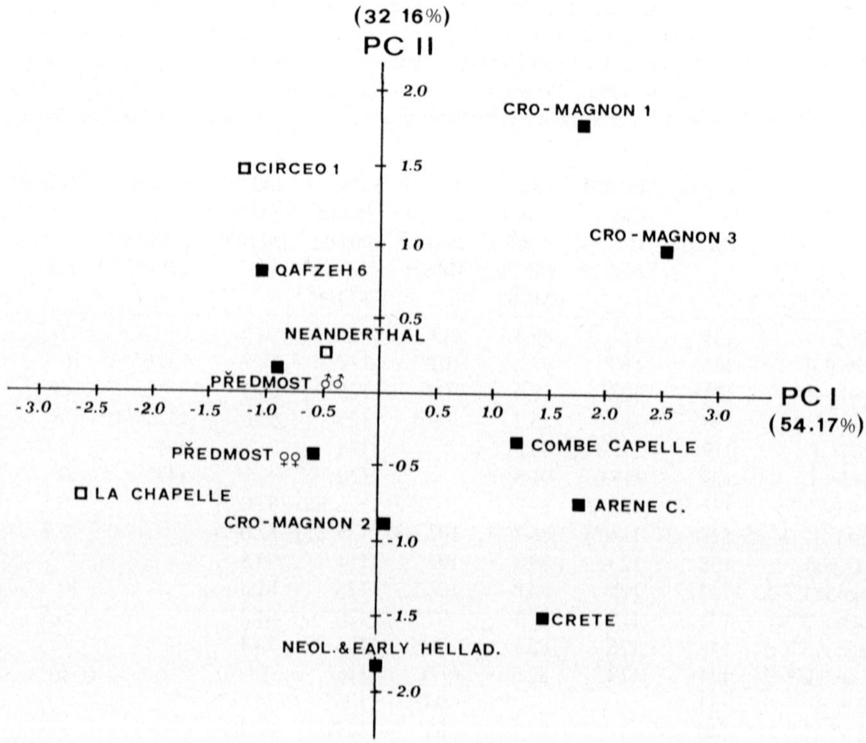

Figure 10. Principal components analysis, based on four frontal variables (frontal arc, frontal chord, min. frontal breadth, max. frontal breadth) in some Neanderthal (empty squares) and 'Upper Palaeolithic' (full squares) samples.

(71.1 and 72.2 respectively) (Fig. 11), the Skhūl-Qafzeh sample (69.7; N=6), and the European Neanderthal average (68.9; N = 5) than to the Předmostí sample (66.6; N = 9)[3]. In the latter sample, the flattening of the tibia is correlated to a certain extent with the subtrocanteric flattening of the femur.

The same pattern of relationships also occur in the bivariate representation of the midshaft index of the humerus and meric index (Fig. 12), where the Crete sample falls near Qafzeh and Neanderthals and quite far from modern humans. The principal components analysis based on 6 variables of the humerus, femur and tibia indicates an intermediate position for the Crete sample between Neanderthals and modern humans (Fig. 13).

The characteristics observed enable us to align the Crete human remains with *Homo sapiens sapiens* specimens which preserve some archaic features. The latter could correspond or be related to anatomically modern *Homo sapiens* as defined by Bräuer (1984a, b).

4 DISCUSSION

In the present state of our knowledge, the human remains reported here represent the most ancient remains from Crete.[4] They are most probably related to some external influence rather

Table 2. Comparative data for the humerus, femur and tibia (measurement in mm). [a] Data from Matiegka (1938), [b] Data from Trinkaus (1976), [c] Data from Trinkaus (1980), [d] Data from de Lumley (1973), p.569, [e] Data from Klaatsch & Hauser (1910), [f] Data from Billy (1969), [g] Data from Trinkaus (1983), [h] Data from Vandermeersch (1981b), * Measure defined by Martin & Saller (1957), rt = right; lt = left.

	Humerus Midshaft diameters		Mid-shaft index	Femur Subtrocanteric diameters		Meric index	Tibia Midshaft diameters		Mid-shaft index
	max (M5)*	min (M6)*	(M5/ M6)	ant-post (M10)*	transv (M9)*	(M10/ M9)	ant-post (M8)*	transv (M9)*	(M9/ M8)
Crete rt ♀	-	-	-	-	-	-	28.3	20.0	70.7
lt	-	18.3	79.5	24.5	30.2	81.1	-	-	-
Předmost III rt[a] ♂	22	18	81.8	24	38	63.1	37.7	20.5	54.4
lt	21	18	85.7	24	38	63.1	38	19	50
Předmost IV rt[a] ♀	22	16	72.7	25	35.8	69.8	31.6	20	63.3
lt	20	20	100	25	34	73.5	31	19	61.3
Předmost IX rt[a] ♂	19	18	94.7	23	33	69.7	27	20	74.0
lt	17	16	94.1	23	33	69.7	28	20	71.4
Předmost X rt[a] ♀	20	17	85	22.6	35	64.6	28	20	71.4
lt	20	16	80	23	35.3	65.1	29.5	18.5	62.7
Předmost XIV rt[a] ♂	21	17	80.9	22.5	33	68.2	30.5	19.5	63.9
lt	20	17	85	23.4	35	66.8	30.5	21	68.8
Amud 1 lt ♂	-	-	-	29.4[b]	36.9[b]	79.7	38.4[c]	24.3[c]	63.3
La Chapelle 1 rt ♂	-	-	-	29.4[b]	33.0[b]	89.1	35.5[c]	22.8[c]	64.2
lt	-	-	-	-	-	-	35.5[c]	22.8[c]	64.2
La Ferrassie 1 rt ♂	25.0[e]	17.5[e]	70.0	29.9[b]	38.0[b]	78.7	35.9[c]	22.7[c]	63.2
lt	23.0[e]	15.0[e]	65.2	29.5[b]	36.0[b]	81.9	34.5[c]	22.7[c]	65.8
La Ferrassie 2 rt ♀	20.0[e]	14.0[e]	70.0	27.6[b]	33.4[b]	82.6	28.1[c]	20.9[c]	74.4
lt	-	-	-	27.9[b]	32.0[b]	87.2	28.6[c]	21.9[c]	76.6
Combe Capelle rt[e] ♂	20.5	15	73.2	25	29	86.2	36	21	58.3
lt	24	15	62.5	26	30	86.6	37	25	67.5
Chancelade rt[f] ♂	25.5	18.5	72.5	25	35	71.4	-	-	-
lt	22?	17?	77.3	24	(37)	64.8	-	-	-
Shanidar 4 rt[g] ♂	26.8	19.0	70.8	-	-	-	-	-	-
lt	-	-	-	(31.0)	(36.5)	(84.9)	-	-	-
Shanidar 6 rt[g] ♀	20.0	14.4	72	-	-	-	-	-	-
lt	-	-	-	24.6	30.5	80.7	29.5	21.3	72.2
Qafzeh IX rt[h] ♀	23	18.5	80.4	27.3	32.4	84.2	-	-	-
lt	-	-	-	27.1	32.5	83.4	-	-	-
Neanderthal rt[h] ♂	25	19.5	78	30	34	88.2	-	-	-
lt	-	-	-	29.5	35.5	83.1	-	-	-

than to local evolution since no earlier human forms are known from the Island. In any case, the question of how humans crossed from the continent to Crete remains. Today, the island is separated from Greece and Turkey by two straits, 100 km long and with a depth varying from 100 to 600 m. The straits are dotted with several islands. However, different conditions existed in the past. From the beginning of the Pleistocene, mammalian faunas on Crete have a clearly insular character, manifested in different patterns of faunal association, all characterized by a

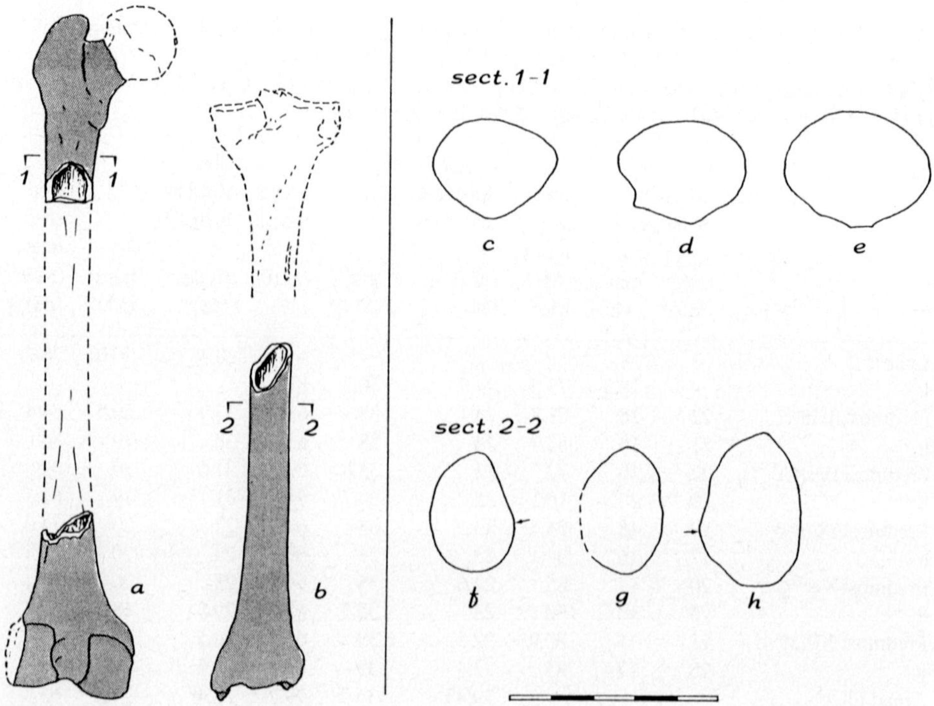

Figure 11. Posterior views (schematics) of the left femur (a) and the right tibia (b) of the Crete human remains. Diaphyseal cross sections at the middle of the gluteal tuberosity of the femora from Crete (left) (c), Shanidar 6 (left) (d) and Shanidar 1 (right) (e). Diaphyseal cross section at midshaft of the tibiae of Crete (right) (f) Shanidar 6 (right) (h) and Shanidar 2 (left) (h). The sections are viewed in a proximal-distal direction; anterior is above and the arrows indicate the positions of the interosseous crests. Scale in centimeters.

small, but varying, number of endemic species. During the Lower and Middle Pleistocene periods there were different migratory waves. The exact nature of these are not completely agreed upon by all researchers.

Some authors think that Crete was never connected to the Continent during the Pleistocene (Dermitzakis & Sondaar 1979). According to these workers, the appearance of Elephas, Hippopotamus, Cervidae, etc., on the island would have taken place by swimming via the sea or by passive transport (Sondaar 1986). Other researchers, however, assume that continental bridges existed even though they were rather discontinuous (Azzaroli 1982, Malatesta 1985, Caloi et al. 1986, Palombo 1986). These bridges were formed during periods of strong tectonic foldings that occurred during the course of the Lower and early Middle Pleistocene (Angelier 1977, Angelier & Le Pichon 1980, Angelier et al. 1982). Such neotectonic phenomena, probably continued in the Upper Pleistocene, are still active as shown by the crustal block movements from holocenic shorelines in Crete, Karpathos and Antikythira (Pirazzoli et al. 1982, Keraudren & Sorel 1984).

Glacial-eustatismic phenomena concomitant with strong horst/graben activity (documented for the Aegean area) then followed (Angelier et al. 1982). This activity would have made the shore line fluctuate, forming connections betwen Crete and the Continent at various times during the Lower, Middle and Upper Pleistocene, perhaps in successive stages (Kotsakis et al.

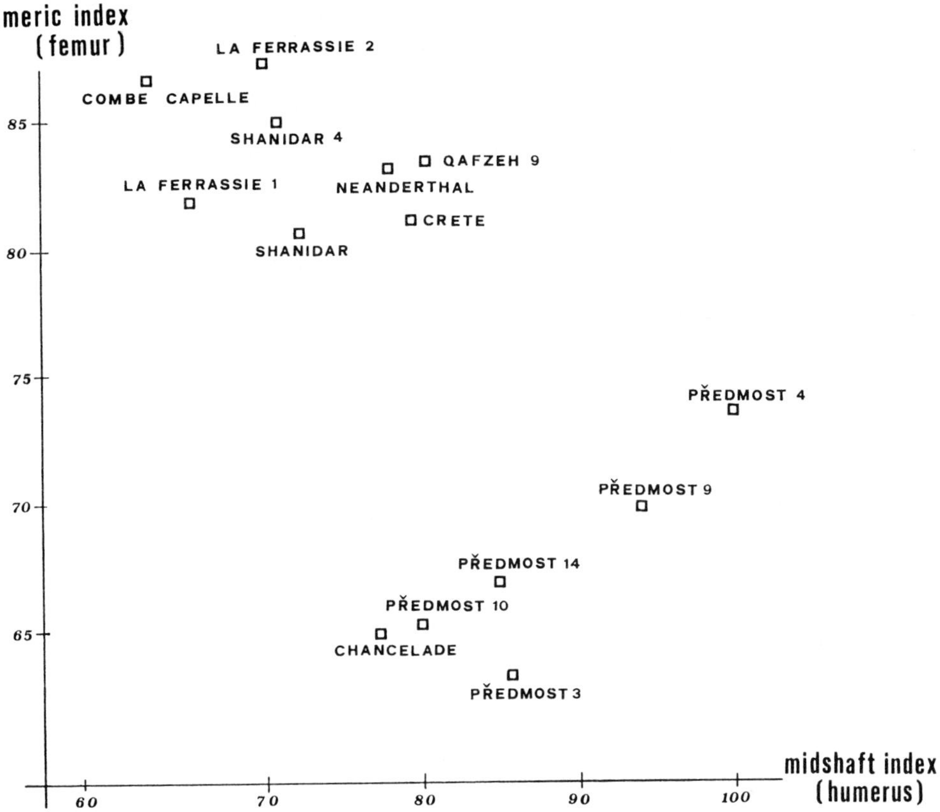

Figure 12. Bivariate representation of midshaft index of the humerus and platymeric index of the femur.

1979, Palombo 1986). In this way, the entry of the present types of faunal species (as well as that of humans) into the Cretan area might be explainable (Kotsakis et al. 1979, Kotsakis 1987). This could have happened during the Würm, coinciding with a humid period dated back to approximately 50 000 years ago and favoured the entry of humans.

The problem of the provenience of Crete modern type humans and, generally speaking, the problem of *Homo sapiens sapiens* origins in Europe and in the Near East must be considered. At the time when the Crete specimen lived, Neanderthals were widespread all over Europe, but modern humans already existed in the Near East. In fact, on the basis of recent dates proposed for the Qafzeh remains, modern humans appear to date to approximately 90 000 years ago in the Levant (Valladas et al. 1988). At present, the oldest *Homo sapiens sapiens* representatives found in Europe are dated to approximately 35 000-30 000 years ago (Smith 1984). Therefore the modern human form that lived in Crete approximately 50 000 years ago represents a rather new element, since it seems to push back the date of the presence of *Homo sapiens sapiens* in Europe and to suggest an origin from neighbouring regions of the eastern Mediterranean. Without excluding the possibility of an evolution of modern humans from Neanderthals in some eastern Europe regions, some authors think that anatomically modern humans might have

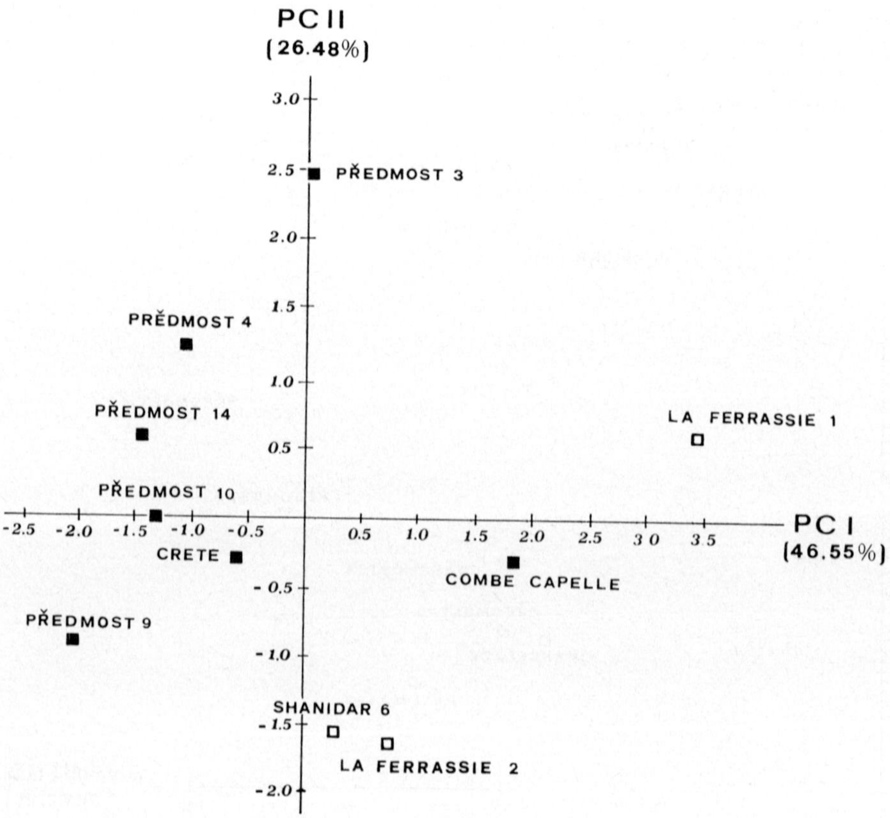

Figure 13. Principal components analysis based on 6 variables of the humerus, femur and tibia (max. and min. midshaft diameters of the humerus; ant.-post. and transv. subtrocanteric diameters of the femur; ant.-post. midshaft diameters of the tibia).

reached Europe from the Near East. This agrees with the greater antiquity attributed to *Homo sapiens sapiens* in this region, where they most likely evolved from *H. erectus* through intermediate stages, as the El Zuttiyeh skull would suggest. On the other hand it could be related to the arrival of already modern forms coming from Africa, where their presence in the eastern regions dates to at least 100 000 years ago.

Concerning the relationships between *Homo sapiens sapiens* forms of the Near East and Europe, different opinions have been expressed. Ferembach (1985, 1986) suggests that Cromagnoids would have evolved from an Aterian ancestor who had come from Africa during a sea regression, while the Combe Capelloids would have been derived from the Qafzeh type that spread into Europe from the Near East. Vandermeersch (1978, 1982) asserts that Cromagnoid forms could derive from proto-Cromagnoid elements of the Near East of which Qafzeh would be the most ancient representative. Qafzeh, in turn, would have locally evolved from *H. erectus* through the archaic *Homo sapiens* forms, as El Zuttiyeh (Vandermeersch 1982, 1989). Above all, the Afro-European *sapiens* hypothesis suggested by Bräuer (1984a, b) must be mentioned. This model states that anatomically modern forms evolved in sub-Saharan Africa around 100 000 years ago and then spread to Asia and Europe. Furthermore, Howells (1988), thinks that early moderns moved along the Nile Valley and reached the Middle East between 40 and 70 000 years ago, a point of view which agrees with the Afro-European *sapiens* hypothesis. According to Trinkaus (1986), a modern sub-Saharan population slowly expanded towards the North around 50 000 years ago.

According to these views, new populations spread into Europe and replaced the Neanderthals, thanks to some physical or cultural advantages, but with a certain degree of interbreeding between these groups (Vandermeersch 1984, Bräuer 1989). But, according to Jelinek (1976), *Homo sapiens sapiens* in central Europe was genetically connected with the previous populations of *Homo sapiens neanderthalensis*; in regions of eastern Europe several Neanderthal populations were also contemporary with the early populations of *Homo sapiens sapiens* and interbred with them. Smith (1982, 1984) also thinks that a morphological continuum exists between *Homo sapiens neanderthalensis* and *Homo sapiens sapiens* in south-central Europe and that there is no convincing evidence to indicate a migration of fully evolved modern *Homo sapiens* into south-central Europe. Consequently the archaic-to modern-*Homo sapiens* transition in these regions was indigenous.

The Crete remains suggest that a modern form might have reached Crete through the Hellenic peninsula or up the eastern Mediterranean coasts. In our opinion a migration of *Homo sapiens sapiens* from Near East to south-eastern Europe must be accepted; nevertheless, the replacement of *Homo sapiens neanderthalensis* in some regions may not have been complete. In conclusion, the *Homo sapiens sapiens* remains from Crete could be one of the most ancient representatives of modern humans or, at least, of his most ancient radiation in south-eastern Europe.

The diffusion of modern forms into the European regions, possibly from the Near East, might have taken place more gradually than we think. There are reasons, therefore, for excluding the possibility that Neanderthals were violently or, in any case, quickly supplanted by modern humans. *Homo sapiens* and Neanderthals might have lived side-by-side for a long time, at least some millenia, in Europe and appear to have interbred, as Upper Paleolithic forms from eastern Europe seem to suggest.

ACKNOWLEDGEMENTS

We are grateful for the suggestions received from Günter Bräuer, Yves Coppens, Denise Ferembach, Roger Saban and Fred Smith.

NOTES

1 The species recognized by Prof. S.I. Di Geronimo are: Mollusca: *Haliotis lamellosa* Lamarck; *Diodora graeca* (L.); *Patella lusitanica* Gmelin; *Patella sp.; Monodonta turbinata* (Born); *Tricolia pullus* (L.); *Tricolia* sp. (operculum); *Littorina neritoides* (L.); *Truncatella subcylindrica* (L.); *Rissoa* sp.; *Rissoina bruquierei* (Payraudeau); *Bittium reticulatum* (Da Costa); *Cerithium* sp. (juv.); *Epitonium*? sp.; *Erosaria spurca* (L.); *Cypreidae* sp.; *Tonna galea* (L.) (fragm.); *Columbella rustica* L.; *Hinia* sp. (fragm.); *Conus mediterraneus* Hwass; *Chiton* sp.; *Barbatia barbata* (L.); *Arcanoae* L.; *Glycymeris* sp. (juv.); *Mytilaster* sp.; *Mytilus* sp. (fragm.); *Modiolus barbatus* (L.) (juv.); *Modiolus* sp. (interior model); *Chlamys* sp. (fragm.); *Anomia*? sp.; *Ctena decussata* (Costa); *Loripeslacteus (L.); Cardium* sp.; *Acanthocardia tuberculata* (L.). Bryozoa: *Margaretta cereoides* (Ellis & Solander). Foraminifera: Ammonia beccarii (L.); *Elphidium crispum* (L.); *Miliolidae*. Echinodermata: aculeus.
No climatic differentiating elements as regards to present age have been observed in the identified species (Facchini et al. 1989).

2 According to Accorsi & Bandini Mazzanti (in press) the pollen spectra of samples A and B reveal a floral landscape marked by sclerophyllae mediterranean wood with some aspects of high bush, some pine-woods, and deciduous woods. Among the phanerophytae and arboreal/shrubby plants (approximately 2/3 of each spectrum) can be counted *Quercus ilex, Quercus coccifera* type, *Olea, Pistacia, Phyllyrea, Arbutus, Erica, Juniperus, Pinus* (mostly *Pinus brutia*), *Q. pubescens* type, *Carpinus betulus, C. orientalis, Ostrya, Castanea, Juglans, Alnus, Platanus, Betula* etc.; among the erbaceous plants appear *Gramineae, Cichorioideae, Artemisia, Helianthemum, Asphodelus, Umbelliferae, Chenopodiaceae*, etc. Sample C is characterized by a sharp regression of both pines mediterranean wood elements with diffused deciduous broadleaves.
The spectrum is dominated by arboreal species (a foresting rate of 73%) such as *Carpinus orientalis, Ostrya, Quercus, Juglans, Corylus, Salix, Platanus, Pinus, Cupressus*, etc. Scarce erbaceous plants denote the presence of aquatic environments. The vegetational picture reflects a climatic situation cooler and more humid than the previous one.
The method of detaching bones from the calcareous breccia have found out to be rather complicated and were restricted to the most important portions in order to avoid damaging the remains (see also Facchini et al. 1989).

3 For Skhūl-Qafzeh, Shanidar, Amud, Spy, European Neanderthals and Early Upper Palaeolithic group data, see Trinkaus (1983), for Cro-Magnon data see Vallois & Billy (1965), for Grimaldi Barma Grande data see Graziosi (1942), for Předmostí data see Matiegka (1938).

4 The presence of humans in Crete together with some Mousterian tools has been reported by Poulianos (1968) in the district of Kania. Specifically in a conglomerate were found bones of human fetus which, owing to the presence of a mental eminence, the author related to the modern form and, consequently, to an epoch more recent than the Mousterian one (approximately 25 000 years ago).

REFERENCES

Accorsi, C.A. & M. Bandini Mazzanti (in press). Analisi polliniche correlate a una calotta cranica di Homo sapiens proveniente dall'Isola di Creta. *Com. International Conference: Early Man in Island Environments*. Oliena (Sardegna) 1988.

Angel, J.L. 1944. A racial analysis of the ancient Greeks: An essay on the use of morphological types. *Am. J. Phys. Anthropol.* 2: 329-376.

Angelier, J. 1977. Essai sur la néotectonique et les derniers stades tarditectoniques de l'arc égéen et de l'Égée méridionale. *Bull. Soc. géol. France* XIX-3: 651-662.

Angelier, J. & X. Le Pichon 1980. La subduction hellenique et l'expansion égéenne: reconstruction cinematique et interpretation dynamique. *C.R. somm. Soc. géol. Fr.* fasc. 5: 158-161.

Angelier, J., N. Lybéris, X. Le Pichon, E. Barrier & P. Huchon 1982. The tectonic development on the hellenic arc and the sea of Crete: A synthesis. In X. Le Pichon, S.S. Angustithis & J. Mascle (eds), Geodynamics of the Hellenic arc and Trench. *Tectonophysics* 86: 159-196.

Azzaroli, A. 1982. Insularity and its effects on terrestrial vertebrates: Evolutionary and biogeographical aspects. In E. Montanaro Gallitelli (ed.), Palaeontology, Essent. of Hist. Geology. *Modena:* pp. 193-213.

Billy, G. 1969. Le squelette post-cranien de l'Homme de Chancelade. *L'Anthropol.* 73: 207-246.

Bottema, S. 1980. Palynological investigation on Crete. *Rev. Paleobot. Palynol.* 31: 193-217.

Bräuer, G. 1984a. A Craniological Approach to the Origin of Anatomically Modern *Homo sapiens* in Africa and Implications for the Appearance of Modern Europeans. In. F.H. Smith & F. Spencer (eds), *The Origins of Modern Humans: A World Survey of the Fossil Evidence*. New York: Alan R. Liss, pp. 327-410.

Bräuer, G. 1984b. The 'Afro-European *sapiens*-hypothesis', and hominid evolution in East Asia during the late Middle and Upper Pleistocene. In P. Andrews & J.L. Franzen (eds), The Early Evolution of Man with Special Emphasis on Southeast Asia and Africa. Frankfurt am Main. *Cour. Forsch. Inst. Senkenberg* 69: 145-165.

Bräuer, G. 1989. The evolution of Modern Humans: A Comparison of the African and non-African Evidence. In P. Mellars & C. Stringer (eds), *The Human Revolution: Behavioural and Biological Perspectives on the Origins of Modern Humans*. Edinburgh: Edinburgh University Press, pp. 123-154.

Caloi, L., T. Kotsakis & M.R. Palombo 1986. La fauna a vertebrati terrestri del Pleistocene delle Isole del Mediterraneo. *Geologica Romana* 25: 235-256.

Chamla, M-C. 1980. Étude des variations métriques des couronnes dentaires des nord-africains, de l'Épipaléolithique à l'époque actuelle. *L'Anthropol.* 84: 254-271.

Coon, C.S. 1962. *The origin of races.* New York: Knopf. It. Transl. L'origine delle razze. Milano: Bompiani, 1970.

Day, M.H. & C.B. Stringer 1982. A reconsideration of the Omo Kibish remains and the *erectus-sapiens* transition. *Pretirage 1er Congr. Int. Palaeont. Hum. Nice*, pp. 814-846.

de Lumley-Woodyear, M.A. 1973. *Anténéandertaliens et Néandertaliens du bassin méditerraneen occidental européen.* Paris, CNRS.

Dermitzakis, M.D. & P.Y. Sondaar 1979. The importance of fossil mammals in reconstructing palaeogeography with special reference to the Pleistocene Aegean archipelago. *Ann. Geol. Pays Hellen.* 28(1978): 808-840.

Facchini, F., G. Giusberti, C.A. Accorsi, M. Bandini, S.I. Di Geronimo & Y. Yokoyama 1989. Restes de *Homo sapiens sapiens* provenant de l'ile de Crète. In G. Giacobini (ed.), *Hominidae. Proc. 2nd Intern. Congr. of Hum. Palaeont. Turin*, Sept. 28-Oct. 3.1987. Milan: Jaca Book, pp. 467-474.

Ferembach, D. 1985. On the origin of the Iberomaurusians (Upper Palaeolithic: North Africa). A New Hypothesis. *J. Hum. Evol.* 14: 393-397.

Ferembach, D. 1986. Les hommes du Paléolithique supérieur autour du bassin méditerranéen. *L'Anthropol.* 90: 579-587.

Graziosi, P. 1942. L'uomo fossile della Barma Grande ai Balzi Rossi nel Museo di Mentone. *Archiv. Antrop. Etnol.* 72: 22-50.

Hillson, S. 1986. *Teeth.* Cambridge: Cambridge University Press.

Howell, F.C. 1951. The place of Neanderthal Man in human evolution. *Am. J. Phys. Anthropol.* 9: 379-416.

Howells, W.W. 1973. Cranial Variation in Man. *Peabody Museum Papers* 67.

Howells, W.W. 1988. The meaning of the Neanderthals in human evolution. In Fondation Singer-Polignac (ed.), *L'evolution dans sa realité et ses diverses modalités.* Paris: Masson, pp. 221-239.

Jelinek, J. 1976. The *Homo sapiens neanderthalensis* and *Homo sapiens sapiens* relationship in Central Europe. *Anthropol. (Brno)* 14: 79-81.

Jelinek, A.J. 1982. The Tabun Cave and Palaeolithic Man in the Levant. *Science* 216: 1369-1375.

Keraudren, B. & D. Sorel 1984. Relations entre sédimentation, tectonique et morphologie dans le Plio-Pleistocène de Karpathos (Grèce); mouvements verticaux et datations radiométriques. *L'Anthropol.* 88: 49-61.

Klaatsch, H. & O. Hauser 1910. *Homo aurignacensis* Hauseri; ein paläolitischer Skeletfund aus dem unteren Aurignacien der Station Combe-Capelle bei Montferrand (Perigord). *Praehist. Zeits.* I Heft 3/4: 273-338.

Kotsakis, T. 1987. Elementi di paleobiogeografia dei micromammiferi della Grecia durante il Cenozoico. In P. Crucitti (ed.), *Atti del convengo di Zoologia Ellenica Soc. Romana di Sc. Nat. Roma* 1986, pp. 20-24.

Kotsakis, T., C. Petronio & G. Sirna 1979. The quaternary vertebrates of the Aegean Islands: Palaeogeographical implications. *Ann. Géol. Pays Hellen.* 30: 31-64.

Leroi-Gourhan, Arl. 1988. Le cadre naturel et les Néandertaliens. In M. Otte (ed.), *L'Homme de Néanderthal,* vol. 2, L'Environnement. Liège, *ERAUL* 29: 95-100.

Lovejoy, C.O., A.H. Burnstein & K.G. Heiple 1976. The Biomechanical Analysis of Bone Strength: A Method and Its Application to Platycnemia. *Am. J. Phys. Anthropol.* 44: 489-506.

Lovejoy, C.O. & E. Trinkaus 1980. Strength and Robusticity of the Neanderthal Tibia. *Am. J. Phys. Anthropol.* 53: 465-470.

Malatesta, A. 1985. *Geologia e paleobiologia dell' era glaciale.* Roma: Nuova Italia Scientifica.

Martin, R. & K. Saller 1957. *Lehrbuch der Anthropologie.* Stuttgart: Fischer Verlag.

Matiegka J. 1934. *Homo Předmostensis: Fosilní Člověk Z Předmostí na Morave I. Lebky (L'Homme fossile de Předmostí en Moravie [Tchécoslovaquie] I. Les crânes).* Prague: Cěské Akademie Věd a Umění.

Matiegka, J. 1938. *Homo Předmostensis. Fosilní Člověk z Předmostí na Morvae II. Ostatni Casti Kostrove (L'Homme Fossile de Předmostí en Moravie (Tchécoslovaquie) II. Autres parties du squelette).* Prague: Cěské Akademie Věd a Umění.

Palombo, M.R. 1986. I grandi mammiferi pleistocenici delle isole del Mediterraneo: tempi e vie di migrazione. *Boll. Soc. Paleont. It.* 24(1985): 201-224.

Pirazzoli, P.A., J. Thommeret, Y. Thommeret, J. Laborel & L.F. Montappioni 1982. Crustal block movements from Holocene shorelines: Crete and Antikythira (Greece). *Tectonophysics* 86: 27-43.

Poulianos, A.N. 1968. Palaeolithic find on Crete? *Proceed. Hellenic Anthrop. Soc. Athens.* 51-53.

Saban, R. 1984. *Anatomie et évolution des veines méningées chez les hommes fossiles.* Paris: E.N.S.B. – C.T.H.S.

Sergi, S. 1974. *Il cranio Neandertaliano del Monte Circeo (Circeo I).* Roma: Acc. Naz. Lincei.

Sergi, S., R. Parenti & G. Paoli 1974. Il giovane paleolitico della caverna delle Arene Candide. *Mem. Ist It. Paleont. Um.* II: 13-38.

Simonelli, V. 1897. *Candia – Ricordi di escursione.* Parma: Battei.

Smith, F.H. 1976. A Fossil Hominid Frontal from Velika Pécina (Croatia) and a Consideration of Upper Pleistocene Hominids from Yugoslavia. *Am. J. Phys. Anthropol.* 44: 127-134.

Smith, F.H. 1982. Upper Pleistocene Hominoid Evolution in South-Central Europe: A Review of the Evidence and Analysis of Trends. *Curr. Anthropol.* 23: 667-703.

Smith, F.H. 1984. Fossil Hominids from the Upper Pleistocene on Central Europe and the Origin of Modern Europeans. In. F.H. Smith & F. Spencer (eds), *The Origins of Modern Humans: A World Survey of the Fossil Evidence.* New York: Alan R. Liss, pp. 137-209.

Sondaar, P.Y. 1986. The Island Sweepstakes. Why did pigmy elephants, dwarf deer, and large mice once populate the Mediterranean? *Nat. Hist.* 95: 50-57.

Stringer, C.B., J.-J. Hublin & B. Vandermeersch 1984. The Origin of Anatomically Modern Humans in Western Europe. In F.H. Smith & F. Spencer (eds), *The Origins of Modern Humans: A World Survey of the Fossil Evidence.* New York: Alan R. Liss, pp. 51-135.

Trinkaus, E. 1976. The evolution of the hominid femoral diaphysis during the Upper Pleistocene in Europe and the Near East. *Z. Morph. Anthropol.* 67: 291-319.

Trinkaus, E. 1980. Sexual Differences in Neandertal Limb Bones. *J. Hum. Evol.* 9: 377-397.

Trinkaus, E. 1983. *The Shanidar Neandertals.* New York: Academic Press.

Trinkaus, E. 1986. Les néandertaliens. *La Recherche* 180: 1040-1047.

Valladas, H., J.L. Reyss, J.L. Joron, G. Valladas, O. Bar-Yosef & B. Vandermeersch 1988. Thermoluminescence dating of Mousterian 'Proto-Cro-Magnon' remains from Israel and the origin of modern man. *Nature* 331: 614-616.

Vallois, H.V. & G. Billy 1965. Nouvelles recherches sur les hommes fossiles de l'Abri de Cro-Magnon. *L'Anthropol.* 69: 47-74, 249-272.

Vallois, H.V. & B. Vandermeersch 1972. Le crâne mousterien de Qafzeh (*Homo* VI): étude anthropologique. *L'Anthropol. 76: 71-96.*

Vandermeersch, B. 1978. Quelques aspects du problème de l'origine de l'Homme moderne. In Fondation Singer-Polignac (ed.), *Les origines humaines et les époques de l'intelligence.* Paris: Masson, pp. 153-157.

Vandermeersch, B. 1981a. Les premiers *Homo sapiens* au Proche Orient. In D. Ferembach (ed.), *Les Processus de l'Hominisation.* Paris: CNRS, pp. 97-100.

Vandermeersch, B. 1981b. Les hommes fossiles de Qafzeh (Israel). *Cah. Paleontol.* (*Paleoanthropol.*) Paris: CNRS.

Vandermeersch, B. 1982. The first *Homo sapiens sapiens* in the Near East. In A. Ronen (ed.), The Transition from Lower to Middle Palaeolithic and the Origin of Modern Man. Oxford: *Brit. Arch. Rep. Int. Series.* S1 51: 297-299.

Vandermeersch, B. 1984. A propos de la découverte du squelette Néanderthalien de Saint-Césaire. *Bull. Mém. Soc. Anthropol.* 3: 191-196.

Vandermeersch, B. 1989. L'origine de l'Homme moderne: Le point de vue du paléanthropologue. In G. Giacobini (ed.), *Hominidae. Proc. 2nd Intern. Congr. of Hum Palaeont. Turin,* Sept. 28-Oct.3.1987. Milan: Jaca Book, pp. 415-421.

Wolpoff, M.H. 1980. Cranial Remains of Middle Pleistocene European Hominids. *J. Hum. Evol.* 9: 339-358.

The evolution of *Homo sapiens* as indicated by features of the postcranium

Gail E. Kennedy
Department of Anthropology, University of California, Los Angeles, USA

1 INTRODUCTION

Until recently it has been assumed by most workers that the postcranium of early members of the genus *Homo*, particularly *Homo erectus*, was essentially like that of modern man. This belief was due to several factors: The generally poor preservation of early hominine postcranial elements, the very rare association between early hominine cranial and postcranial elements and, perhaps most importantly, the discovery in 1892 (Dubois 1893) of the complete Trinil femur from geological layers then thought to contain cranial remains of early hominines. This specimen, while grossly pathological, appeared to differ in no significant way from the femora of modern man.

However, as early as 1941 Weidenreich attempted to correct this notion by describing the few and fragmentary postcranial elements associated with cranial and dental remains of 'Sinanthropus' recovered at Zhoukoudian. In the 7 femoral shaft fragments (only I and IV are complete enough for study), he noted thickened cortical bone with pronounced medullary stenosis, a medial convexity in the proximal shaft and a low position of minimum shaft breadth along with several other non-modern human features. In 1971 Day published on an associated femoral shaft and os coxae (OH 28) from Bed IV, Olduvai Gorge: The femoral shaft showed the same complex of features found at Zhoukoudian, and Day therefore attributed OH 28 to the taxon *Homo erectus*.

My own recent work has focused on the 2 best known postcranial elements of early hominines, the femur and os coxae. Earlier publications, using femoral materials and employing phenetic analysis, have shown close groupings between 'Sinanthropus', OH 28, KNM ER 737 and 1481 (Kennedy 1973, 1983a, b, 1984).

The analysis reported here has been based on the philosophies and techniques of phylogenetic systematics (Hennig 1966). Although not without fault in hominid systematics (e.g. Tobias 1985) the approach recently has been widely applied to issues of hominine evolution (e.g. Stringer 1987, Chamberlain & Wood 1987, and many others). The outgroup used here consisted of the African pongids, which have generally been acknowledged to be monophyletic with hominids (e.g. Kluge 1983).

2 THE OS COXAE

This bone is reported for *H. erectus* from OH 28 (Day 1971) and KNM WT 15000 (Brown et al. 1985); KNM ER 3228, Arago XLIV and SK 3155b have tentatively been assigned to the taxon (Rose 1984, Sigmon 1982, Brain et al. 1974 respectively). The os coxae is also known for the robust australopithecines (SK 50), for the gracile australopithecines (STS 14,65; MLD 7,25; TM 1605; AL 288-1ao) and for a number of Neanderthals. The os coxae is represented in the African 'archaic *H. sapiens*' only by Kabwe.

The African pongid sample shows a vertically oriented ilium without lateral flare, no pronounced vertical buttresses and no forward protuberance of the anterior superior iliac spine. The australopithecines (both robust and gracile), OH 28, KNM ER 3228, Arago and SK 3155b all show a derived pattern relative to the pongids in having a craniocaudally shortened ilium with a lateral flare and a pronounced forward protuberance of the anterior superior iliac spine. KNM WT 15000 also shows a 'remarkable degree of iliac flare' (Brown et al. 1985:791); the condition of the anterior iliac spine is not yet reported. Kabwe also shows indications of strong lateral iliac flare (Stringer 1986, and pers. observ.). Iliac flare is somewhat difficult to judge in the Neanderthals because of the incompleteness of most specimens, but the complete Kebara 2 pelvis shows only 'ordinary' lateral flare (Rak & Arensburg 1987).

The australopithecines show a double buttress to the ilium: A posteriorly placed acetabulocristal buttress and an anteriorly placed acetabulospinous buttress, with the latter buttress being the most developed (Robinson 1972). The buttresses are not strongly developed in the juvenile australopithecines (MLD 7,25 and TM 1605), but they are present (pers. observ.). The Pan ilium demonstrates neither buttress strongly; but Mednick (1955) has shown, using split line technique, that there is increasing bone density in the areas where the buttresses occur in early bipeds. In the preserved fossil hominine ossa coxae (OH 28, KNM ER 3228, Arago and SK 3155b, the Neanderthals and Kabwe) the acetabulospinous buttress has been greatly reduced or lost, and the acetabulocristal buttress (the 'pillar') is retained and is highly developed, often with an associated and highly developed iliac tubercle. The posterior buttress and tubercle

Table 1. Os coxae.

Sample	Acetabulo-cristal buttress		Medial rotation of ischium		Reference
	Present	Absent	Present	Absent	
Gorilla		x		x	Kennedy (on orginal)
Pan		x		x	Kennedy (on orginal)
amHs		x		x	Kennedy (on orginal)
Africa					
OH 28	x		x		Kennedy (on cast)
KNM ER 3228	x		x		Rose (1984)
KNM WT 15000		x	?		Brown et al. (1985)
Kabwe	x		?		Kennedy (on orginal)
STS 14	x		?		Kennedy (on orginal)
AL 288	x		?		Kennedy (on cast)
Europe					
La Chapelle	x		x		Kennedy (on orginal)
La Ferrassie I	x		x		Kennedy (on orginal)
Krapina 207	x		x		Kennedy (on orginal)
Krapina 208	?		x		Kennedy (on orginal)

persist in anatomically modern *Homo sapiens* (amHs) but in a much less developed state. Rose (1984) reports that for one sample of amHs the range of thickness of the acetabulocristal buttress is 9-19 mm (n=40); the buttress measures 19 mm in OH 28 and 21 mm in KNM ER 3228, thus in both fossils the thickness is either at the limit or beyond the range for modern man. A buttress is not present on KNM WT 15000, a condition probably related to its juvenile status (Brown et al. 1985) (Table 1).

Stringer (1986) has reported the probable presence of a strong iliac pillar in Kabwe (E 719). Although much of the superior portion of the ilium is missing in this specimen, the base of the acetabulocristal buttress is very strongly developed. The cortical thickness on the gluteal surface of the preserved ilium (ca. 21 mm above the acetabulum) is great: z=9.7 above Stringer's mean for his modern comparative sample and well above that range. The presence of thickened cortical bone on the gluteal surface of the OH 28 ilium is confirmed by X-rays and personal observation of the original.

Neanderthal ilia (La Chapelle, La Ferrassie I and Tabun C1 originals examined by author) all show strongly developed iliac pillars and pronounced thickening on the gluteal surface of the ilium. Krapina 207, even though sub-adult, shows a very strongly developed pillar; and the iliac crest, while still not fully fused, shows a well developed iliac tubercle. The more complete Kebara 2 shows a massive iliac buttress (Rak & Arensburg 1987). Lateral cortical thickening of the gluteal surface is also present on La Chapelle, La Ferrassie I, Tabun C1 and Kabwe (pers. observ.).

Also characteristic of *Homo erectus* (OH 28, KNM ER 3228, Arago and SK 3155b) is a medial torsion to the ischium (Day 1971, Rose 1984, Sigmon 1982, pers. observ., respectively) a character which is absent in the African pongids, the gracile australopithecines and amHs. SK 50 appears to show considerable torsion to the ischium; this might, however, be exaggerated by post mortem distortion (pers. observ.). Strong ischial torsion like that found in *Homo erectus* is also present on La Chapelle, La Ferrassie I and Krapina 207 and 208 (pers. observ.); the ischial orientation cannot definitively be determined on Kabwe but some torsion may be present. The tuberosity is not sufficiently preserved in Tabun C1 and has not been discussed for Kebara. In the outgroup, the gracile australopithecines and in amHs the ischial tuberosity lacks this rotation and faces posterolaterally. Until additional robust australopithecine ossa coxae are reported, ischial torsion can be considered derived for *Homo erectus* and for the Neanderthals. It seems likely that the lateral facing acetabulum and medial ischial torsion present in *Homo erectus* and the Neanderthals represents a biomechanical complex. As the acetabulum is brought forward by the anterior and lateral flare of the ilium, the ischium must also move in the same plane to maintain the origin of the hamstrings in an optimal biomechanical relationship. The SK 50 ilium also shows very strong anterior and lateral flare and a similar biomechanical adjustment might also be in operation here.

It is unfortunate that the superior pubic ramus is so rarely preserved in fossil hominids, since the Neanderthals show a very distinctive pattern. In that group, the pubis is flattened and there is acetabulo-symphyseal elongation (Trinkaus 1976). Rak & Arensburg (1987) have recently pointed out that in the Kebara 2 pelvis the great length of the superior pubic ramus is due to external or lateral rotation of the ilia. The character state is clearly derived in Neanderthals relative to both pongids and amHs but its relationship to the fossils under study here is not known.

The very absence in the fossil record of the superior pubic rami in early hominines may suggest that in these groups, too, this area of the os coxae was thin and fragile. This speculation must await testing on more complete specimens.

3 FEMUR

In the traits investigated here the pongids, gracile australopithecines (the femur is poorly identified for the robust australopithecines) and amHs share a morphological pattern. These shared traits include lack of pronounced AP curvature (bowing) and small cortical thickness (Table 2) both absolutely and relative to shaft diameters (Kennedy 1973, 1983a). In the cortical variable included here the African pongids and the gracile australopithecines both differ from modern males at the .05 level of confidence.

Derived traits for the genus *Homo* include a round (rather than flattened) femoral neck and 'waisting' of the femoral shaft (Kennedy 1973, 1983a, b). The derived traits for the genus *Homo* may also include lateral flaring to the greater trochanter, but the slight flare present on AL 333-3 and 333w-40, attributed to *A. afarensis* by Johanson & White (1978), makes the polarity of this trait unclear.

The femora of *H. erectus* are derived in having pronounced cortical thickness, both abso-

Table 2. Cortical thickness in the femur. Medial cortex, mid shaft. *All data by Kennedy on X-rays, **Krapina data taken directly at most distal point remaining on shaft..

		Measurement	Standard deviation
amHs			
Males	n=57	7.9	1.42
Females	n=43*	7.1	1.11
Gorilla	n=3	6.1	1.37
Pan	n=3	6.1	0.65
Neanderthal			
Neanderthal (type)		8.3	
Krapina 213		8.5**	
Krapina 214		10.5	
	mean	9.1	1.21
Kabwe 689		7.2	
Omo I		6.8	
Homo erectus			
Peking I		11.5	
Peking IV		10.3	
OH 28		9.2	
KNM ER 1481		10.2	
KNM ER 737		10.5	
	mean	10.3	0.82
Australopithecus			
KNM ER 993		4.6	
SK 82		7.1	
SK 97		6.8	
	mean	6.1	1.36

lutely and relative to external shaft dimensions, which results in medullary stenosis, medial convexity in the proximal shaft and a low position of minimum shaft breadth (Weidenreich 1941, Kennedy 1983b, 1985). It has been argued elsewhere that the additional cortical bone in the *H. erectus* femur is endosteal, not sub-periosteal, in origin (Kennedy 1985).

Differences between *H. erectus* and the comparative males in most of the cortical variables and indices is significant at $p = .05$ (Kennedy 1983a). The cortical variable included here is significant at the $p = .01$ level ($t = -3.71$). Although sample size is small, the Neanderthal cortical values are consistently above those for modern males. However, the probability values are not statistically significant ($t = -1.43$, 58 df). The Arago femora are characterized by pronounced cortical thickening (Lovejoy 1982), although data for these specimens have not been published.

Since very few fossil femora are complete, shaft/articular ratios are difficult to determine; however, a few data are available. KNM ER 1481, a complete *H. erectus* femur (contra Trinkaus 1984) gives a bicondylar breadth/maximum length ratio of 17.08, $z = -0.92$ below the comparative males (amHs data from Pearson & Bell 1919, fossil data, pers. observ. on original). KNM ER 1472 is also complete; and, while it has been tentatively assigned to *Homo* (Leakey 1973), it demonstrates the straight shaft and lack of trochanteric flare usually found in the australopithecines. This specimen gives a ratio of 16.32, $z = -1.86$ against control males. Neanderthals (the 'type', Spy and La Chapelle), however, show very high ratios: $\bar{x} = 20.5$, s.d. 0.97, $p = .005$ against control males. Kabwe, using a reconstructed length of 475 mm, gives a bicondylar breadth/length ratio of $z = +1.07$ against the control males. While recognizing the inadequacy of the fossil sample, it seems that the australopithecines, *H. erectus* and amHs show similar articular/length ratios while the Neanderthals are autapomorphic in the intra-limb proportions of the femur.

The major derived feature of the *H. erectus* femur, greatly increased cortical thickness with medullary stenosis, may be secondarily reflected in other distinctive femoral traits in this taxon. For example, the area of greatest cortical thickness in the *H. erectus* femur is in the medial aspect of the proximal shaft; this is the area occupied externally by the medial convexity. The increased thickness may also be reflected in the low position of minimum shaft breadth. It is only at this point, in the distal 1/4 of the shaft, that compressive biomechanical forces of the shaft shift toward the multifocal forces of the articular areas; thus in the distal 1/4 of the shaft the cortex begins to thin as the medullary canal fills with the cancellous tissue of the articular areas.

The Neanderthal femora show a number of autapomorphs in addition to the derived traits shared with *H. erectus:* These autapomorphs include pronounced AP bowing and the large shaft/articular indices described above. Low Crural Indices reflecting relative shortening of the tibia (to the femur) are also distinctive in the Neanderthals ($\bar{x} = 78.9$) but can be duplicated in at least one modern population (Trinkaus 1981). Published mean Crural Indices for amHs range from 83.0 to 88.8 with a full sample mean of 84.79 (Schultz 1937). KNM WT 15 000, the only *H. erectus* complete enough to determine relative limb lengths, does not show relative shortening of the tibia (A. Walker, pers. comm.), although its Crural Index is not yet published. Kabwe also does not show shortening of the distal limb element: The Kabwe tibia (E691) is complete and is very long (409 mm); mean European Neanderthal tibial length is 325.4 mm, $n = 5$ (Austin 1983). A femur (E689) appears to be from the same Kabwe individual (Pycraft et al. 1928, Kennedy 1984) and has a reasonable reconstructed length of 475 mm, giving a Crural Index for Kabwe of about 86 (see Kennedy 1984). This value is within the modern range. Thus, although Kabwe retains primitive features in the os coxae, its inter- and intra-lower limb proportions are within the amHs range.

Omo-Kibish I has an extensive but comminuted and largely unreported postcranium; data on a reconstructed femur, however, have been published (Kennedy 1984). In every feature which can be examined in this specimen, radiographically and directly, including cortical diameters, Omo-Kibish I is well within the range of modern *H. sapiens* (Kennedy 1984).

4 DISCUSSION

Postcranial data of fossil hominines are extremely limited, and only the os coxae and femur are known across enough groups for the information to be phylogenetically useful; very limited data are available for the tibia. In the os coxae, medial ischial torsion does not occur in the outgroup, the modern *H. sapiens* controls or the gracile australopithecines. Its existence in the robust australopithecines is questionable, but it does occur in both *H. erectus* and the Neanderthals. Both derived features may also occur in Kabwe, the only African 'archaic *sapiens*' for which the bone is well enough preserved to judge.

It seems likely that these 2 features may be interrelated and are complex in their origin. As pointed out, medial ischial torsion may be related to maintaining hamstring efficiency in a laterally flaring and anteriorly rotated ilium. The iliac pillar is highly developed in *H. erectus*, the Neanderthals and Kabwe, and this is accompanied by increased cortical thickness on the gluteal surface of the ilium. The femoral traits included here are quantitative and may all be manifestations of cortical hyperostosis. Data presented here are limited to a single cortical variable, the thickness of the medial cortex at mid shaft; although other data, published elsewhere, show a similar pattern (Kennedy 1973, 1983a, b, 1985). *H. erectus* shows a consistent pattern of cortical hyperostosis absolutely and relative to shaft diameters (Kennedy 1983b, 1985). The Neanderthal sample is small, and while that sample shows a similar pattern of greater cortical diameters (relative to the outgroup, the australopithecines and amHs), the differences are not statistically significant at the $p = .05$ level, the accepted significance level for a 'new' character state in this study. It is obvious that while this level of significance may be statistically defensible, it is biologically arbitrary.

A pattern of cortical hyperostosis of other tubular bones is also present in a *H. erectus* ulna (OH 36, Kennedy, n.d.) and the Neanderthal and Kabwe tibiae (Austin 1983). Thus in postcranial cortical diameters it is clear that the primitive polarity is for thin cortex (both absolutely and relative to total shaft diameters) and that the primitive state is shared by the outgroup, the australopithecines (both robust and gracile), certain of the late Middle or early Upper Pleistocene African post *H. erectus* hominids (see below) and modern man.

Work already published (e.g. Weidenreich 1943 and many others) and in preparation (this author) indicates that hyperostosis of the cranial bitabular compacta also characterizes *H. erectus* and, again, the Neanderthals. In the cranial data, however, the Neanderthal vault thickness at euryon shows high significance from male controls ($p = .005$) (Kennedy, in prep.). Thus, again, *H. erectus* and the Neanderthals share a derived character. Hyperostosis of the cranial bitabular compacta is also found in many other Middle and early Upper Pleistocene hominids whose attributions vary: e.g. Bodo, Petralona and Arago.

It is clear, however, that within the very heterogeneous post *H. erectus* hominids in sub-Saharan Africa there is a sub-group that shows a complex of fully modern features. Omo-Kibish I (Kennedy 1984), Border Cave 1 and 5 (de Villiers 1973, Beaumont et al. 1978, Rightmire 1979) and Klasies River Mouth (various specimens) (Singer & Wymer 1982) have, unequivocally, been attributed to amHs. Of interest, however, is that all of these sub-Saharan specimens appear to pre-date 100 ky (Butzer et al. 1978, Butzer 1969, and Singer & Wymer 1982, respectively). This date, falling within the Eemian, is equivalent to or may actually precede the appearance of Neanderthals in Europe.

The presence of cortical/tabular hyperostosis in *H. erectus*, the Neanderthals and other pre-modern specimens (Kennedy, in prep.) suggests the appearance of a new genetic state, perhaps reflecting change at a single regulatory gene locus, at the point of origin of *H. erectus*. From the data considered here it is possible to argue that the origin of the *H. erectus* clade began with and is characterized, in part, by a genetic event which coded for either increased endosteal apposition or decreased endosteal resorption, either of which would lead to the characteristic cortical hyperostosis and medullary stenosis in the tubular bones and thickness of the bitabular compacta. The pronounced iliac pillar in these groups may also be due, in part, to a pattern of increased cortical bone development.

However, the phylogenetic relationship between various groups of Middle and early Upper Pleistocene hominids and of these to modern humans is difficult to investigate. The Neanderthals as individuals are relatively well known because of their practice of burial while other European and sub-Saharan hominids are fewer in number, often more fragmentary and the postcrania are poorly known.

It seems clear that such diverse groups of European and African hominids cannot all be accommodated within the modern species, and yet it is not generally agreed what the definitive features of the modern species may be. A proposal provided by Day & Stringer (1982) requires further investigation. Hypothetically if absence of hyperostosis is considered as a definitive modern sapiens trait, then individuals within both the Neanderthals and some post *H. erectus* hominids cannot be attributed to amHs.

In a previous paper (Kennedy 1985) I have stressed the uniqueness of cortical/tabular hyperostosis: Among terrestrial mammals only *H. erectus* and, to a lesser extent, the Neanderthals and certain of the 'archaic *sapiens*' appear to demonstrate such profound generalized cortical/tabular hyperostosis. The adaptive significance of this trait is not clear, but it may be related to the maintenance of steady serum clacium levels when dietary sources of calcium vary widely (see Kennedy 1985 for a full discussion of this hypothesis). This particular hypothesis is difficult (impossible?) to test, but inferential support may be provided by the archaeological record (Isaac 1981) and by microwear on teeth (Walker 1981). A major dietary shift is unusual in animals, and perhaps the hyperostosis is a physiological reflection of a significant shift toward carnivory away from our long primate history as plant consumers. This hypothesis has the fault of not explaining the later loss of hyperostosis in modern *H. sapiens* unless it is hypothesized that pronounced cortical/tabular hyperostosis did not characterize the direct ancestor of amHs. Thus, our relatively thin cortical/tabular bone could be either a retained primitive trait or a reversion. An alternative hypothesis to explain cortical/tabular hyperostosis may involve neoteny or paedomorphosis. In juvenile modern humans bone apposition occurs primarily on the endosteal surface (Garn 1972). If a high rate of internal apposition continued, without concordant sub-periosteal resorption, then hyperostosis and medullary stenosis might occur. The presence of heterochrony and particularly neoteny have been suggested by many authors as an explanation for certain hominine characters (see Gould 1977 for a review). A fault with this model is that modern humans, highly neotenous in other characters, do not demonstrate hyperostosis. The most parsimonious explanation (but not necessarily the most correct) is that modern *H. sapiens* did not pass through a hyperostotic phase.

5 CONCLUSION

Both *H. erectus* and the Neanderthals show similar derived traits in the femur and os coxae: Among these are medial torsion to the ischium and a massively developed iliac pillar. Kabwe, the only post-*H. erectus* African hominid which has adequate postcranial representation certainly shows the latter trait and possibly the former. Relative and absolute cortical hyperostosis, which is probably also reflected in the iliac pillar, is found in the femur and in other known tubular bones such as the ulna and tibia. The cortical hyperostosis found in the postcranium is very likely reflected in the hyperostosis of the cranial bitabular compacta in *H. erectus*. Somewhat reduced, but statistically significant, vault hyperostosis characterizes the Neanderthals and a number of late Middle or early Upper Pleistocene African specimens: e.g. Kabwe, Florisbad, Saldanha.

One theoretical, if not practical, advantage of cladistics is that only a single trait is required to demonstrate affinity, if homoplasy and reversion can be ruled out. Of the postcranial traits considered here, cortical hyperostosis is a derived trait in *H. erectus* and, to a lesser extent, the Neanderthals and most of the 'archaic *H. sapiens*' . Hyperostosis must be considered a derived character since it is not known to have existed in any other hominoid group; therefore, *a priori*, homoplasy and reversion can be ruled out. This is not to say that there are no other autapomorphs in *H. erectus* or the Neanderthals, but within the protocols of phylogenetic systematics a single, derived trait is sufficient to identify a new clade, and this single feature is sufficient for ordering the hominids. However, it is also clear that there is a sub-set within the sub-Saharan post *H. erectus* hominids (e.g. Omo-Kibish I, Border Cave and Klasies River Mouth) which do not show such hyperostosis. Within the structures of phylogenetic systematics these individuals (and the populations they represent) may represent early members of our own species.

There are several implications of these conclusions. One evolutionary scenario suggested is that some *H. erectus* populations in Europe and the circum-Mediterranean evolved there directly into the Neanderthals. Indeed, individuals such as Arago appear to represent evidence of late *H. erectus*–Neanderthal populations. In sub-Saharan Africa ecology and distance seem to have resulted in a more complex evolution of later Pleistocene hominids. While there is growing evidence of early (> 100 ky) modern populations, other individuals (e.g. Kabwe, Saldanha) appear to represent late survivors of more primitive populations. This scenario presents at least 2 problems. First, other explanations would have to be found for the Upper Pleistocene-age eastern European (e.g. Předmost) and Near Eastern (e.g. Skhūl) populations often regarded as 'hybrids'. Second, amHs autapomorphs would have to characterize that sub-set of sub-Saharan populations but would have to be absent in the Neanderthals. A test of this scenario obviously rests on defining the modern human autapomorphs and as yet there is no consensus what the modern autapomorphs may be. A final issue of the early sub-Saharan origin of modern *H. sapiens* is that members of this group were not associated with an advanced Upper Paleolithic blade tool technology as were later modern humans in western Europe. This problem would be solved by the decoupling of cultural and morphological evolution. Nevertheless, using other data several workers have similarly proposed an early emergence of modern humans in sub-Saharan Africa (Cann et al. 1987, Bräuer 1989, Hublin 1986). A final issue which requires investigation is the desirability of basing a classification on a single trait as it is possible to do in cladistics.

ACKNOWLEDGEMENTS

I would like to thank those people who allowed me to study fossils in their care: M.H. Day, St. Thomas's Hospital Medical School; C. Stringer, British Museum (Natural History); J.J. Hublin, Université Paris VI; R. Leakey, National Museums of Kenya; R. Brain, Transvaal Museum; and J. Radovčić, Croatian Natural History Museum. I would also like to thank J. Austin for the use of her unpublished data.

REFERENCES

Austin, J. 1983. Investigations in Mid and Upper Pleistocene Hominid Tibiae. Unpublished Thesis: University of California, Los Angeles.

Beaumont, P., H. De Villiers & J. Vogel 1978. Modern man in sub-Saharan Africa prior to 49 000 years BP: A review and evaluation with particular reference to Border Cave. *S. Afr. J. Sci.* 74: 409-419.

Brain, R., E. Vrba & J. Robinson 1974. A new hominid innominate bone from Swartkrans. *Ann. Trans. Mus.* 29: 55-66.

Bräuer, G. 1989. The evolution of modern humans: A comparison of the African and non-African Evidence. In P. Mellars & C.B. Stringer (eds), *The Human Revolution: Behavioral and Biological Perspectives on the Origin of Modern Humans.* Chicago: University of Chicago Press, pp. 124-155.

Brown, F., J. Harris, R. Leakey & A. Walker 1985. Early *Homo erectus* skeleton from West Lake Turkana, Kenya. *Nature.* 316: 788-92.

Butzer, K. 1969. Geological interpretation of two Pleistocene hominid sites in the Lower Omo Basin. *Nature* 222: 1133-1135.

Butzer, K., W. Beaumont & J. Vogel 1978. Lithostratigraphy of Border Cave, Kwazulu, South Africa: A Middle Stone Age sequence beginning c. 195 000 BP. *J. Archaeol. Sci.* 5: 317-341.

Cann, R., M. Stoneking & A. Wilson 1987. Mitochondrial DNA and human evolution. *Nature* 325: 39-44.

Chamberlain, A. & B. Wood 1987. Early hominid phylogeny. *J. Hum. Evol.* 16: 119-133.

Day, M.H. 1971. Post cranial remains of *Homo erectus* from Bed IV, Olduvai Gorge, Tanzania. *Nature* 232: 383-387.

Day, M.H. & C. Stringer 1982. A reconsideration of the Omo Kibish remains and the *erectus-sapiens* transition. *Proc. 1er Cong. International de Palaeontologie Humaine, Nice.* Pretirage: 814-846.

Dubois, E. 1893. *Verslag van het Nijnuezan, No. 5.* Batavia.

Garn, S. 1972. *The Earlier Gain and Later Loss of Cortical Bone.* Springfield: C. Thomas.

Gould, S.J. 1977. *Ontogeny and Phylogeny.* Cambridge: Belknap.

Hennig, W. 1966. *Phylogenetic Systematics.* Urbana: University of Illinois Press.

Hublin, J. 1986. Some comments on the diagnostic features of *Homo erectus. Anthropos (Brno)* 23: 175-87.

Isaac, G. 1981. Archaeological tests of alternative models of early hominine behavior: Excavations and experiments. *Phil. Trans. R. Soc. Lond. B.* 292: 177-188.

Johanson, D. & T. White 1978. A systematic assessment of early African hominids. *Science* 202: 321-330.

Kennedy, G.E. 1973. The Anatomy of the Middle and Lower Pleistocene Hominid Femora. Ph.D. Dissertation. University of London.

Kennedy, G.E. 1983a. A morphometric and taxonomic assessment of a hominine femur from the Lower Member Koobi Fora, Lake Turkana. *Am. J. Phys. Anthropol.* 61(4): 429-36.

Kennedy, G.E. 1983b. Some aspects of the evolution of fossil and recent hominine femora. *J. Hum. Evol.* 12: 587-616.

Kennedy, G.E. 1984. The emergence of *Homo sapiens:* The post cranial evidence. *Man* 19: 94-110.

Kennedy, G.E. 1985. Bone thickness in *Homo erectus. J. Hum. Evol.* 14: 699-708.

Kluge, A. 1983. Cladistics and the Classification of the Great Apes. In R. Ciochon & R. Corruccini (eds), *New Interpretations of Ape and Human Ancestry.* Plenum: New York, pp. 151-180.

Leakey, R. 1973. Evidence for an advanced Plio-Pleistocene hominid from East Rudolf, Kenya. *Nature* 242: 447-450.

Lovejoy, O. 1982. Diaphyseal biomechanics of the locomotor skeleton of Tautavel Man with comments on the evolution of skeletal changes in Late Pleistocene man. *Proc. 1er Cong. International de Paléontologie Humaine, Nice*. Pretirage, pp. 447-470.

Mednick, L. 1955. Evolution of the human ilium. *Am. J. Phys. Anthropol.* 13: 203-215.

Pearson, K. & J. Bell 1919. A Study on the Long Bones of the English Skeleton. *Drapers Co Mem., Biometrica Series XI, Pt. 1, The Femur.* Cambridge: Cambridge Univ. Press.

Pycraft, W., G. Elliott-Smith, M. Yearsley, J. Carter, R. Smith, A. Hopwood, D. Bate & W. Swindon 1928. *Rhodesia man and associated remains*. London: Clowes.

Rak, Y. & B. Arensburg 1987. Kebara 2 Neanderthal pelvis: First look at a complete inlet. *Am. J. Phys. Anthropol.* 73: 227-231.

Rightmire, P. 1979. Implications of Border Cave skeletal remains for later Pleistocene human evolution. *Curr. Anthropol.* 20: 23-35.

Robinson, J. 1972. *Early Hominid Posture and Locomotion.* Chicago: University of Chicago Press.

Rose, M. 1984. A hominine hip bone, KNM ER 3228, from East Lake Turkana, Kenya. *Am. J. Phys. Anthropol.* 63: 371-378.

Schultz, A. 1937. Proportions, variability and asymmetries of the long bones of the limbs and the clavicles in man and apes. *Hum. Biol.* 9: 281-328.

Sigmon, B. 1982. Comparative morphology of the locomotor skeleton of *Homo erectus* and the other fossil hominids with special reference to the Tautavel innominate and femora. *Proc. 1er Cong. International de Paléontologie Humaine, Nice*. Pretirage, pp. 422-446.

Singer, R. & J. Wymer 1982. *Human Skeletal Remains in the Middle Stone Age at Klasies River Mouth*. Chicago: University of Chicago Press.

Stringer, C. 1986. An archaic character in the Broken Hill innominate E. 719. *Am. J. Phys. Anthropol.* 71: 115-120.

Stringer, C. 1987. A numerical cladistic analysis for the genus *Homo*. *J. Hum. Evol.* 16: 135-146.

Tobias, P. 1985. Single characters and the total morphological pattern redefined: The sorting effected by the selection of morphological features of the early hominids. In E. Delson (ed.), *Ancestors. The Hard Evidence*. New York: R. Liss, pp. 94-101.

Trinkaus, E. 1976. The morphology of European and Southwest Asian Neanderthal pubic bones. *Am. J. Phys. Anthrop.* 44: 95-104.

Trinkaus, E. 1981. Neanderthal limb proportions and cold adaptation. In. C.B. Stringer (ed.), Aspects of Human Evolution. *Symp. for the Study of Human Biology,* Vol. 21, pp. 187-224.

Trinkaus, E. 1984. Does KNM ER 1481 establish *Homo erectus* at 2.0 my BP? *Am. J. Phys. Anthropol.* 64: 137-138.

Villiers, H. de 1973. Human skeletal remains from Border Cave, Ingwavuma District, Kwazulu, South Africa. *Transv. Mus.* 28: 229-256.

Walker, A. 1981. Diet and teeth. *Phil. Trans. R. Soc. Lond. B.* 292: 57-64.

Weidenreich, F. 1941. The extremity bones of *Sinanthropus pekinensis*. *Palaeont. Sin. N.S.D.* 5: 1-150.

Weidenreich, F. 1943. The skull of *Sinanthropus pekinensis:* a comparative study on a primitive hominid skull. *Palaeont. Sin. N.S.D.* 10: 1-484.

Paleolithic archaeology, ancient behavior, and the transition to modern *Homo*

Francis B. Harrold
Department of Sociology, Anthropology and Social Work, University of Texas at Arlington, Arlington, USA

The pattern of the transition (to modern Homo) is as equivocal as its timing and at present it is not possible to decide among several alternatives. The very abundant archaeological record is of little help in resolving these problems. Pilbeam (1986:336)

1 INTRODUCTION

The origin of modern *Homo sapiens* has in recent years become one of the most disputed topics in paleoanthropology. At issue are very different interpretations of recent human evolution. Some physical anthropologists favor a monocentric model in which anatomically modern humans (or AMH) evolved originally in a restricted area, probably Africa, and then spread to replace archaic populations elsewhere (e.g. Stringer & Andrews 1988). Other propose a multiregional model, wherein modern form evolved essentially independently in different regions (e.g. Wolpoff 1989). Between these extremes still others advance models with important roles for gene flow. Bräuer's (1984a, 1989) 'hybridization and replacement' model proposes an African origin for AMH, and subsequent population replacement combined with modern-archaic gene flow in complex, regionally variable patterns. Smith (1985) and Trinkaus (1986) suggest different processes of transition in different parts of the world; regional continuity, augmented by gene flow, is envisioned in some areas.

Proponents of each model attempt to show that the patterning in various lines of evidence – fossil, archaeological, and molecular[1] – supports their proposal better than others. As an archaeologist, I will not attempt here to evaluate these competing models of what might be called the 'what, where and when' of recent human evolution. I will instead discuss our ideas about why – about the selective pressures thought to have shaped this evolution (whenever and wherever it occurred), and about their archaeological implications.

2 MORPHOLOGY, BEHAVIOR AND ARCHAEOLOGY

Despite their disputes, researchers are in broad agreement that the most striking aspect of the skeletal transition from archaic to modern *Homo sapiens*[2] was the reduction in robusticity. They also seem in accord that the differences between archaic and modern humans are, at least in large part, genetically based rather than developmental (Trinkaus & Smith 1985:326).

Table 1. Inferred links between morphological changes and archaeological evidence in the emergence of modern *Homo sapiens* (based primarily on Trinkaus & Smith 1985, Smith 1985, Trinkaus 1986, 1987).

Morphological changes	Behavioral implications	Suggested cultural developments	Proposed archaeological correlates
1. Reduced anterior dentition, facial prognathism/massiveness, and nuchal and related structures	Less use of anterior teeth as tools Less forceful and repetitive chewing	a) More efficient, task-specific tools (especially tools with better leverage) b) Improved food processing facilities and tools	a) Greater typological and technological diversity (especially stone blades, bone tools, hafted implements) b) More efficient hearths (cf. Brace 1979)
2. Reduced upper body massiveness (primarily muscular)	Less strength habitually used in manipulative tasks	a) As in 1a above b) Tools and techniques for 'stand-off' hunting: e.g. projectiles, snares	a) As in 1a above b) Projectile points, atlatls; occurrence in archaeofaunas of species usually snared or trapped
3. Changed thumb phalangeal proportions and less hand robusticity	More use of precision grip, less use of power grip	a) As in 1a above (especially tools requiring fine manipulative ability)	a) As in 1a above (especially bladelets and other small tools)
4. Reduced lower limb robusticity	Lower levels of sustained locomotor activity	More systematic, less opportunistic foraging, based on extensive environmental knowledge, facilitated by (information storage and transmission systems) more complex social organization	a) Evidence from faunal and spatial analysis (inter-site and intra-site) of logistically-organized procurement systems b) Evidence of higher populations c) Notational systems, art; complex burials; artifact style zones
5. Longer legs and longer distal segments	Cultural substitutes for anatomical adaptations to cold (removing constraints on selection for efficient long-distance walking)	Better heat conservation through improved clothing, hearths, shelters	Bone pins; more efficient and complex hearths; more numerous and thermally efficient shelters
6. Higher, rounder cranium (reduced platycephaly)	(If due to accelerated postnatal brain growth [Trinkaus & LeMay 1982]) longer period of learning; greater amount of acquired cultural information (especially if combined with greater longevity [Trinkaus & Thompson 1987])	Information storage and transmission systems; more complex social organization	Notational systems, art; complex burials; artifact style zones

Whatever their evolutionary relationship with later peoples, Neanderthals, for instance, were not simply genetically modern folk who led very hard lives, but populations differing from AMH in the frequencies of a number of genes. Furthermore, the morphological differences between archaic and modern people are interpreted as due mainly to natural selection, as opposed to random genetic drift (e.g. Smith 1985, Trinkaus & Smith 1985, Trinkaus 1986).

When we ask why AMH emerged, there is surprising agreement among those who express an opinion. The selective pressures which remodelled robust archaic form are seen as rooted in cultural changes which resulted in habitually different patterns of human behavior. There seems to be a consensus which goes as follows:

Postcranial robusticity decreased during the emergence of AMH because of decreased selection for strength and massiveness; energetically-expensive heavy morphology was selected against when it was no longer needed. Similarly, large anterior dentition and associated structures began to shrink when no longer selected for. But why did the ancient characteristics of hominid robusticity become superfluous? Because technological and other cultural changes allowed adaptive problems to be solved in ways requiring less direct and repetitive application of physical strength.

Such an explanatory approach is both plausible and in accord with general evolutionary theory (e.g. Lewontin 1974), according to which behavioral adaptations typically precede, and help shape selection for, morphological change.

This approach has been developed over time by a number of authors, especially Brace (1962, 1979), S. Binford (1968), and Brose & Wolpoff (1971, Wolpoff 1980). Recent formulations are fairly specific about the relationship between various morphological changes across the transition, and the behavioral shifts which they imply; furthermore, they specify at least some of the archaeological evidence which the cultural innovations would be expected to produce.

Table 1 recapitulates these proposed links; it is drawn mainly from the work of Trinkaus and Smith, who have been the most explicit about the causal linkages it presents. It lists six morphological changes across the transition; then, for each, its broad behavioral implications; then the cultural innovations suggested to account for the habitual behavior change; and finally the archaeological evidence seen as diagnostic of the cultural innovations. For instance, the first morphological change (reduction of anterior dentition and functionally associated structures), is believed to have been selected for when the characteristic use of the front teeth as tools decreased (and as chewing became less forceful and repetitive). This in turn is attributed to the development of efficient and specialized tools to relieve the dentition of much of its load (as well as better cooking and other means of preparation to make food more easily chewed), thus relaxing selection for the heretofore necessary robust anatomy. Finally, the archaeological indicators of such cultural innovations include more diverse and specialized toolkits, including implements with better leverage characteristics (stone blade tools, bone tools, generally long and less brittle than flint, and especially hafted implements), specialized hearths, and so on.

Several points should be emphasized about Table 1. First, it does not include every morphology-behavior-archaeology (or MBA) link ever proposed, but only those commonly advanced today. One issue conspicuous by its absence concerns the behavioral implications of Neanderthal pubis length. Trinkaus (1986, 1983) has proposed that such pubic form implied a large bony birth canal, which in turn suggested significant Neanderthal-AMH differences in gestation length, the cultural wherewithal to care for very altricial neonates, and the developmental effects of birth at different gestation lengths. However, the recent description of the Kebara pelvis (Rak & Arensburg 1987) casts doubt on earlier inferences about the birth canal. Current explanations of Neanderthal pelvis form relate it simply to functional demands on the pelves of short, massive mothers bearing massive infants (Rosenberg 1988), or to the condition preceding locomotor changes favoring more efficient walking (Trinkaus 1986, and comment in Rosenberg 1988). In any event, there is no longer a causal MBA chain proposed especially for pelvic change over the transition.

Also missing from the table is the suggestion that the transition to modern *Homo* involved reorganized or more complex brains – i.e. that modern people had inherently greater cultural abilities than archaics. Some workers regard this suggestion with distaste as vitalistic or worse, while others find it attractive in explaining the successful spread of modern morphology. Especially intriguing is the suggestion that the ability to use fully symbolic language separates modern from archaic *Homo* (Lieberman 1989, Pilbeam 1986). However, it is widely agreed that we cannot reliably infer linguistic ability from fossil skulls and brain endocasts (Holloway 1985). Furthermore, no one knows how we could archaeologically distinguish between behavioral change due to enhanced cultural capacities, and that due to 'ordinary' innovation or diffusion. Thus the suggestion of archaic-modern differences in cognitive capacities remains untestably moot (Trinkaus 1986:209).

It is also important that the links in Table 1 (especially those involving postcranial traits) are based heavily on studies of Neanderthals, not archaic populations in general. Given the paucity of non-Neanderthal remains, this is unavoidable, but should be kept in mind when 'Out of Africa' models are advanced. Most saliently, for instance, the propositions about limb changes in row 5 of the table are explicitly related to cold adaptations, and thus probably irrelevant to the origin of AMH in tropical regions[3].

A third general point is that Table 1 does not describe a complete theory of the relationship between morphology and archaeology, for none has been developed. At best, it sketches a generalized model of how changes in form are related to behavior and, in turn, to the archaeological record. The current main concerns of workers in this field are the 'where and when' questions about basic patterning in the fossil record, and 'how' questions about phylogenetic relations among ancient populations. The 'why' issues like those in the table are typically addressed secondarily, as authors seek to show that their models of the transition are consistent with evolutionary theory and the archaeological record. The relationships posited are plausible, but they are not the result of rigorous theory-building to link morphological and archaeological evidence.

Probably the most important point to make about Table 1 is that its proposed MBA relationships are not associated with any particular model of the archaic-modern transition, but potentially consistent with any of them. However one argues that the transition happened, the 'reasons why' – the links proposed in the table – seem broadly the same. In a multiregional model, the events in Table 1 presumably took place in many regions, if perhaps at different times and rates. The initial behavioral 'kick' might have been from diffusion of cultural innovations, but the evolutionary process itself was essentially *in situ*. For monocentrists, Table 1's sequence happened *in situ* only once. However, the subsequent spread of AMH would have been due to the competitive advantages their cultural behavior afforded them; people with modern morphology were eventually found everywhere because the cultural behavior favoring that morphology was competitively superior to old patterns of 'paleocultural' behavior, to use Jelinek's (1977) term. In a hybridization-and-replacement model, the gene flow resulting in some 'archaic' alleles in post-transition populations would not change the fact of the overall adaptive superiority of the incoming modern morphology. The same principle would hold true for those who suggest different transitional processes in different areas, and advocate gene flow as an important contributor to at least some regional transitions. In the case of gene flow, incoming alleles would be selected for, not because of innate 'progressiveness,' but because they enhanced fitness in the local environment (Smith 1985:215). And why should these alleles enhance fitness? Because cultural innovations, probably diffused along with the new genes, had relaxed selection for robusticity. Thus the MBA links in Table 1 can be seen to underlie all of the very different accounts of the origins of modern humanity.

The generalized scheme treated here leaves many variables unspecified, such as the rate of the various skeletal changes or their degree of synchronicity. But it does entail one general expectation. Since changes in behavior are seen as catalysts for changing patterns of selection, physical change could not have preceded cultural change. Even if there were complex feedback between skeletal and cultural changes (Trinkaus & Smith 1985), it is logically necessary that some behavioral change occurred as the initial 'kick-off' of the feedback. If culture change preceded physical evolution by only a short timespan (a few centuries?), we might archaeologically perceive the two as contemporary, given the coarse resolution of our dating methods. On the other hand, it is possible that a considerable time lag could elapse between behavioral and morphological change, if a given population lacked the alleles allowing gracilization, and had to 'await' their arrival via gene flow or mutation[4]. In any event, we would expect that morphological could not precede cultural change.

In further examining this model, we will be addressing two questions: (1) Are its expectations met in the archaeological and fossil record? (2) More fundamentally, are these expectations justified?

3 THE MBA MODEL AND THE EVIDENCE

This section will constitute not an exhaustive review, but a brief characterization of the fit between the MBA model and the available evidence concerning the transition to modern *Homo*.

3.1 *Western Europe*

Much remains to be learned, but where sufficient data are available the correlation between morphological change and expected archaeological change in this area is fairly good. Sample sizes are very small, though, for some classes of evidence. The transition to AMH here coincided broadly with the Middle-Upper Paleolithic transition and its considerable change in artifact typology and technology. For many, the transition is most convincingly interpreted as involving population replacement shortly after 35 000 BP (e.g. Stringer et al. 1984, Harrold 1989), though gene flow and Neanderthal genetic contribution to later populations are not ruled out. Others, however, deny abrupt archaeological and/or morphological change (e.g. Clark & Lindly 1989, Simek & Price 1990).

The proposed fit, though, is not apparent in the faunal evidence. Recent studies, especially in Cantabrian Spain, find no important subsistence change until 20 000 BP or so, long after the biological transition (e.g. Clark 1987, Straus 1990, Chase 1989). Indeed, it has been suggested that even this late archaeofaunal shift merely reflects climatic variation rather than new subsistence strategies (Simek & Snyder 1988). It could be replied that our sample of well-analyzed western European archaeofaunas (particularly Mousterian ones) is still small, and that a common methodology and set of analytical conventions has not yet been adopted by zooarchaeologists. However, it is certainly true that no change in human subsistence can yet be demonstrated across the transition in western Europe.

A final point worth making is that while our attention has been understandably focused on the Middle-Upper Paleolithic transition, it is conceivable that earlier cultural changes – during the Mousterian or at its boundary with the Lower Paleolithic – could have led to the selective pressures favoring gracilization. Brace (1979), for example, has proposed that the structure of Mousterian hearths implies cooking practices which reduced selection for massive masticatory apparatus. Archaeologists have tended to downplay the significance of the Lower-Middle Paleolithic distinction (e.g. Jelinek 1977), but significant transitions in human behavior have also been suggested for the Middle Paleolithic/Middle Stone Age (e.g. Clark 1989). However, we currently lack systematic documentation of such important behavioral change.

3.2 *Central Europe*

The data base is not yet as good in this region; dating is sometimes problematic, fossil remains typically fragmentary, and archaeological samples usually rather small. But the evidence we have is broadly consistent with the MBA model. A good case can be made for a more gradual morphological transition here than in the west (Smith 1984), one that had ended by 36 000-38 000 BP. Some, however, argue for population replacement (e.g. Allsworth-Jones 1986). The earliest regional Upper Paleolithic is easily 40 000 years old and apparently exceeds 43 000 years at Bacho Kiro in the southeast of the region (Allsworth-Jones 1986). The dating implies an east-west cline in physical and cultural change perhaps within this region, and certainly between it and western Europe, due to either population movement or diffusion and gene flow. Once again, the clearest fit between the MBA model and the archaeological record is in lithic and bone typology and technology. We know even less than in western Europe about settlement patterns and subsistence, but at least there is nothing overtly in conflict with the MBA model (as is also the case in eastern Europe (Klein 1973, Hoffecker 1988), where human fossils are rare).

3.3 *Southwestern Asia*

The situation in this region is more confusing, particularly in the wake of the dispute over the dating of the AMH remains from Qafzeh. It has been argued that, if the Qafzeh remains are roughly contemporary with the very similar ones from Skhūl, then there was a relatively straightforward transition (whether by evolution or replacement) from Neanderthals to modern humans ending by around 45 000 BP (e.g. Smith 1985, Marks 1990). Recently, though, the case for a greater age for the Qafzeh people has been supported by thermoluminescence dates from that site on the order of 90 000 BP (Valladas et al. 1988). If these dates are correct, then modern people in this region not only long coexisted with Neanderthals (with little indication of gene flow between the populations), but may have preceded them by as much as 30 000 years.

The regional Middle-Upper Paleolithic transition is, on current evidence, a gradual one around 45 000 BP evidenced by shifts in lithic technology and typology (Marks 1988). Evidence for important transformations in human lifeways at this time is underwhelming. Sites are actually fewer than in the Middle Paleolithic (probably due to erosion); settlement pattern changes seem to reflect climatic change. No documentation of more sophisticated procurement systems has yet been made, and the archaeological record is poor in burials, art or notational items, and even bone tools.

Most striking of all is the 'poor fit' between the regional biological and archaeological transitions. The Qafzeh hominids (and those from Skhūl as well) are associated with unremarkable Levantine Mousterian assemblages. If the new TL dates are correct, modern *Homo* had been in this region for over 40 000 years before the apparently modest cultural changes of the Middle-Upper Paleolithic transition. This situation would directly contradict expectations from the MBA model, according to which cultural change precedes morphological change.

Other possibilities do still exist. If the TL dates (which need further confirmation) are wrong and the Qafzeh remains are only 45 000 years old, then the case for roughly synchronous morphological and archaeological change is still alive. Signs of cultural innovation before the transition to the local Upper Paleolithic have also been suggested (e.g. Trinkaus 1986:200). Jelinek (1982) placed both the Skhūl and Qafzeh lithic assemblages at the end of a progressive Levantine trend (for him, a temporal trend) toward greater width/thickness ratios in lithic flakes. One could speculate that this high ratio represents greater technological sophistication, and that its association with modern *Homo* indicates a behavior-morphology link. One could even point to the occasional occurrence in the region of blade-rich industries or facies in Middle Paleolithic times, such as the Amudian (Jelinek 1990) as indicators of cultural innovation. Their association with AMH could be suggested, though it has not been found. All of this is extremely speculative, however; the behavioral significance of the Amudian or of flake width-thickness ratios is unclear. It is unknown whether they represent behavioral advances which could affect selective pressures on anatomy.

3.4 *Africa*

In the fragmentary fossil record of north Africa, archaic or transitional fossils like those from Jebel Irhoud seem reliably older than AMH ones (Bräuer 1984b), and are associated with the Levalloiso-Mousterian. Others, such as Dar-es Soltane 2, are variously characterized as transitional (Smith 1985) or modern (Clark 1983), and are associated with the Aterian. This is an industry of mainly Middle Paleolithic composition, dating back at least 45 000 years, with certain 'advanced' traits like tanged (and presumably hafted) tools (Ferring 1975). More data are badly needed, but it might be maintained that typological/technological change more or less coincides with physical change, consistent with the MBA model (Trinkaus 1986:200).

Dating and the overall amount of information in eastern and southern Africa are problematical, given this area's importance in recent 'Out of Africa' models of AMH origins. But at least some anatomically modern fossils from the area seem to be at least 90 000 years old (Omo, Border Cave, Klasies River Mouth), and perhaps a good deal older (Bräuer 1984a, b, Stringer & Andrews 1988).

It is also apparent that the most striking changes in the archaeological record in this part of the world are associated with the Middle Stone Age-Later Stone Age (MSA-LSA) transition some 30 000-40 000 years ago – a transition reminiscent of and contemporaneous with the Middle-Upper Paleolithic transition elsewhere (Klein 1983, 1987, Volman 1984). However, signs of cultural innovation are apparent in the MSA, which extends some 200 000 years into the past (Clark 1989). There are advances over the Acheulean in lithic technology and typology, most notably in the Howieson's Poort MSA variant, with its well-made backed crescents and blades. There is also some indication of utilization of new food sources (e.g. seals, mollusks). Interpretations of MSA behavior vary widely: At the extremes, Binford (1984) doubts that MSA people were capable of hunting sizeable animals, while Deacon (1989) sees their behavioral capacities as essentially modern.

While the most impressive cultural change in the sub-Saharan African record follows the transition to modern *Homo* by many millennia, it is at least possible to maintain that significant cultural advances preceded or coincided with the transition to AMH. However, there is still little consensus concerning the timing and nature of either morphological or cultural change in this vital region.

For some areas not discussed here (e.g. eastern Asia), we simply have too little information about the fossil and archaeological records to generalize at all. In Africa and the Near East, there is still room to claim congruence between expectation and the record, but also some evidence against the MBA model. I think the case for the posited connections is best in Europe, but far from firmly established. In sum, we face an ambiguous situation. A shortage of data, and diametrically opposed interpretations of existing information, leave the model neither confirmed nor discredited.

4 IS THE MODEL VALID?

That there may be something wrong with the links presented in Table 1 should give us pause. For human paleontologists, the possibility that a significant evolutionary event like the appearance of AMH was unaccompanied by notable behavioral change raises unsettling questions. Why did skeletal remodelling occur, if not through selection related to new behavior patterns? Could it have been due only to genetic drift? For archaeologists, a lack of congruence between evolutionary change and our measures of behavioral change (typological, technological, faunal, etc.) would cast doubt on the validity of these measures. Did the transition really occur without behavioral change? Or are we simply unable to detect and characterize the change with our methods? As an archaeologist, I will leave the first problem to more qualified colleagues, and offer some thoughts about the second. The specifically archaeological implictions of Table 1 are lodged primarily in the relationship between the third and fourth columns – between ancient behavior and archaeological remains.

The epistemological and methodological issues raised in these columns are not new in Paleolithic archaeology. For two decades there have been efforts to replace comfortable old conceptions of the past with more scientifically adequate inferences. We have come to realize that much of what we thought we knew was misleading, wrong, or simply not yet knowable (e.g. Binford 1981, 1984, Conkey 1987, Simek & Rigaud 1987, Straus 1987). New methods in artifact and faunal analysis, emphasis on middle-range theory linking data patterning with behavior, recognition that many non-human factors shape the archaeological record, and respect for the need to justify our reconstructions on grounds more secure than plausibility have come to be widely accepted as prerequisites to real advances in Paleolithic archaeology.

So far, though, new analytical approaches have not 'set right' our understanding of behavioral change at the transition to AMH. They have sometimes led to conflicting findings. For instance, Binford has argued (1981, 1982, 1984) primarily on the basis of faunal analyses that pre-Upper Paleolithic (and pre-LSA) subsistence and cultural behavior were far simpler than in later times. Others have also inferred considerable behavioral differences between archaic and modern humans on the basis of faunal, artifactual, or other evidence (e.g. Klein 1973, 1983, Dibble 1983, Chase and Dibble 1987, Harrold 1989, White 1989, Allsworth-Jones 1986). Even among these broadly convergent studies there is disagreement about how and why later humans behaved differently – e.g. whether population replacement of archaics by AMH is the reason for these differences, or a vitalistic non-explanation.

At the same time, other studies have concluded that cultural continuity predominated during the transition to modern *Homo*, at least regionally (e.g. Clark 1987, Clark & Lindly 1989, Chase 1989, Straus 1987, 1990, Marks 1990, Simek & Price 1990).

What are we to make of these contradictions? A brief reply is that they demonstrate the still-immature state of Paleolithic archaeology. As noted in the Pilbeam quotation at this essay's beginning, the archaeological record is indeed rich, but it has not told us much in unequivocal fashion. More specifically, we need progress on two fronts.

Binford's often-repeated contention (e.g. Binford 1981) that archaeology produces too many post-hoc accommodative arguments and not enough reliable reconstructions based on solidly constructed theory is well taken in the context of this discussion. We need better-established inferential models and procedures. Two examples should suffice to make this point. A whole complex of zooarchaeological methods has come into use over the past decade or so. They variously involve calculation and manipulation of numbers of individual bone specimens (NISP's), minimum numbers of individuals (MNI's), mortality profiles, and indices of faunal diversity and body part utility. When applied to the same data, different approaches may lead to varying conclusions about past subsistence (e.g. Klein 1983, Binford 1984). To what extent are these methodologies conflicting or complementary? Can they be sorted out and integrated for reliable knowledge of past subsistence?

In lithic studies, the significance of the morphological variation monitored in traditional typologies is unclear. That culture-stratigraphic units constructed on this basis (e.g. the Aurignacian) do not correspond to ancient 'cultures' is widely realized, but how do we understand them? Could the shared patterns of artifact manufacture and modification which they imply be related to widespread mating networks on language groupings (cf. Soffer 1985:444-452, Wiessner 1983)? Just what is their relevance for understanding past behavior?

Secondly, the data base needs to be enlarged. While data take on their full significance only in a particular theoretical context, theory (as distinct from sheer speculation) cannot exist in a data vacuum. One reason for the current shortcomings in archaeological inference relative to Table 1 is the spotty Paleolithic archaeological record, which, Rorschach-like, is often sufficiently ambiguous to accomodate various scenarios devised to fit it. I think any archaeologist working on detailed regional problems will agree that the amount of Paleolithic material excavated with modern methods and analyzed by all relevant methodologies is disappointingly meager. Archaeologists' analyses are as plagued as those of humans paleontologists by problems of small and unrepresentative data samples.

Pending advances on both fronts, we will be unable to evaluate fully the soundness of the relationships between behavior and archaeology posited in Table 1.

Finally, a solidly based MBA model will need the integrated efforts of archaeologists and physical anthropologists. Each group has tended in the past to keep to its side of a subdisciplinary divide, too often using uncritically the work of the other (Binford 1987). Without concerted effort the MBA model, which underlies all principle scenarios for the origin of modern *Homo*, will remain open to question.

NOTES

1 I will not discuss molecular evidence, for however valuable it may prove in establishing phylogenetic relationships, it is uninformative about our topic here, the selective factors underlying evolutionary change.

2 Following Trinkaus & Smith (1985), I use the term 'transition' here to refer generally to archaic-to-modern *Homo sapiens* passage, whether global or regional, and whether by *in situ* evolution, population replacement, gene flow, or some combination thereof.

3 There are difficulties regarding the whole issue of lower limbs and walking during the transition to modern *Homo*. If lower limb robusticity was reduced because of less walking (Table 1, point 4) due in turn to more efficient foraging systems, then why would limbs become longer (Table 1, point 5), and pelvis morphology change (e.g. Trinkaus comment in Rosenberg 1988), so as to allow more efficient walking over long distances?

4 I am indebted to Fred Smith for this point.

REFERENCES

Allsworth-Jones, P. 1986. *The Szeletian and the Transition from Middle to Upper Palaeolithic in Central Europe*. Oxford: Oxford University Press.

Binford, L.R. 1981. *Bones: Ancient Men and Modern Myths*. New York: Academic Press.

Binford, L.R. 1982. Comment on White's 'Rethinking the Middle/Upper Palaeolithic Transition.' *Curr. Anthropol.* 23(2): 177-181.

Binford, L.R. 1984. *Faunal Remains from Klasies River Mouth*. Orlando: Academic Press.

Binford, L.R. 1987. The hunting hypothesis, archaeological methods, and the past. *Yearbook of Physical Anthropology* 30: 1-9.

Binford, S.R. 1968. Early Upper Pleistocene adaptations in the Levant. *Am. Anthropol.* 70: 707-717.

Brace, C.L. 1962. Refocusing on the Neanderthal problem. *Am. Anthropol.* 64: 729-741.

Brace, C.L. 1979. Krapina, 'Classic' Neanderthals, and the evolution of the European face. *J. Hum. Evol.* 8: 527-550.

Bräuer, G. 1984a. The 'Afro-European *sapiens* hypothesis', and hominid evolution in East Asia during the late Middle and Upper Pleistocene. *Cour. Forsch-Inst. Senckenberg* 69: 145-165.

Bräuer, G. 1984b. A craniological approach to the origin of modern *Homo sapiens* in Africa and its implications for the appearance of modern Europeans. In F.H. Smith & F. Spencer (eds), *The Origins of Modern Humans: A World Survey of the Fossil Evidence*. New York: Alan R. Liss, pp. 327-410.

Bräuer, G. 1989. The evolution of modern humans: A comparison of the African and non-African evidence. In P. Mellars & C. Stringer (eds), *The Human Revolution: Biological and Behavioural Perspectives on the Origins of Modern Humans*. Edinburgh: University of Edinburgh Press, pp. 123-154.

Brose, D.S. & M.H. Wolpoff 1971. Early Upper Palaeolithic man and late Palaeolithic tools. *Am. Anthropol.* 73: 1156-1194.

Chase, P.G. 1989. How different was Middle Palaeolithic subsistence? In P. Mellars & C. Stringer (eds), *The Human Revolution: Biological and Behavioural Perspectives on the Origins of Modern Humans*. Edinburgh: University of Edinburgh Press, pp. 321-337.

Chase, P.G. & H.L. Dibble 1987. Middle Palaeolithic symbolism: A review of current evidence and interpretations. *J. Anthropol. Archaeol.* 6: 263-296.

Clark, G. 1987. From the Mousterian to the Metal Ages: Long-term change in the human diet of Northern Spain. In O. Soffer (ed.), *The Pleistocene Old World: Regional Perspectives*. New York: Plenum, pp. 293-316.

Clark, G. & J.M. Lindly 1989. The case for continuity: Observations on the biocultural transition in Europe and Western Asia. In P. Mellars & C. Stringer (eds), *The Human Revolution: Biological and Behavioural Perspectives on the Origins of Modern Humans*. Edinburgh: University of Edinburgh Press, pp. 626-676.

Clark, J.D. 1983. The significance of culture change in the early Later Pleistocene in Northern and Southern Africa. In E. Trinkaus (ed.), The Mousterian Legacy. Oxford: *Br. Archaeol. Rep.* S164: 1-12.

Clark, J.D. 1989. The origins and spread of modern humans: A broad perspective on the African evidence. In P. Mellars & C. Stringer (eds), *The Human Revolution: Biological and Behavioural Perspectives on the Origins of Modern Humans*. Edinburgh: University of Edinburgh Press, pp. 565-588.

Conkey, M. 1987. Interpretive problems in hunter-gatherer regional studies: Some thoughts on the European Upper Palaeolithic. In O. Soffer (ed.), *The Pleistocene Old World: Regional Perspectives*. New York: Plenum, pp. 63-77.

Deacon, H.J. 1989. Late Pleistocene palaeoecology and archaeology in the Southern Cape, South Africa. In P. Mellars & C. Stringer (eds), *The Human Revolution: Biological and Behavioural Perspectives on the Origins of Modern Humans*. Edinburgh: University of Edinburgh Press, pp. 547-564.

Dibble, H.L. 1983. Variability and change in the Middle Palaeolithic of Western Europe and the Near East. In E. Trinkaus (ed.), The Mousterian Legacy. Oxford: *Br. Archaeol. Rep.* S164: 53-72.

Ferring, C.R. 1975. The Aterian in North African prehistory. In F. Wendorf & A.E. Marks (eds), *Problems in Prehistory: North Africa and the Levant.* Dallas: SMU Press, pp. 113-126.

Harrold, F.B. 1989. Mousterian, Chatelperronian, and Early Aurignacian: Continuity or discontinuity? In P. Mellars & C. Stringer (eds), *The Human Revolution: Biological and Behavioural Perspectives on the Origins of Modern Humans.* Edinburgh: University of Edinburgh Press, pp. 677-713.

Hoffecker, J.F. 1988. Early Upper Palaeolithic settlement on the Russian plain. In J.F. Hoffecker & C.A. Wolf (eds), The Early Upper Palaeolithic: Evidence from Europe and the Near East. Oxford: *Br. Archaeol. Rep.* S437. In press.

Holloway, R.L. 1985. The poor brain of *Homo sapiens neanderthalensis:* See what you please. In E. Delson (ed.), *Ancestors: The Hard Evidence.* New York: Alan R. Liss, pp. 319-324.

Jelinek, A.J. 1977. The Lower Palaeolithic: Current evidence and interpretations. *Ann. Rev. Anthropol.* 6: 11-32.

Jelinek, A.J. 1982. The Tabun cave and Palaeolithic man in the Levant. *Science* 216: 1369-1375.

Jelinek, A.J. 1990. The Amudian in the context of the Mugharan tradition at the Tabun cave (Mount Carmel), Israel. In P. Mellars (ed.), *The Emergence of Modern Humans.* Edinburgh: University of Edinburgh Press, pp. 81-90.

Klein, R.G. 1973. *Ice-Age Hunters of the Ukraine.* Chicago: University of Chicago Press.

Klein, R.G. 1983. The Stone Age prehistory of Southern Africa. *Ann. Rev. Anthropol.* 12: 25-48.

Klein, R.G. 1987. Reconstructing how early people exploited animals: Problems and prospects. In M. Nitecki & D. Nitecki (eds), *The Evolution of Human Hunting.* New York: Plenum, pp. 11-45.

Lewontin, R. 1974. *The Genetic Basis of Evolutionary Change.* New York: Columbia University Press.

Lieberman, P. 1989. The origins of some aspects of human language and cognition. In P. Mellars & C. Stringer (eds), *The Human Revolution: Biological and Behavioural Perspectives on the Origins of Modern Humans.* Edinburgh: University of Edinburgh Press, pp. 391-414.

Marks, A.E. 1990. The Middle and Upper Palaeolithic of the Near East and the Nile Valley: The problem of cultural transformation. In P. Mellars (ed.), *The Emergence of Modern Humans.* Edinburgh: University of Edinburgh Press, pp. 56-80.

Pilbeam, D.R. 1986. The origin of *Homo sapiens:* The fossil evidence. In B. Wood, P. Martin & P. Andrews (eds), *Major Topics in Primate and Human Evolution.* Cambridge: Cambridge University Press, pp. 331-38.

Rak, Y. & B. Arensburg 1987. A new Neanderthal pelvis: First look at a complete inlet. *Am. J. Phys. Anthropol.* 46: 29-44.

Rosenberg, K.R. 1988. The functional significance of Neanderthal pubic length. *Curr. Anthropol.* 29(4): 595-617.

Simek, J.F. & H.A. Price 1990. Chronological change in Perigord lithic assemblage diversity. In P. Mellars (ed.), *The Emergence of Modern Humans.* Edinburgh: University of Edinburgh Press, pp. 243-261.

Simek, J.F. & J.-P. Rigaud 1987. 'Arms too short to box with God': Problems and prospects for Palaeolithic prehistory in Dordogne, France. In O. Soffer (ed.), *The Pleistocene Old World: Problems and Prospects.* New York: Plenum, pp. 47-61.

Simek, J.F. & L.M. Snyder 1988. Assemblage diversity in Perigord archaeofaunas. In H.L. Dibble & A. Montet-White (eds), *The Pleistocene Prehistory of Western Eurasia.* Philadelphia: University of Pennsylvania Press, pp. 321-332.

Smith, F.H. 1984. Fossil hominids from the Upper Pleistocene of central Europe and the origin of modern Europeans. In F.H. Smith & F. Spencer (eds), *The Origins of Modern Humans: A World Survey of the Fossil Evidence.* New York: Alan R. Liss, pp. 137-209.

Smith, F.H. 1985. Continuity and change in the origin of modern *Homo sapiens. Z. Morph. Anthropol.* 75: 197-222.

Soffer, O. 1985. *The Upper Palaeolithic of the Central Russian Plain.* Orlando: Academic Press.

Straus, L.G. 1987. Paradigm lost: A personal view of the current state of Upper Palaeolithic research. *Helinium* 27:157-171.

Straus, L.G. 1990. The Early Upper Palaeolithic of Southwest Europe: Cro-Magnon adaptations in the Iberian peripheries, 40 000-20 000 BP. In P. Mellars (ed.), *The Emergence of Modern Humans.* Edinburgh: University of Edinburgh Press, pp. 276-302.

Stringer, C.B. & P. Andrews 1988. Genetic and fossil evidence for the origin of modern humans. *Science* 239: 1263-68.

Stringer, C.B., J.-J. Hublin & B. Vandermeersch 1984. The origin of anatomically modern humans in Western Europe. In F.H. Smith & F. Spencer (eds), *The Origins of Modern Humans: A World Survey of the Fossil Evidence*. New York: Alan R. Liss, pp. 51-136.

Trinkaus, E. 1983. *The Shanidar Neandertals.* New York: Academic Press.

Trinkaus, E. 1986. The Neandertals and modern human origins. *Ann. Rev. Anthropol.* 15: 193-218.

Trinkaus, E. 1987. Bodies, brawn, brains and noses: Human ancestors and human predation. In M. Nitecki & D. Nitecki (eds), *The Evolution of Human Hunting*. New York: Plenum, pp. 147-175.

Trinkaus, E. & M. LeMay 1982. Occipital bunning among later Pleistocene hominids. *Am. J. Phys. Anthropol.* 57: 27-35.

Trinkaus, E. & F.H. Smith 1985. The fate of the Neandertals. In E. Delson (ed.), *Ancestors: The Hard Evidence*. New York: Alan R. Liss, pp. 325-33.

Trinkaus, E. & D.D.Thompson 1987. Femoral diaphyseal histomorphometric age determinations for the Shanidar 3, 4, 5 and 6 Neandertals and Neandertal longevity. *Am. J. Phys. Anthropol.* 72: 123-29.

Valladas, H., J.L. Reyss, J.L. Joron, O. Bar-Yosef & B. Vandermeersch 1988. Thermoluminescence dating of Mousterian 'Proto-Cro-Magnon' remains from Israel and the origins of modern man. *Nature* 331: 614-16.

Volman, T. 1984. Early prehistory of southern Africa. In R.G. Klein (ed.), *Southern African Prehistory and Palaeoenvironments*. Rotterdam: Balkema, pp. 169-220.

White, R. 1989. Production complexity and standardization in early Aurignacian bead and pendant manufacture. In P. Mellars & C. Stringer (eds), *The Human Revolution: Biological and Behavioural Perspectives on the Origins of Modern Humans*. Edinburgh: University of Edinburgh Press, pp. 366-390.

Wiessner, P. 1983. Style and social information in Kalahari San projectile points. *Am. Antiq.* 48: 253-276.

Wolpoff, M.H. 1980. *Palaeoanthropology.* New York: Knopf.

Wolpoff, M.H. 1989. Multiregional evolution: The fossil alternative to Eden. In P. Mellars & C. Stringer (eds), *The Human Revolution: Biological and Behavioural Perspectives on the Origins of Modern Humans*. Edinburgh: University of Edinburgh Press, pp. 62-108.

Neanderthal cognition and the Middle to Upper Paleolithic transition

Jan F. Simek
Department of Anthropology, University of Tennessee, Knoxville, USA

1 INTRODUCTION

In recent years, the idea has been revived that archaic European *Homo sapiens* (the Neandertals) could not have been our biological or cultural ancestors. Primarily because of new data on mitochondrial DNA variation in modern human populations (e.g. Cann et al. 1987) and a few reinterpretations of the human fossil record (Stringer & Andrews 1988), it has become rather fashionable to view the Neandertals as an evolutionary dead end, primitive and untalented when compared with ourselves (e.g. Putnam 1988).

In archaeology, a number of studies have appeared that seek and explain differences between 'their' Middle and 'our' Upper Paleolithic 'Cultures' in vitalistic terms. These studies see distinct, biologically-based capacities for culture between Neandertals and anatomically-modern *Homo sapiens* which implicate migration and replacement as the primary cause of culture change (Harrold 1989, Mellars 1973, 1989a, b). Such a typological approach was first applied to the study of artifact technology and subsistence forms (e.g. Mellars 1973), but correlations between hominids and behavior were difficult to find in these realms (Clark & Lindly 1989, Freeman 1973, Simek & Price 1990, Simek & Snyder 1988, Straus 1977). Indeed, the relation between physical and behavioral adaptations is complex to say the least. In western Asia, modern human fossils have been found with Middle Paleolithic tools (Brose & Wolpoff 1971, Bar-Yosef this volume). Neandertal skeletons have been found with Upper Paleolithic tools in western Europe (Leveque & Vandermeersch 1980, Stringer et al. 1984), and, according to some scholars (e.g. Minugh-Purvis 1988, Wolpoff 1988), what may be an anatomically modern fossil was recovered at Krapina from a Mousterian cultural layer (Simek 1988).

Because of these complications, those that seek biological causes for the Middle-Upper Paleolithic Transition turn more and more to the cognitive realm for support (e.g. Chase & Dibble 1987, Mellars 1989b). In these kinds of studies, simple criteria for recognizing 'human' cognitive behavior are usually defined based on modern human analogues, and comparisons are made between dichotomized Middle and Upper Paleolithic 'Cultures'. Middle Paleolithic artifacts are usually found wanting in terms of criteria for human cognition. Observed differences are related to extant chronologies, (which, in western Europe, see the transition as occurring some 35 000 years ago, roughly coincident with the disappearance of the Neandertals), and change is deemed too rapid to have involved direct evolution or even biological or cultural mixture between modern and archaic *Homo sapiens* groups.

In this brief paper, I cannot deal with all aspects of culture change across the Middle-Upper

Paleolithic boundary. Instead, I will limit myself to a discussion of a few dimensions of the archaeological record often used in assessing the cognitive content of Middle Paleolithic culture. Recently, three such dimensions have been emphasized: Cognition expressed in mundane lithic materials (Chase & Dibble 1987, Dibble 1989, Mellars 1989a, White 1982), symbolic behavior expressed in terms of personal adornment (White 1986, 1989), and spectacular paleolithic 'art works' (Chase & Dibble 1987). In my discussion, I will closely examine empirical data relevant to these criteria. I hope to show that, even in a region where migration and replacement are probable (western Europe), the process of culture change was by no means clear, especially in the realm of cognition[1], and biological causation is unlikely. Before examining data for cognition, evidence for migration by late Pleistocene modern humans into western Europe will be presented.

2 EVIDENCE FOR REPLACEMENT

It is in western Europe that perhaps the best documentation for replacement of archaic by modern humans can be found (Leroi-Gourhan 1984, Leroyer & Leroi-Gourhan 1983, Smith et

Table 1. Cultural stratigraphy of Le Piage (after Champagne & Espitalie 1981).

Level	Cultural attribution
A	Sterile
B	Sterile rocky layer
CE	Solutrean/Early Magdalenian
F	Aurignacian
F1	Chatelperronian
G	Aurignacian
H	Sterile
I	Aurignacian
J	Aurignacian
K	Aurignacian
L	Substratum

Table 2. Cultural stratigraphy of Roc de Combe (after Bordes & Labrot 1967).

Level	Cultural attribution
1	Evolved Perigordian
2	Upper Perigordian V
3	Upper Perigordian V
4	Upper Perigordian IV
5	Evolved Aurignacian
6	Aurignacian II
7	Aurignacian I
8	Chatelperronian
9	Aurignacian
10	Chatelperronian

al. 1989). This evidence hinges on the association of fully modern human fossils with early Upper Paleolithic Aurignacian tools at Cro-Magnon (Lartet & Christy 1868) and of a Neandertal skeleton with an early Upper Paleolithic Chatelperronian assemblage at St. Cesaire (Leveque & Vandermeersch 1980). A few less complete finds, including teeth from the Chatelperronian at Arcy-sur-Cure (Leroi-Gourhan & Leroi-Gourhan 1964) and the infant cranium fragments from the Early Aurignacian at Le Piage (Champagne & Espitalie 1981) may support these associations. Other 'discoveries', like the Combe Capelle skeleton, are unconvincing for a variety of reasons. Reasonable chronological estimates place both the Chatelperronian and Aurignacian traditions (and, thus, both hominid forms) in western Europe at about 35-38 000 years ago (Leroi-Gourhan 1984, Leroyer & Leroi-Gourhan 1983), indicating temporal coincidence of the archaic and modern forms of *Homo sapiens*.

Evidence for Aurignacian and Chatelperronian contemporaneity is more dramatically expressed at several sites in southwestern France.

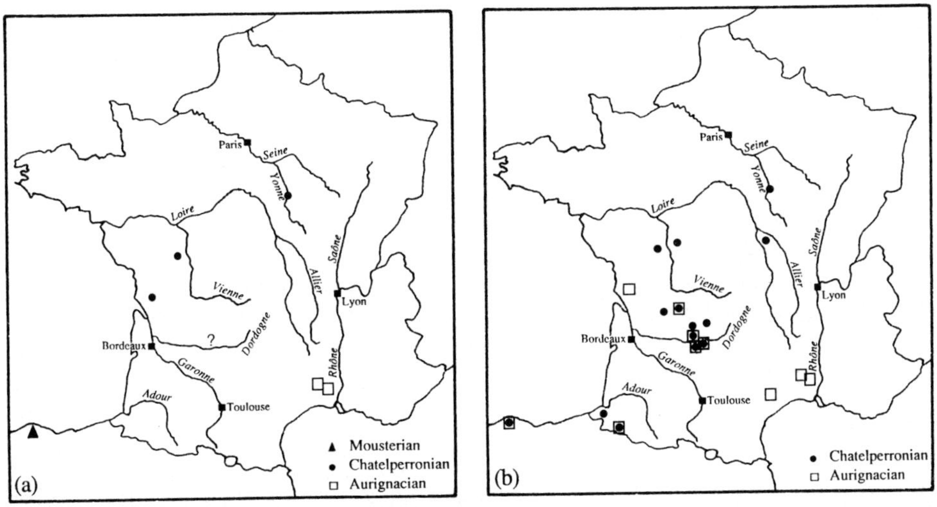

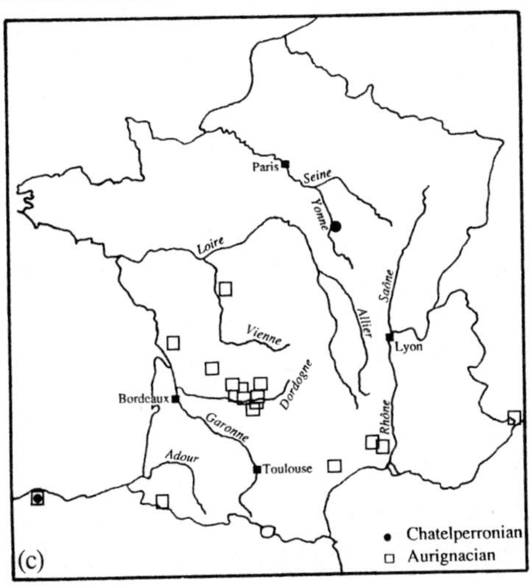

Figure 1. Spatial relations between Mousterian, Chatelperronian and Aurignacian sites in France and northern Spain at three points in time: (a) Cottes/Hengelo, (b) Laville's Recent Wurm Phase 1, (c) Laville's Recent Wurm Phase 2 (after Leroyer & Leroi-Gourhan 1983, Harrold 1987).

Table 1 shows the stratigraphy from Le Piage, where Aurignacian levels appear above and below a well defined Chatelperronian assemblage found in level F.1 (Champagne & Espitalie 1981). At Roc de Combe, Bordes & Labrot (1967) found a similar pattern of Aurignacian/Chatelperronian interstratification (Table 2). These sites indicate that the makers of both assemblage traditions were contemporary, at least in a limited area for a limited time.

Contemporaneity in western Europe is also evident at a more regional scale. Based on pollen profile correlations among sites, Leroyer & Leroi-Gourhan (1983) show that Aurignacian assemblages spread into western Europe from the east. Appearing first along western European coasts, the Aurignacian then moves inland where Chatelperronian assemblages are found at the same time (Fig. 1a). The two traditions co-exist for a time, as seen at Le Piage and Roc de Combe, although the Chatelperronian is circumscribed in the northern part of its previous range (Fig. 1b). After about 3000 years, the Aurignacian has replaced the Chatelperronian over most of western Europe, with the latter assemblages found only in a few sites at the periphery of the area (Fig. 1c).

Present data on assemblage systematics and chronology, therefore, strongly suggest that migration by modern humans and replacement of Neandertals occurred in western Europe beginning around 38 000 years ago. But did this replacement involve the cultural 'overwhelming' of a mentally inferior people by a culturally superior one, as some would argue? Or was the change a more subtle part of a larger evolutionary process?

3 TECHNOLOGICAL AND INDUSTRIAL SOPHISTICATION

One focus of inquiry concerning Neandertal cognitive capabilities is lithic technology. Generally, Middle Paleolithic industries are seen as culturally impoverished when compared to Upper Paleolithic traditions (Chase & Dibble 1987, Mellars 1989a). In Middle Paleolithic lithics, evidence for cognitive complexity has been sought by reference to several criteria. Chase and Dibble argue that one indication for true cognition might be redundant morphological pattern-

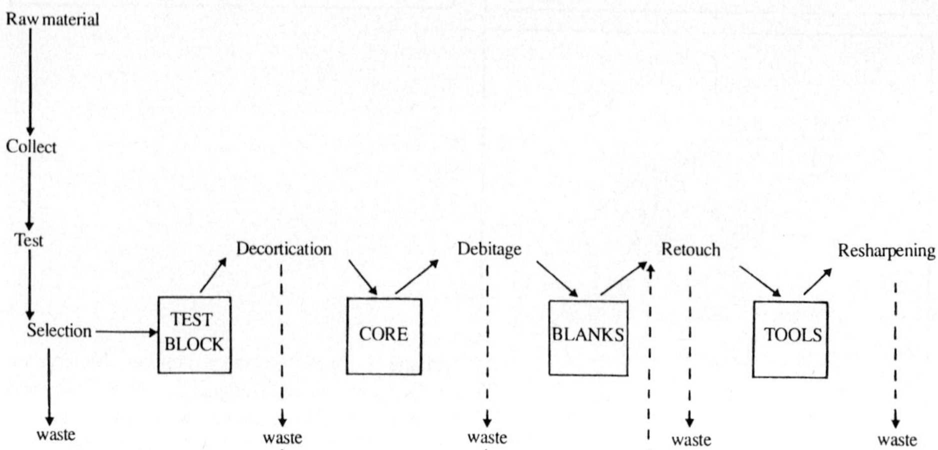

Figure 2. Reduction sequence defined by Chadelle for Upper Paleolithic blade technologies (after Chadelle 1983).

ing in lithic artifact form; they suggest that such patterning reflects 'categories defined within the culture of the makers' (1987:266). Chase and Dibble go on to argue that, although a sequence of both varying and redundant artifact morphologies existed since the very beginnings of the Upper Paleolithic, little formal change is apparent in Middle Paleolithic artifacts, indicating that earlier cultures may not have involved much symbolic behavior (Chase & Dibble 1987:267, Dibble 1989). Thus, one criterion for recognizing human cognition seems to be the presence of standardized artifact types. By implication, lack of such patterning reflects a lack of cultural categorization (Dibble 1989).

Another kind of evidence for Mousterian cognitive inferiority concerns the complexity and sophistication of their lithic assemblages. Middle Paleolithic tool variation has been characterized by White as reflecting, 'mechanical redundancy...that may imply more programmed behavior than that of later people' (1986:17), and that, 'the number of stages between conception of a tool and its completion was much greater in the Upper Paleolithic than in the Mousterian' (1986:35). In contrast to Middle Paleolithic materials, Upper Paleolithic assemblages are seen as 'highly standardized yet technologically complex' (Chase & Dibble 1987:267). These arguments imply two more criteria for identifying cognition in lithic assemblages over and above the presence of standardized tool types. If Upper Paleolithic blades assemblages are more sophisticated (which seems to mean they involve more effort, more preparation), they should have a more complicated reduction sequence than Middle Paleolithic industries. Upper Paleolithic assemblages are also seen as more varied and diverse, i.e. involving more tool types.

I do not believe that any of these criteria are met when data are examined. It is true that Bordes' Middle Paleolithic typology contains only 63 types while the Upper Paleolithic type list has now over 100 classes. But this in no way means that the Middle Paleolithic types are somehow less well defined. Using either typology involves certain analytical decisions, but in both schemes, the types themselves are defined in very similar ways based on aspects of tool technology (Bordes 1950, 1961, Sonneville-Bordes 1960, Sonneville-Bordes & Perrot 1956). In fact, Middle Paleolithic types are just as 'standard' as Upper Paleolithic ones. (Or, perhaps, Upper Paleolithic classes are just as 'sloppy' as those in the Middle Paleolithic list). And while some Upper Paleolithic types may seem to our modern eyes more 'loaded with style' than any Middle Paleolithic ones, this still remains an undemonstrated impression that cannot be used as the basis for serious interpretation.

The notion that types reflect mental categories on the part of the makers is somewhat more disturbing as used in this context. Dangers inherent in this line of argument were pointed out long ago (e.g. Binford 1962, Dunnell 1971). Many factors (like style, function, raw materials, geological processes after deposition) might have affected artifact form. The physical effects of some of these factors can be assessed today using a variety of techniques (microwear analysis, for example), but past human intentions are far more difficult to observe. We might want to believe that certain artifact forms were designed according to mental templates. However, such mentalist explanations are untestable with archaeological data and thus difficult to support by normal scientific analyses. Both Middle and Upper Paleolithic tool types are equally affected by a variety of factors producing formal variation, and mental categories are as difficult to identify for one as for the other. Strictly mentalist criteria are difficult to defend on theoretical grounds, and are probably useless anyway in assessing the cognitive content of lithic artifacts (e.g. Sackett 1982).

The notion that Upper Paleolithic technologies were more sophisticated than Middle Paleolithic ones is outmoded in light of recent work based on refitting and replicative experiments. Distinct but equally complicated reduction sequences have been defined for Upper Paleolithic

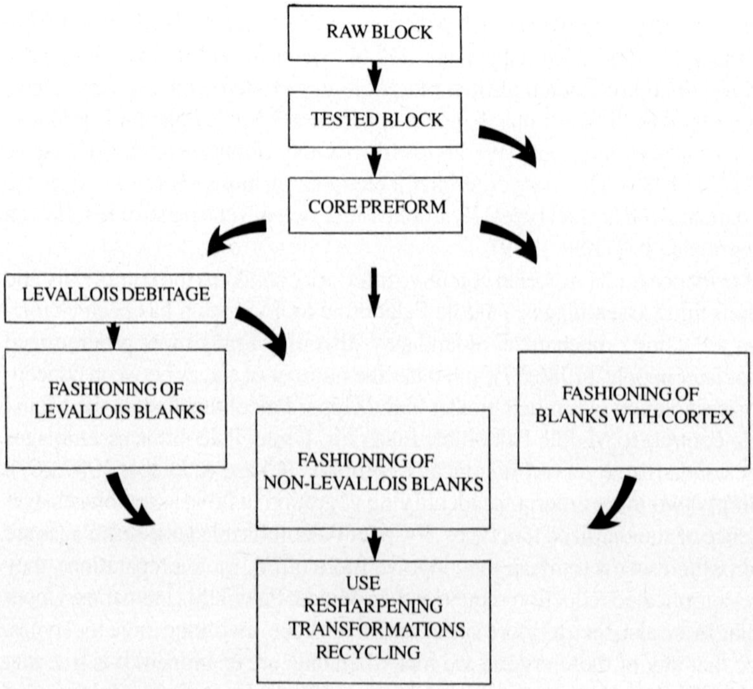

Figure 3. Reduction sequence defined by Geneste for Middle Paleolithic flake technologies (after Geneste 1985).

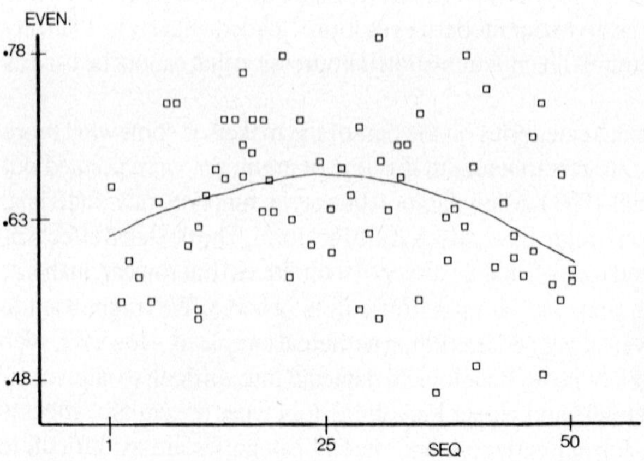

Figure 4. Lithic assemblage evenness for paleolithic stone tool assemblages from the Perigord. Line describes significant polynomial regression through these points (after Simek & Price 1990).

blade technologies by Chadelle (1983), who defined five reduction stages resulting in 21 classifiable artifact forms (Fig. 2); and Geneste (1985), who found four stages and 26 separate technological products produced by Middle Paleolithic technologies (Fig. 3). Moreover, one

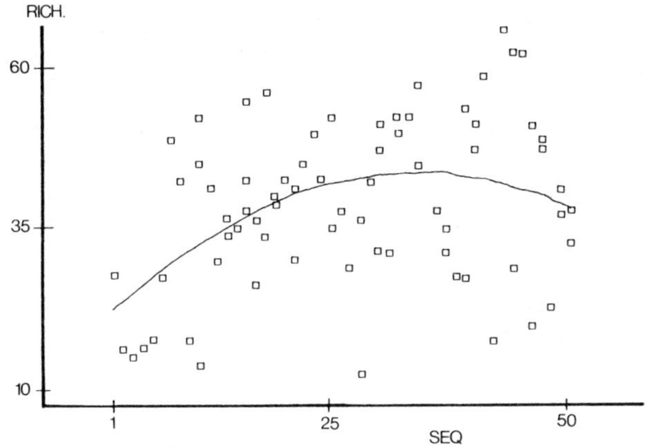

Figure 5. Lithic assemblage richness for paleolithic stone tool assemblages from the Perigord. Line describes significant polynomial regression through these points (after Simek & Price 1990).

production system can grade into the other, as Marks & Volkman (1983) discovered at Boker Tachtit. It is simply judgmental to see one technology as 'more advanced' than the other because the products have more pleasing forms.

The idea that Upper Paleolithic assemblages are more diverse than Middle Paleolithic ones is not supported by examination of the data. Price and I have demonstrated (using a set of 235 paleolithic assemblage inventories) that statistical measures of assemblage diversity[2] do show a significant, if weak, pattern of change over time in the Perigord region of southwestern France (Simek & Price 1990). However, the pattern does not seem to highlight the Middle-Upper Paleolithic Transition as a time of radical change. When diversity is expressed as Evenness (Fig. 4), diverse lithic tool assemblages do not arise until late in the Upper Paleolithic (during the Upper Perigordian phase), when site sizes and numbers also increase. Similar change patterns are observed when diversity is expressed as assemblage Taxonomic Richness (Fig. 5). While it is true that the Upper Paleolithic type list contains over 40 more types than the Lower and Middle Paleolithic typology, obviously no single assemblage has all of the types present. When assemblages are grouped according to archaeological tradition, no Upper Paleolithic tradition has all types in the list. In fact, a maximum of 85 types occurs (when all the Gravettian and Magdalenian assemblages are grouped into their respective general traditions). Of course, these traditions appear later in the Upper Paleolithic sequence; Early Upper Paleolithic traditions (those usually associated with the Middle-Upper Paleolithic Transition) have essentially the same assemblage richness as the preceding Middle Paleolithic. I might also note that maximum individual site values show very little variation over the entire sample time range.

Assertions that Upper Paleolithic lithic assemblages are more standardized, are technologically more sophisticated, and are more complicated than Mousterian assemblages have no empirical basis. If, as is argued, these criteria reflect cognitive abilities (which, it should be stressed, remains undemonstrated), then there is no radical difference between the Middle and Upper Paleolithic.

4 PERSONAL ADORNMENT

In several recent works, White has argued that items of personal adornment (perforated shell, bones, ivory, etc., interpreted as beads) 'explode' onto the Paleolithic scene with the Aurig-

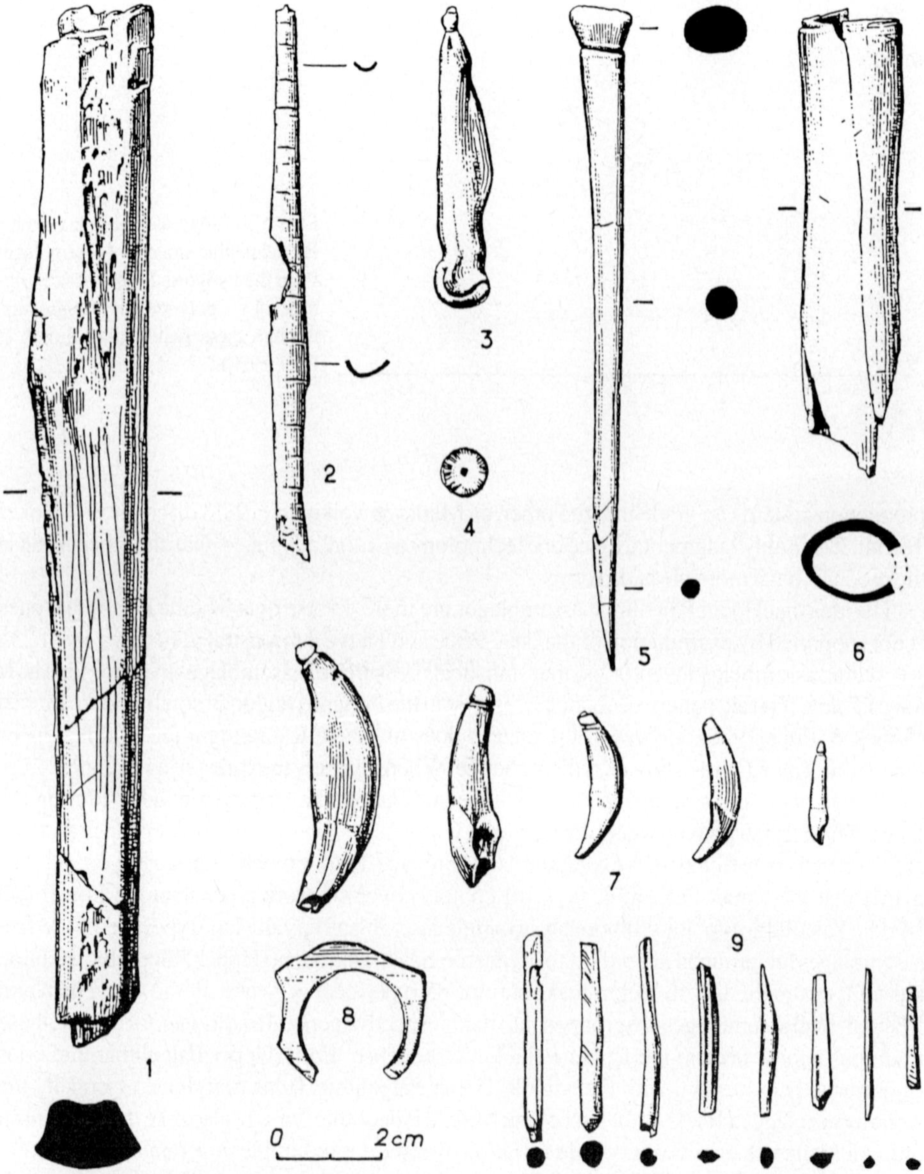

Figure 6. Artifacts from the Chatelperronian of the Grotte de Renne at Arcy-sur-Cure (Yonne) (after Leroi-Gourhan & Leroi-Gourhan 1964).

nacian and reflect a more-complex adaptation linked to the emergence of anatomically modern humans (White 1986, 1989). White sees this jewelry, which is indeed abundant in western Europe, as representative of increased self-definition and social display among moderns; he argues that, 'the absence of both (definition and display functions) among Neandertals must be seen as evolutionarily significant' (White 1989:375).

While it is clear that items of personal adornment are an important part of Aurignacian material culture, I do not agree with White that there is necessarily any coincident evolutionary implication of Neandertal cultural poverty. On the one hand, Aurignacian beadwork has a rather distinctive distribution. In western Europe for example, a very few sites (most within a single area, the Vallon de Castelmerle) have provided much of White's sample, a fact he himself notes (1989:375). The majority of Aurignacian sites lack jewelry in significant amounts, and although early excavation techniques certainly biased the present sample (Rigaud & Simek 1987), the fact is that bead concentrations are few and far between. Moreover, the hundreds of beads that have been found can rather easily be explained by a much smaller number of necklaces, bracelets, or decorated clothes that, one assumes, would have been the actual units of definition or display. All in all, I think, this suggests a complex internal social context for bead production and use that may say more about the Aurignacian and its makers than about other cultures or the evolution of culture itself.

Perhaps equally important, there is evidence suggesting that Neandertals may themselves have made and worn items of personal adornment. In the Grotte de Renne at Arcy-sur-Cure, Andre Leroi-Gourhan (Leroi-Gourhan & Leroi-Gourhan 1964) discovered several rich Chatelperronian layers (Couches VIII, IX and X). Within these layers, an abundant bone and antler tool industry was recovered, consisting of shaped and decorated baguettes, bone awls, modified bird bones, and at least eight artifacts, including carved and grooved deer and fox teeth, that must be considered articles of personal adornment (Fig. 6). Found in association with over 500 Chatelperronian knives, the context of these artifacts cannot be doubted (Leroi-Gourhan & Leroi-Gourhan 1964:38-49). It is interesting to observe that the Aurignacian assemblages at the Grotte de Renne are impoverished in modified bone with only two pieces that might represent items of personal decoration (Leroi-Gourhan & Leroi-Gourhan 1964:49-50).

As was discussed above, the only secure hominid fossil association for the Chatelperronian is the St. Cesaire Neandertal (Leveque & Vandermeersch 1980). In addition, teeth recovered from the Grotte de Renne are statistically closer in size to Neandertals than modern humans (Hunt, n.d.). Thus, Neandertals might well have adorned their persons for self definition and social display in certain contexts, although the frequency of this behavior may have been less than for the Aurignacian. Nevertheless, it would seem difficult to attribute the development of personal decoration to any innate, modern human ability to use symbols not shared by contemporary archaic *Homo sapiens*.

5 SPECTACULAR ART WORK

When considering more obvious evidence for symbolic behavior (art), discussions usually entail presence/absence arguments. Nearly always, the Middle Paleolithic is characterized as lacking artifacts that we today recognize as art: 'In many ways, the most striking difference between Middle and Upper Paleolithic is the contrast between the rich and highly developed art found in the latter period and the almost complete lack of it in the former' (Chase & Dibble 1987:280). In comparison, the Upper Paleolithic is characterized as rich in this kind of material; from the beginning, it is argued, fully modern humans created visual symbols on their walls, on rocks around them, even on themselves.

Again, close examination of the relevant data indicates that the evolution of art was a more complex process than the prevalent view allows. Middle Paleolithic peoples almost certainly did use symbols and create art. Early Upper Paleolithic peoples made far less art than in more

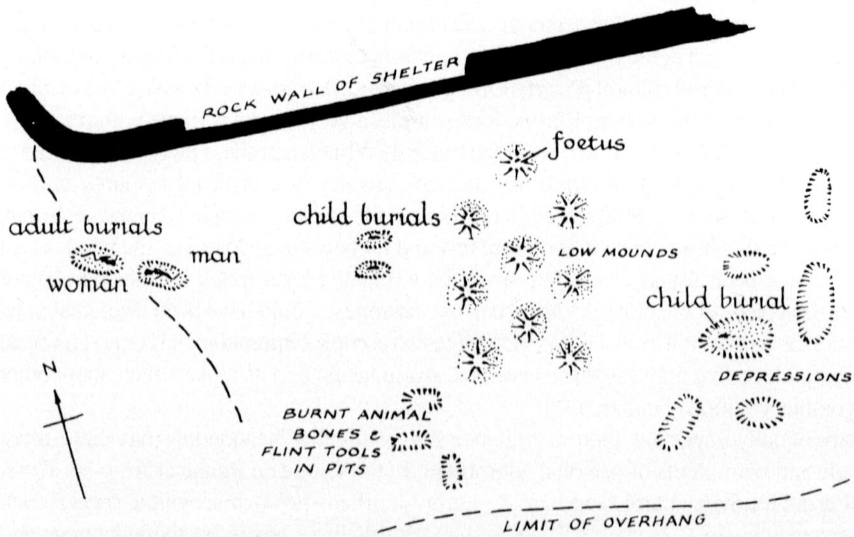

Figure 7. Map of burial complex at La Ferrassie (after Wymer 1982).

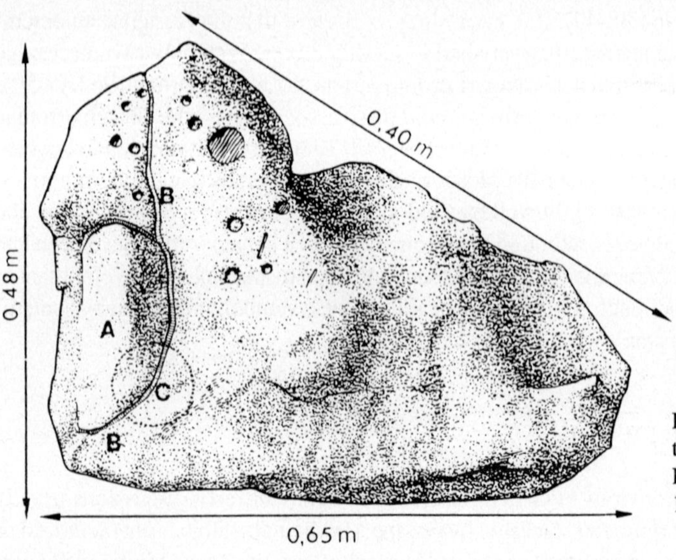

Figure 8. Stone cap found on the grave of La Ferrassie Hominid no. 6 (after Heim 1976).

recent time periods. If an artistic explosion occurred, it was later during the Upper Paleolithic and probably had little to do with the biological shift from Neandertals to modern humans.

Perhaps the best-known (but least referenced) use of art in Mousterian times comes from the large rockshelter site, La Ferrassie. The shelter was excavated for over thirty years by Denis Peyrony (from 1896 to 1929) and has been the subject of numerous publications (e.g. Peyrony 1934, Delporte 1984, Delporte & Maziere 1977, Heim 1976, 1982a, 1982b). Between 1909 and 1921, Peyrony discovered the remains of seven Neandertals buried in sedimentary units C and D during the earliest (Mousterian) occupations at the site (Peyrony 1934:24-36). Despite recent, uninformed assertions to the contrary (Gargett 1989), most archaeologists familiar with the data believe that these bodies were found organized into a mortuary complex that included graves, associated non-grave features, and grave goods.

Two adults and five young children, were intentionally buried in six graves (Fig. 7). Nine other pits were located nearby that, according to Peyrony, contained concentrations of fragmented bone (1934:30-31). Nine mounds, easily distinguished from the surrounding matrix on stratigraphic grounds, formed a symmetrical pattern in the center of the shelter and were associated with one of the infant burials (Peyrony 1934:31-33). All of the graves contained finely-worked Mousterian tools, according to Peyrony, and two graves, holding La Ferrassie Hominids no. 2 and no. 6, were capped with stones and contained artifacts that, considering what is classified as Aurignacian art work, must be called art.

Because he understood the import of his discoveries at La Ferrassie, Peyrony published verbatim long passages from his excavation notes. The grave containing Hominid no. 2 yielded an engraved bone fragment, and Peyrony was clear in reporting the association.

> *We soon uncovered three flat stones about 20 cm to a side, one placed where we would find the cranium and others above the arms. Over the entire surface of the skeleton, in the sandy reddish-brown sediments of the archaeological level, there were, in much greater numbers than elsewhere in the same deposit, large splinters of animal bone having used areas on their external surfaces around 2 to 3 cm (flakers). One of these bones shows a series of fine intentional scratches recalling the notched bones from the aurignacian levels.* (1934:24)

Even Peyrony had difficulty distinguishing between this piece of Middle Paleolithic 'art' and that he had found in the overlying Aurignacian levels.

The capstone for the grave containing Hominid no. 6 is particularly interesting (Fig. 8), since it has a series of pits and deep grooves worked into its surface. In virtually all respects, the artifact resembles objects considered as art when found in Aurignacian artifacts. Again, Peyrony was very precise in attributing this piece to the Middle Paleolithic/Neandertal grave. In addition to verbal descriptions, he provided a profile of the burial pit showing that the capstone was found well within the grave fill (Fig. 9). In form, this piece must be considered as equivalent in style and sophistication to the earliest Upper Paleolithic art works.

The nature and frequency of Early Upper Paleolithic art has also been oversimplified in recent accounts (e.g. Mellars 1989b:362).

White is correct when he states that the number of ivory and shell beads increases during the Aurignacian, although (as we have seen) jewelry cannot be considered an Aurignacian innovation. However, other forms of art (sculptures, engravings, etc.) are rare in the Aurignacian of western Europe. In his 1971 synthesis of Paleolithic art in France, Leroi-Gourhan lists a total of 19 Aurignacian art pieces, a tiny number when compared with the hundreds that appear during the later Badegoulian and Magdalenian periods (Leroi-Gourhan 1971).

Trajectories of change during the Upper Paleolithic are even more evident when only wall art, a western European phenomenon, is considered. A recent complete catalog of parietal art in France (Ministry of Culture 1987) confirms the general paucity of Aurignacian art, showing a great difference between early and later Upper Paleolithic traditions in frequencies of wall art works (Fig. 10). When three large engraved stones found detached from cave walls are classified as parietal art, the Aurignacian only has three pieces of wall art; later time periods show marked increases in this medium. It is, of course, the late Upper Paleolithic (Solutrean and

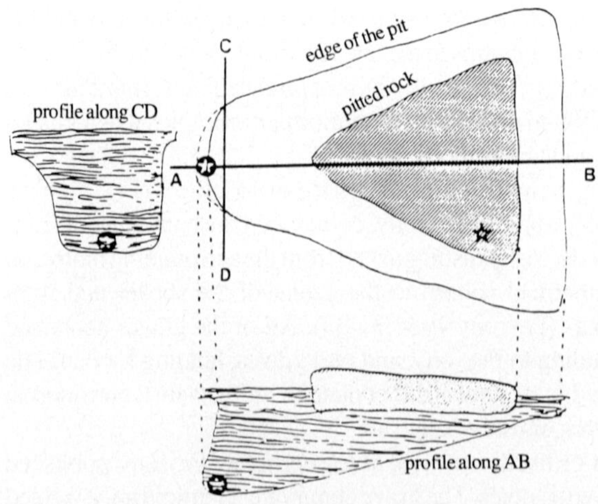

Figure 9. Profile of La Ferrassie Hominid no. 6 grave showing relation of engraved stone to grave fill (after Peyrony 1934).

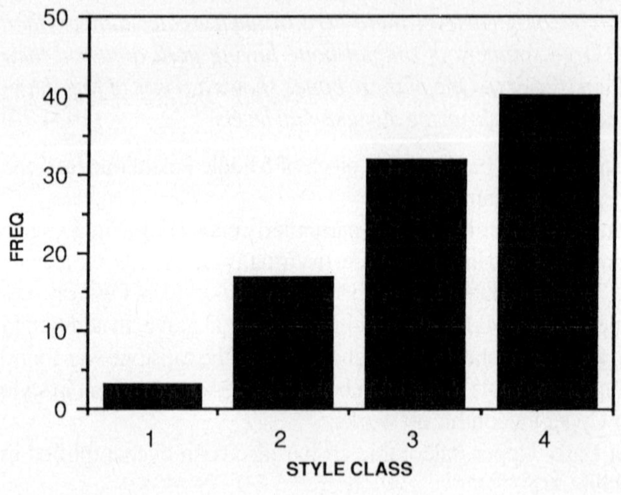

Figure 10. Histogram of wall art frequencies over time. Data are from Leroi-Gourhan (1971).

Magdalenian) that produced the great bulk of paleolithic wall decoration. Clearly, there is a variety of change trajectories even within art or symbolic expression during the Paleolithic. Nevertheless, the Middle Paleolithic and Chatelperronian, while certainly not rich in symbolic objects, cannot be viewed as lacking any evidence for artistic expression. Thus, artistic behavior was not a uniquely modern human characteristic.

6 CONCLUSION

It has not been possible to cover all aspects of the cultural changes that occurred during the late Pleistocene and which are frequently interpreted as reflecting biological differences between

Neandertals and fully modern humans. I hope, however, that I have been able to show that several of the criteria often employed to distinguish modern human cognition from Neandertal cognitive poverty do not display the radical change patterns predicted by this view. Evolutionary scenarios that see modern human intellectual superiority as an integral part of the Middle-Upper Paleolithic Transition process, are not supported by the very data proponents of such perspectives like to use. Other archaeological dimensions, subsistence for example, show equally complex and distinct change trajectories (e.g. Simek & Snyder 1988, Straus 1977). Close examination of archaeological data from western Europe, where migration and replacement were probable change parameters, suggests that the effects of such parameters on culture was anything but direct.

If we are to understand the evolution of culture as we know it, then we must resist simple vitalistic explanations. Neandertal culture was certainly rich, and it took several thousands of years of cultural evolution for the Upper Paleolithic to replace it, even in western Europe. That interaction and exchange took place between Neandertals and modern humans occupying the same region seems certain, perhaps producing the Chatelperronian as a hybrid culture in western Europe (an idea first proposed by Bordes 1972). And in the presence of such cultural exchanges, it is indeed very hard to imagine how biological interaction did not occur.

ACKNOWLEDGEMENTS

The author acknowledges the support of the National Science Foundation (BNS 8507856 and BNS 8606536) for some of the research reported here.

NOTES

1 It should be noted right away that most of the data I discuss were discovered and published long ago. Yet, they are still germane to present debates and should be brought back into the discussion.

2 In these diversity studies, we use two standard measures of assemblage diversity, Evenness (Fig. 4) and Richness (Fig. 5). Evenness is calculated from frequency table data as Pielou's *J* statistic (Pielou 1975):

$$J = \frac{H}{H_{max}}$$

Where *H* is the common Shannon-Weaver information statistic (Shannon & Weaver 1949). The value measures how a series of discrete classes are filled and varies between 0 (only one class represented) and 1 (all classes have equal frequencies). Richness is measured as the number of classes in a frequency table having values greater than zero. Both measures have been used and evaluated in the archaeological literature (Grayson 1984, Kintigh 1984, Rhode 1988).

REFERENCES

Binford, L.R. 1962. Archaeology as anthropology. *American Antiquity* 28: 217-25.

Bordes, F. 1950. Principes d'une méthode d'ètude des techniques de débitage et de la typologie du paléolithique ancien et moyen. *L'Anthropologie* 54: 19-34.

Bordes, F. 1961. *Typologie de Paléolithique Ancienne et Moyenne*. Bordeaux: Delmas.

Bordes, F. 1972. Du paléolithique moyen au paléolithique supérieur: continuité ou discontinuité? In F. Bordes (ed.), *The origin of Homo sapiens*. Paris: UNESCO, pp. 211-18.

Bordes, F. & J. Labrot 1967. La stratigraphie de Roc de Combe (Lot) et ses implications. *Bull. Soc. Préhist. Franc.* 64: 15-28.

Brose, D. & M.L. Wolpoff 1971. Early upper paleolithic man and late middle palaeolithic tools. *Am. Anthropol.* 73: 1156-94.

Cann, R.L., M. Stoneking & A.C. Wilson 1987. Mitochondrial DNA and human evolution. Nature 325: 31-36.

Chadelle, J.-P. 1983. *Technologie et Utilization du Silex au Périgordien Supérieur.* Memoire de EHESS, Université de Toulouse.

Champagne, F. & R. Espitalie 1981. *Le Piage: Site Préhistorique de Lot.* Paris: Société Préhistorique Française.

Chase, P.G. & H.L. Dibble 1987. Middle paleolithic symbolism: A review of current evidence and interpretations. *J. Anthropol. Arch.* 6: 263-93.

Clark, G.A. & J.M. Lindly 1989. The case of continuity: Observations on the biocultural transition in Europe and Western Asia. In P. Mellars & C.B. Stringer (eds), *The Human Revolution: Behavioural and Biological Perspectives on the Origins of Modern Humans*, vol. I. Edinburgh: Edinburgh University Press, pp. 626-76.

Delporte, H. (ed.) 1984. *Le Grand Abri de la Ferrassie: Fouilles 1968-73*. Paris: Université de Provence.

Delporte, H. & G. Maziere 1977. L'Aurignacien de la Ferrassie: observations préliminaires a la suite des fouilles recentes. *Bull. Soc. Préhist. Fran.* 74: 343-61.

Dibble, H.L. 1989. The implications of stone tool types for the presence of language during the lower and middle palaeolithic. In P. Mellars & C.B. Stringer (eds), *The Human Revolution: Behavioural and Biological Perspectives on the Origins of Modern Humans*, vol. I. Edinburgh: Edinburgh University Press, pp. 415-32.

Dunnell, R.C. 1971. *Systematics in Prehistory.* New York: Free Press.

Freeman, L.G. 1973. The significance of mammalian faunas from palaeolithic sites in cantabrian Spain. *Am. Antiq.* 384-44.

Gargett, R.H. 1989. The evidence for Neanderthal burial. *Curr. Anthropol.* 30: 157-90.

Geneste, J.-M. 1985. Analyse Lithique d'Industries Mousteriennes du Périgord: Une Approche Technologique du Comportement des Groupes Humains au Paléolithique Moyen. Thèse de Doctorat. Bordeaux: University of Bordeaux I.

Grayson, D.K. 1984. *Quantitative Zooarchaeology: Topics in the Analysis of Archaeological Faunas.* New York: Academic Press.

Harrold, F.B. 1989. Mousterian, Chatelperronian, and Early Aurignacian in western Europe: Continuity or discontinuity? In P. Mellars & C.B. Stringer (eds), *The Human Revolution: Behavioural and Biological Perspectives on the Origins of Modern Humans*, vol. I. Edinburgh: Edinburgh University Press, pp. 677-713.

Heim, J.-L. 1976. Les hommes de la Ferrassie I. Le gisement, les squelettes d'adultes: crane et squelettes de tronc. *Archives de l'Institute de Paléontologie Humaine* 35: 1-368.

Heim, J.-L. 1982a. Les hommes de la Ferrassie II. Les squelettes d'adultes: squelettes de membres. *Archives de l'Institute de Paléontologie Humaine* 38: 1-272.

Heim, J.-L. 1982b. *Les Enfants Neanderthaliens de la Ferrassie*. Paris: Masson.

Hunt, D. n.d. The Neandertal-modern *Homo sapiens* transition: A study of the fossil remains of Arcy-sur-Cure and Saint-Cesaire and their association to Chatelperronian-Aurignacian hominids. Unpublished manuscript, Department of Anthropology, University of Tennessee.

Kintigh, K.W. 1984. Measuring archaeological diversity by comparison with simulated assemblages. *Am. Antiq.* 49: 44-54.

Lartet, E. & A. Christy 1868. *Reliquiae Aquitanicae, Being Contributions to the Archaeology and Palaeontology of southern France*. London: Rupert Jones.

Leroi-Gourhan, A. 1971. *Treasures of Prehistoric Art*. New York: Adams.

Leroi-Gourhan, A. & Ar. Leroi-Gourhan 1964. Chronologie des grottes d'Arcy-sur-Cure (Yonne). *Gallia Préhistoire* 7: 1-64.

Leroi-Gourhan, Ar. 1984. La place du neandertalien de St.-Cesaire dans la chronologie würmienne. *Bull. Soc. Préhist. Fran.* 81: 196-98.

Leroyer, C. & Ar. Leroi-Gourhan 1983. Problemes de chronologie: le castelperronien et l'aurignacien. *Bull. Soc. Préhist. Fran.* 80: 41-44.

Leveque, F. & B. Vandermeersch 1980. Découverte de restes humains dans un niveau castelperronien a Saint-Césaire (Charante-Maritime). *Comptes Rendues de l'Academie des Sciences de Paris*, series 2, 291: 187-89.

Marks, A.E. & P.W. Volkmen 1983. Changing core reduction strategies: A technological shift from the middle to upper palaeolithic. In E. Trinkaus (ed.), *The Mousterian Legacy: Human biocultural change in the Upper Pleistocene*. Oxford: BAR, pp. 35-51.

Mellars, P. 1973. The character of the Middle-Upper palaeolithic transition in southwest France. In C. Renfrew (ed.), *The Explanation of Prehistoric Culture Change*. London: Duckworth, pp. 255-76.

Mellars, P. 1989a. Technological changes at the Middle-Upper Palaeolithic transition: Economic, social and cognitive perspectives. In P. Mellars & C.B. Stringer (eds), *The Human Revolution: Behavioural and Biological Perspectives on the Origins of Modern Humans*, vol. I. Edinburgh: Edinburgh University Press, pp. 338-65.

Mellars, P. 1989b. Major issues in the emergence of modern humans. *Curr. Anthropol.* 30: 349-85.

Ministry of Culture of France 1987. *L'Art des Cavernes*. Paris: Ministry of Culture.

Minugh-Purvis, N. 1988. Interpreting immature craniofacial morphology at Krapina: Problems and possibilities. Paper presented at the 12th ICAES meeting, Zagreb, Yugoslavia, July 1988.

Peyrony, D. 1934. La Ferrassie-Mousterien, Périgordien, Aurignacien. *Préhistoire* 3: 1-92.

Pielou, E.C. 1975. *Ecological Diversity*. New York: Wiley.

Putnam, J. 1988. The search for modern humans. *Nat. Geo.* Oct.: 438-77.

Rhode, D. 1988. Measurement of archaeological diversity and the sample size effect. *Am. Antiq.* 53: 708-16.

Rigaud, J.Ph. & J.F. Simek 1987. Arms too short to box with God: Problems and prospects for palaeolithic archaeology in Dordogne, southwest France. In O. Soffer (ed.), *The Pleistocene Old World: Regional Perspectives*. New York: Plenum, pp. 47-61.

Sackett, J. 1982. Approaches to style in lithic archaeology. *J. Anthropol. Arch.* 1: 59-112.

Shannon, C.E. & W. Weaver 1949. *The Mathematical Theory of Communication*. Urbana: University of Illinois Press.

Simek, J.F. 1988. Stone Tools from Krapina (Croatia, Yugoslavia). Paper presented at the 12th ICAES meeting, Zagreb, Yugoslavia, July 1988.

Simek, J.F. & H.A. Price 1990. Chronological change in Perigord lithic assemblage diversity. In P. Mellars (ed.), *The Emergence of Modern Humans. Edinburgh: Edinburgh University Press*, pp. 243-261.

Simek, J.F. & L.M. Snyder 1988. Changing assemblage diversity in Perigord Archaeofaunas. In H. Dibble & A. Montet-White (eds), *Upper Pleistocene Prehistory of Western Eurasia*. Philadelphia: University of Pennsylvania Press, pp. 321-32.

Smith, F.H., J.F. Simek & M.S. Harrill 1989. Geographic variation in supraorbital torus reduction during the later Pleistocene (c. 80 000-15 000 BP). In P. Mellars & C.B. Stringer (eds), *The Human Revolution: Behavioural and Biological Perspectives on the Origins of Modern Humans*, vol. I. Edinburgh: Edinburgh University Press, pp. 172-93.

Sonneville-Bordes, D. 1960. *Le Paléolithique Supérieur en Périgord*. Bordeaux: Delmas.

Sonneville-Bordes, D. & J. Perrot 1956. Lexique typologique du paléolithique supérieur. *Bull. Soc. Préh. Fran.* 53: 408-12.

Straus, L.G. 1977. Of deerslayers and mountain men: Palaeolithic faunal exploitation in cantabrian Spain. In L.R. Binford (ed.), *For Theory Building in Archaeology*. New York: Academic Press, pp. 40-76.

Stringer, C.B. & P. Andrews 1988. Genetic and fossil evidence for the origin of modern humans. *Science* 239: 1263-68.

Stringer, C.B., J.-J. Hublin & B. Vandermeersch 1984. The origin of anatomically modern humans in western Europe. In F.H. Smith & F. Spencer (eds), *The Origins of Modern Humans: A World Survey of the Fossil Evidence*. New York: Alan R. Liss, pp. 51-136.

White, R. 1982. Rethinking the middle to upper paleolithic transition. *Curr. Anthropol.* 23: 169-92.

White, R. 1986. *Dark Caves, Bright Visions*. New York: W.H. Norton.

White, R. 1989. Production complexity and standardization in early aurignacian bead and pendant manufacture: Evolutionary implications. In P. Mellars & C.B. Stringer (eds), *The Human Revolution: Behavioural and Biological Perspectives on the Origins of Modern Humans*, vol. I. Edinburgh: Edinburgh University Press, pp. 366-90.

Wolpoff, M.H. 1988. The dental remains from Krapina. Paper presented at the 12th ICAES meeting, Zagreb, Yugoslavia, July 1988.

Wymer, J.J. 1982. *The Palaeolithic Age*. New York: St. Martins.

Social transformations at the Middle to Upper Paleolithic transition: The implications of the European record

Olga Soffer
Department of Anthropology, University of Illinois, Urbana, USA

1 INTRODUCTION

Between some 100 000 and 30 000 years ago a new type of hominid appeared in the northern latitudes of the Old World. This time slice saw *Homo sapiens sapiens* replace *Homo sapiens neanderthalensis* and a transition from Middle to Upper Paleolithic tool industries and adaptations. The significance and synchrony of the changes in both the morphology of the hominids and in the nature of the archaeological record they left behind has been a subject of much recent debate in the literature (Mellars & Stringer 1989, Trinkaus 1983a, b, 1987, Smith & Spencer 1984, White 1982). This chapter examines the record of this transition from regions often considered only anecdotally in these discussions – the East European or Russian Plain and Crimea – and, synthesizing it with the record from the rest of Europe as well as with paleoanthropological data on the hominid remains, argues that the transition involved major socio-cultural transformations.

2 THE RECORD FROM EUROPEAN USSR

2.1 *The environmental given*

Two physiogeographic features of the East European Plain and Crimea significantly affected Late Pleistocene hominid occupation there. First, the Plain is a vast flat expanse of land rimmed by north-south running mountain chains of varying heights which restrict vertical biotic differentiation to the western and eastern margins of the Plain. The second is that Crimea, which today is a peninsula, was an island during Interglacial and some warmer interstadial times, and a part of the Plain during stadial times (Gerasimov & Velichko 1982). This suggests that Crimea would have been peopled during glacial times, and its archaeological record implies that this occurred during the last, Valdai, glaciation (Boriskovskij 1984, Velichko & Praslov 1978).

Paleoenvironmental reconstructions indicate that climatic conditions during the last (Mikulino) Interglacial and subsequent interstadial periods were differentially distributed accross Europe, and that the East European Plain and Crimea witnessed significantly more precipitation and pronounced warming than the remainder of Europe, which resulted in their forestation (Musil 1985, Velichko & Pecsi, in press).

At the glacial maximum (20 000 to 18 000 BP), and in a similar although less dramatic fashion during preceeding stadial episodes, the Plain changed to an open landscape covered by a periglacial steppe – one with a dramatic reduction in latitudinal biotic differentiation and diversification (Soffer 1985 with references, Gerasimov & Velichko 1982). The only exception to this resultant stadial hyperzonality on the mainland was found in the Dnestr-Prut region where vertical differentiation in elevations permitted biotic diversification. The other exception was Crimea whose physiogeography features a flat plain in the north-northwest and three parallel chains of low mountain ranges in the southeast. Thus the plains, foothills, and mountain ranges near the Dnestr-Prut and in southeastern Crimea created a number of ecotones with more complex, diverse, and productive biotic communities during both stadial and interstadial times (Musil 1985, Soffer 1989 with references).

2.2 *Hominid remains*

Hominid remains dating to the period under discussion have been surprisingly sparse in this area and almost exclusively restricted to Crimea (Boriskovskij 1984, Day 1976, Praslov & Rogachev 1982, Soffer 1985, Oakley et al. 1971). Sites on the Plain have yielded a total of one molar from the early Middle Paleolithic site of Rozhok on the coast of the Sea of Azov, and a recently discovered incomplete skeleton of an infant found at Kostenki 12 in a layer dated to about 33 000 BP (Anikovich 1987). The former has been described as consistent with Neanderthal dentition while the latter's developmental age may preclude any morphological assignment. Layers with Middle Paleolithic inventories in Crimea contained 11 individuals – nine of whom are classified as Neanderthals and 2 as early anatomically modern humans (EAMH) (Kolosov 1983, 1986, Trinkaus 1984). Three anatomically modern individuals came from late Upper Paleolithic layers probably dating considerably younger than 20 000 BP (Boriskovskij 1984, Day 1976, Oakley et al. 1971).

2.3 *The archaeological record*

2.3.1 *Distribution of the sites*

Late Middle and Early Upper Paleolithic inventories on the Russian Plain come predominantly from open air locations while those in Crimea are from rock shelters. In the Dnestr-Prut area these inventories usually come from stratified deposits with a number of Middle Paleolithic layers which, after a depositional hiatus, are often overlaid by Upper Paleolithic ones (Boriskovskij 1984, Soffer 1989 with references). Stratification of occupation is likewise in evidence at the Kostenki-Borshchevo sites along the Don. Here, however, only Upper Paleolithic occupations are represented (Praslov & Rogachev 1982, Soffer 1989). The central part of the Plain, by contrast, lacks stratified sites dating to either the Middle or the Upper Paleolithic and contains no sites where the two periods are found superimposed. Middle Paleolithic sites here are extremely sparse while the Upper Paleolithic ones all date younger than 26 000 BP (Soffer 1985, 1989).

The Crimean record is somewhat different. First, both known Middle and Upper Paleolithic sites are restricted to the southeastern mountainous part of the peninsula (Boriskovskij 1984, Kolosov 1983, 1986, Velichko & Praslov 1978). This area contains over 800 archaeologically investigated rock shelters with 33 primarily single-layer Middle Paleolithic occupations, an absence of identified Early Upper Paleolithic sites, and 13 Late Upper Paleolithic sites, less than half of which are stratified. While both Middle and Upper Paleolithic sites here are known from rock shelters, Kolosov (1988) has recently argued that this bias is a result of archaeological research and not indicative of past hominid preferences.

What is significant about the Middle Paleolithic record from the study area is that it is precisely those regions with the greatest vertical differentiation in elevations which contain evidence for continuous hominid occupation throughout the last glaciation. I have interpreted this regionally disparate data for the Middle Paleolithic to indicate that ecotonal areas with the greatest biotic diversification may have seen more continuous hominid occupation while other more homogeneous parts of the Plain witnessed a pattern of sporadic and discontinous colonization and abandonment (Soffer 1989).

Regions occupied during the Early Upper Paleolithic are, likewise, localized in two disparate parts – the middle Dnestr and the middle Don. The appearance of Upper Paleolithic sites in the Kostenki-Borshchevo area marks a recolonization of the plain, one which, in contrast to previous aborted attempts, expands through time. After around 26 000 BP this region includes increasing numbers of single layer sites on the eastern, central and southern parts of the Plain. Two coeval early Upper Paleolithic cultures have been identified here, both of which show ties to the south-southeastern steppe zone and Crimea. The Spitsyn culture features prismatic cores, worked bone and a predominant use of superior exotic flint which came from up to 300 km to the south-southwest (Boriskovskij 1984, Soffer 1989 with references, Praslov, pers. comm.). The Streletskaya, with strong Middle Paleolithic elements, exclusive use of local raw materials, and absence of bone tools, art and items of personal adornment, bears some stylistic affinities to Crimean Middle Paleolithic bifacial industries (Boriskovskij 1984, Kolosov 1983, 1986, Soffer 1989).

2.3.2 *Food management strategies*

I have argued elsewhere that faunal data from the late Middle and early Upper Paleolithic sites on the Russian Plain indicate more regionally circumscribed opportunistic subsistence strategies during the Middle Paleolithic; a change to foraging adaptations which involved mapping onto resources over wider areas and entailed seasonal group mobility is evident in the Upper Paleolithic (Soffer 1989). I also argued that data from the sites indicate that this change neither occurred overnight at the Middle to Upper Paleolithic rubicon nor was necessarily associated with different types of hominids. Faunal remains from Spitsyn sites, for example, are better explained by seasonal scheduling and group mobility while those from Streletskaya sites appear to indicate regionally circumscribed opportunistic food procurement.

Although they have been selectively used in the past to purportedly show Neanderthal specialization of procurement, Crimean data unfortunately cannot be used in such discussions. Faunal remains from Starosel'ye, the site used in these arguments where over 98% of identified specimens came from wild asses, were found both with a very late Middle Paleolithic inventory and, more importantly, with remains of EAMHs (Alekseev 1976, Formozov 1958). All of the published Crimean faunal remains come from rock shelters which also contain an uncomfortable abundance of carnivores who were, more than likely, responsible for generating some of the faunal materials. Thus 'who did what to whom' must be first sorted out before this problematic data set can be used for cultural constructs.

2.3.3 *Inventories, features, and site types*

Evidence for raw material procurement and use and for the utilization of space at the sites similarily indicate changes between Middle and Upper Paleolithic adaptations. Middle Paleolithic inventories on the Russian Plain and in Crimea, as elsewhere, contain a paucity of bone tools and no art of items of personal adornment. They show a redundant use of local lithic raw materials, regardless of their quality, which suggests a very localized settlement system while early Upper Paleolithic sites contain varying but redundant amounts of superior exotics (Soffer 1989 with references).

Relatively small sized oval concentrations of cultural remains, measuring 30 to 50 m^2 in area, sometimes with shallow hearths and often at least partially over-lapping one another, are characteristic of the known Middle Paleolithic sites and probably resulted from repeated palimpsest occupations. Elaborate features reported at some sites, mammoth bone dwellings at Molodova I and IV for example, are in all likelihood fanciful present day constructs (Soffer 1989). The organization of space and the structure of features is far clearer at the Upper Paleolithic sites. This clear patterning, like its muted Middle Paleolithic predecessor, is redundant from site to site and more recognizable though ethnographically known examples.

3 DIFFERENCES ACCROSS THE MIDDLE TO UPPER PALEOLITHIC BOUNDARY

3.1 *Archaeological*

The differences noted between the inventories and site structures at Middle and Upper Paleolithic sites discussed above have been noted for other parts of Europe as well (for raw material procurement see Geneste 1988, Kozlowski 1989, but also Roebroeks et al. 1988, for subsistence behavior see Binford 1985, Straus 1985, White 1982 with discussion; for worked bone, art and items of personal adornment see Conkey 1983, Chase & Dibble 1987, Marshack 1988, White 1982, 1989, for site structure and size see Clark & Lindly 1989, Simek 1987, White 1982, 1985). These differences have recently led a number of archaeologists to argue that the changes between the Middle and Upper Paleolithic involved unspecified but major socio-cultural transformations (Binford 1985, Whallon 1989, White 1982, 1989).

Given the disparate archaeological data sets from the Russian Plain and from Crimea which pattern in a similar manner, I have suggested that the transition from Middle to Upper Paleolithic adaptation involved 1) a qualitative change in both the perception and utilization of nature, 2) that it entailed a change in food management strategies and in the accompanying settlement patterns and 3) that this change clearly did not happen overnight nor at a particular chronological or morphological rubicon (Soffer 1989). The lack of a synchronic fit betwen cultural and morphological changes noted in Crimea, which is likewise in evidence in other parts of the Old World (e.g. Neanderthal with early Upper Paleolithic industry at St. Cesaire, EAMHs with Middle Paleolithic tools at Qafzeh and Skhūl) suggests that we are not dealing with innate differences in the capacities of the two types of hominids but, more likely, just differences in performance.

3.2 *Morphological and thanatological change*

In considering what these performance or habitual behavior differences may have been, I next turn to some salient paleoanthropological differences noted between the Neanderthals and EAMHs. First, Neanderthals are considered to have been much more robust and stronger hominids adapted to significantly greater physical exertion (Trinkaus 1983b, 1987, Trinkaus & Smith 1985, Jacobs 1985a, b). Second, their survival to ages only in the late 30's - early 40's indicates that their longevity was considerably shorter than that of EAMHs who lived to post-reproductive age (Skinner 1981, Trinkaus 1986, 1987 with refs.). Much note has been made in the literature of the changes in crania and in dentition. The consensus of opinion sees the reduction in tooth size (especially of anterior dentition) and accompanying changes in masticatory musculature and cranial configurations of EAMH attributable to reduction in dental loading (i.e. in paramasticatory use of dentition) (Frayer 1984, Smith 1983, Trinkaus 1987, Trinkaus & Smith 1985, Wolpoff 1979, 1980). As this volume as well as chapters in Mellars & Stringer (1989) demonstrate, no consensus exists now on whether these changes reflect population replacement, regional evolution, or a mix of gene flow and population movement.

Frayer's (1986) intra- as well as inter-populational comparisons between the sexes have brought to light even more interesting variability. Specifically, he has argued for differential gracilization rates for EAMHs – with females gracilizing much earlier than males. Using an admittedly small data set from central Europe, Frayer has argued that while EAM females dating before the glacial maximum were considerably more gracile than their Neanderthal equivalents and little distinct from their Late Upper Paleolithic descendants, EAM males were very robust and Neanderthal-like in many morphological features and did not become more gracile until after the glacial maximum. Frayer's observations on asynchronic gracilization rates suggest that whatever behavioral changes occurred at the time, their impact was far more dramatic on the females who were no longer subject to the same amounts of physical exertion as were Neanderthal females. EAM males, however, continued to exhibit much the same degree of muscular hypertrophy (cf. Ben-Itzhak et al. 1988). Since females are more vulnerable than males to reproductive penalties of stress, this reduction in stress would have benefited them, their young, and their reproductive partners as well (Jochim 1988 with references, Zihlman 1981).

Little attention has been paid in the past to morphological evidence for stress in late Pleistocene hominids. Brothwell (1963) considered evidence for dental hypoplasia and, combining both Neanderthal and EAMHs, argued for an increase in this indicator of stress between the late Pleistocene and early Holocene. More recent studies of specific late Pleistocene hominid groups are refuting this conclusion. Molnar & Molnar (1985) noted a much higher incidence of hypoplasia among Neanderthals from Krapina than observed among their Holocene Hungarian sample. Skinner (1981) reported a decrease in dental attrition in immature Upper Paleolithic hominids when compared to the Neanderthals. In the most inclusive study to date Brennan (1987, 1988) examined skeletal remains of Neanderthal and EAMH populations in south-western France for non-specific indicators of stress. Her results indicate a dramatic decrease in such evidence as Harris lines and dental hypoplasia between the Neanderthals and EAM groups living before the glacial maximum. Thus these recent studies also indicate a reduction in stress between Neanderthal and EAMHs, a reduction which, as argued above, would have had more impact on the female and the more vulnerable immature segment of the population.

The age structure of the hominid remains themselves also underscores the benefits of whatever it was that happened at the transition for the sub-adult segment of the population. Table 1 compares the proportion of juveniles (defined at < 12 years of age) among the Pleistocene hominids of northern Eurasia.[1] The significant factor in this table is the decrease in juvenile mortality that can be noted across the transition. While 43.1% of the Neanderthal minimum number of individuals (MNI) sample is under 12 years of age (40.3% of the maximum number of individuals – MaxNI – sample), only 29.7% (of MNI, 28.3% of MaxNI) of EAMH are subadults. Since the sizes of the sampled aged populations are roughly comparable, disparities

Table 1. Pleistocene hominid remains from northern Eurasia. [a] Krapina MNI counts as per Wolpoff (1988, pers. comm.), [b] Estimated MNI values for Changyang of 1-2 individuals tabulated at 1.5.

	Middle Pleistocene	Neanderthals[a] MNI	MaxNI	EAMH MNI	MaxNI
Sample size					
Total number found	85.5[b]	215	298	166	201
Number aged	85.5	202	238	155	180
Juveniles ($\leq$ 12 years)					
number	22	87	96	46	51
%	25.7	43.1	40.3	29.7	28.3
Adults ($\geq$ 13 years)					
number	63.5	115	142	109	129
%	74.3	56.9	59.7	70.3	71.7

Middle Pleistocene Hominids vs Neanderthals:
1)MNI: X^2 value = 7.84; d.f. = 1; $0.001 < p < 0.01$; G statistic = 8.12
X^2 with continuity correction = 7.11; d.f. = 1; $0.001 < p < 0.01$
2)MaxNI: X^2 value = 5.94; d.f. = 1; $0.01 < p < 0.02$; G statistic = 6.16
X^2 with continuity correction = 5.32; d.f. = 1; $0.02 < p < 0.05$

Middle Pleistocene Hominids vs EAMHs:
1)MNI: X^2 value = 0.46; d.f. = 1; $0.50 < p < 0.70$; G statistic = 0.46
X^2 with continuity correction = 0.28; d.f. = 1; $0.50 < p < 0.70$
2)MaxNI: X^2 value = 0.22; d.f. = 1; $0.50 < p < 0.70$; G statistic = 0.22
X^2 with continuity correction = 0.11; d.f. = 1; $0.70 < p < 0.80$

in sample sizes cannot be invoked as explanations for the differences. Furthermore, both samples were primarily derived from comparable cave contexts. Finally, differential preservation favoring adults also does not account for the observed differences, because, if anything, it would produce a steady increase in the proportion of juveniles in a more recent population – a situation which is clearly as expected when we compare Middle Pleistocene hominids[2] (MNI: 25.7% juveniles and 74.3% adults) to the Neanderthals (MNI: 43.1% juveniles and 56.9% adults; MaxNI: 40.3% juveniles and 59.7% adults) and the reverse for the Neanderthals vs EAMHs (MNI:29.7% juveniles and 70.3% adults; MaxNI 28.3% juveniles and 71.7% adults).

The data set used in Table 1 is admittedly a very problematic one. Firstly, it is based on data on northern Eurasian hominid remains published by Oakley et al. (1971) which I have updated to include more recent discoveries (see Note 1 for selection criteria and updating). Since Oakley et al. (1971) do not report estimates for the numbers of individuals represented at most sites, I used their specimen listings to calculate both MNI estimates per age category (counting as individuals only those clearly aged in Oakley et al. 1971) as well as the maximum number of individuals (MaxNI) (assigning each individually numbered specimen to a different individual and ageing them where this is indicated). Secondly, this problematic data set is also a very ambiguous indicator of past mortality because there is no way of reliably factoring out potential biasing cultural practices. Wolpoff (1979), for example, has argued that the Krapina remains which constitute a sizeable component of the Neanderthal sample and which have no children < 3 and almost no adults > 20 years old represented, do not reflect a living population and implied that cultural practices are implicated in the observed demographic imbalance. Both

Table 2. Late Pleistocene hominid remains from northern Eurasia. Legend: HSN no. 1 – Neanderthal sample without Krapina remains: MNI: 59 juveniles, 78 adults; MaxNI: 68 juveniles, 105 adults. HSN no. 2 – Neanderthal sample with Wolpoff's (1988, pers. comm.) estimates for Krapina of 28 juveniles, 37 adults. HSN no. 3 – Neanderthal sample with White's (1988) estimates for Krapina of 10 juveniles, 18 adults. HSN no. 4 – Neanderthal sample with Trinkaus (1988) estimates for Krapina of 5 juveniles, 10 adults, 5-10 unaged individuals.

	HSN no. 1		HSN no. 2		HSN no. 3		HSN no. 4	
	MNI	MaxNI	MNI	MaxNI	MNI	MaxNI	MNI	MaxNI
Sample size								
Total number	150	233	215	298	178	261	170-175	253-258
Number aged	137	173	202	238	165	201	152	188
Juveniles (≤ 12 years)								
number	59	68	87	96	69	78	64	73
%	43.1	39.3	43.1	40.3	41.8	38.8	42.1	38.8
Adults (≥ 13 years)								
number	78	105	115	142	96	123	88	115
%	56.9	43.1	56.9	59.7	58.2	61.2	57.9	61.2

X^2 values: d.f. = 1

1. EAMH vs HSN no. 1
A. MNI: $X^2 = 5.66$, $0.01 < p < 0.02$, G statistic = 5.66, X^2 with continuity correction = 5.09, $0.02 < p < 0.05$
B. MaxNI: $X^2 = 4.75$, $0.02 < p < 0.05$, G statistic = 4.76, X^2 with continuity correction = 4.27, $0.02 < p < 0.05$

2. EAMH vs HSN no. 2
A. MNI: $X^2 = 6.73$, $0.001 < p < 0.01$, G statistic = 6.80, X^2 with continuity correction = 6.17, $0.01 < p < 0.02$
B. MaxNI: $X^2 = 6.48$, $0.01 < p < 0.02$, G statistic = 6.55, X^2 with continuity correction = 5.96, $0.01 < p < 0.02$

3. EAMH vs HSN no 3.
A. MNI: $X^2 = 5.12$, $0.02 < p < 0.05$, G statistic = 5.14, X^2 with continuity correction = 4.60, $0.02 < p < 0.05$
B. MaxNI: $X^2 = 4.65$, $0.02 < p < 0.05$, G statistic = 4.68, X^2 with continuity correction = 4.19, $0.02 < p < 0.05$

4. EAMH vs HSN no. 4
A. MNI: $X^2 = 5.16$, $0.02 < p < 0.05$, G statistic = 5.17, X^2 with continuity correction = 4.63, $0.02 < p < 0.05$
B. MaxNI: $X^2 = 4.54$, $0.02 < p < 0.05$, G statistic = 4.08, X^2 with continuity correction = 4.07, $0.02 < p < 0.05$

Harrold's (1980) recent and Binford's (1968) earlier analyses of Middle and Upper Paleolithic burial data, however, show no special or preferential treatment of juveniles during the Middle Paleolithic.

The Krapina sample is also problematic because no consensus of opinion exists on the minimum number of individuals represented by these highly fragmented remains (cf. Trinkaus 1988, White 1988, Wolpoff 1979). Table 2 compares Neanderthal remains to EAMH at four levels of exclusion or inclusion of the Krapina sample. The observed X^2 values for all relevant permutations consistently show a greater proportion of juveniles among Neanderthal remains. Given that the data set used here is very problematic, I neither claim that it is demographically accurate nor that it unambiguously indicates a major change in the pattern of juvenile mortality. I offer these data as a heuristic device which, together with other indicators discussed above, appears to suggest a reduction of stress on juveniles accross the Middle to Upper Paleolithic transition. I also suspect, but cannot demonstrate because of the ambiguity inherent in sexing fragmentary skeletal remains, that the same decrease in stress should be apparent in the EAM females and be reflected in a decrease in female mortality across the transition as well.

4 DISCUSSION AND CONCLUSIONS

Explanations found in the paleoanthropological literature for the changes at the Middle to Upper Paleolithic transition have focused on inter-populational differences between the Neanderthals and EAMHs. They have pointed to such changes in technology as the transition from tooth use to tool use, or from flake to blade tools, or the advent of insert technology and long distance kills (Frayer 1981, 1984, Smith 1983, Trinkaus 1983a, b, 1986, 1987). These explanations are clearly particularistic and insufficient to account for all the changes noted above, as well as for the mosaic nature of these changes (e.g. male vs. female, EAMHs before and after the glacial maximum). Explanations that invoke 'Deus ex machina' – the arrival from elsewhere of AMHs, whether in person or through gene flow, or some combination of the two – are equally unsatisfactory for they neither account for the non-synchrony between morphological forms and cultural behavior nor, more importantly, specify what it is that the AMHs did that was more adaptive than what the Neanderthals did. Finally, archaeological explanations which propose changes in socio-cultural behavior at the transition likewise leave much to be desired for, by and large, they avoid dealing with the morphological and thanatological data on the hominids themselves.

Given both the morphological and archaeological data on hand, I suggest that the differences between the Middle and Upper Paleolithic adaptations in northern Eurasia lay not in blade vs. teeth but in a dramatic change in economic and social relationships – changes which were immediately felt by the females and the young. Admittedly problematic though each data set I presented is, their patterning in the same direction suggests that the time has come for us to abandon our latent but unproven and unprovable, assumptions that bi-parental provisioning of the young, division of labor, and food sharing between their progenitors goes back to the australopithecines (Lovejoy 1981, Isaac 1983, but cf. Zihlman 1981). I suggest instead the possibility that such reorganization in both subsistence behavior and ensuing social organization took place much later in time, and that the gender based separation and division of labor as we know it from the ethnographic record occurred around the Middle to Upper Paleolithic transition.

I furthermore argue that the small site size we see in the Middle Paleolithic record coupled with the use of local raw materials, opportunistic exploitation of biotic resources, a general absence of clear cut site types, and a regionalization of the sites, suggests the existence of very small, stable, co-residential units who moved within very restricted geographic space. This, in turn, together with evidence for muscle hypertrophy in both sexes, dimorphic feeding ranges postulated for male and female early and middle Pleistocene hominids (Frayer & Wolpoff 1985), and an absence of evidence for the transport of meat indicative of food sharing (Binford 1985), suggests the existence of small sized foraging and co-residential units of mothers with their immature young plus, perhaps, a small number of related adult females, as well as small sized all-male units. Studies of dimorphism in feeding behavior among the primates tentatively suggest that Neanderthal females and their young may have had more restricted day ranges than males. This in turn implies that males may have fed on a greater number of larger herbivores and consumed more large migratory taxa than females who would have ingested more locally abundant small sized species. This hypothesis carries clearly testable propositions and predicts a greater difference in bone chemistry between Neanderthal males and females than between the sexes among EAMHs.

This social construct is just a first approximation and I obviously do not preclude some situational larger sized bisexual aggregations. I offer these scenarios to underscore that, given the data, we should begin thinking about Neanderthal cultural behavior in a non-nuclear family mode – a mode which does not take the !Kung as a base to be somewhat simplified and used as a prehistoric cookie cutter for understanding hominid adaptations in the past at all points of Pleistocene and Pliocene time. The data on adult to juvenile ratios in the Middle Pleistocene sample (74.3% adults to 25.7% juveniles) which are more like the one for EAMH than the Neanderthal ones suggest possibilities of yet other forms of social organization further back in time such as, perhaps, combined group foraging.

Finally, while I have speculated what may have happened, I have not examined possible reasons for why this occurred. Here, again, I have no clear cut answers, just some further speculations. First, based on data from eastern Europe, I strongly suspect that the continent was not fully packed with Neanderthals from the Atlantic to the Urals, but that they were localized in discrete regional patches and continuously present only in regions with vertical differentiation and resource diversification. Benevolent environmental conditions, as found during interglacial and interstadial times, accompanied possibly by modest increases in regional populations, probably led to repeated but unsuccessful attempts to colonize spatially undifferentiated environments. Deteriorating climatic conditions, on the other hand, probably brought about latitudinal population shifts. What they also brought about was an increase in temporal fluctuations in the availability of food resources, in the package size of prey resources, and an increase in the non-synchrony of acquisition among individuals – all features of northern latitudes which select for and promote food sharing (Jochim 1981 with references).

Since, however, we are also dealing with two hominid forms each of which is not exclusively associated with specific cultural behavior, it remains unclear if the change to increased division of labor combined with bi-parental provisioning and food sharing was exclusively a feature of EAMHs. Wolpoff (1979, 1980) for example, has noted an increase in deciduous tooth size among late Neanderthals and interpreted this to mean earlier weaning and a possible reduction in birth spacing. This may suggest either the advent of male provisioning or perhaps, female specific technological advances in the preparation of soft weaning foods. If the former was the case, as it may have been since female gracilization noted for the Early Upper Paleolithic would not have happened overnight but would have taken time to develop, then clearly we are not dealing with innate differences in the capacity for such behavior but rather just with the habitual practice of that behavior – a practice which, I suggest, became entrenched during the Early Upper Paleolithic and generated the kind of an archaeological record we recognize as structurally similar to ethnographically known cases.

ACKNOWLEDGEMENTS

I am grateful to Fred Smith and Günter Bräuer for their original invitation to think on these matters transitional. This chapter has greatly benefited from the advice, criticism and discussion with a number of people, most notably E. Delson, C. Gamble, L. Binford, D. Frayer, L. Klepinger, J. Speth and R. Whallon. The initial unpublished dental data set on Late Pleistocene hominids which I used for an earlier version of this paper was generously provided by M. Wolpoff and D. Frayer to whom I am sincerely grateful. I thank M. Wolpoff not only for his unpublished estimates for Krapina MNIs but also for countless stimulating hours of discussion about Middle and early Upper Paleolithic hominids and their adaptations during which he unfailingly tried to change 'the error of my ways'. Research assistance was provided by Jolee West. Portions of this work, primarily those concerned with Central and Eastern Europe, was supported by the National Geographic Society, Fulbright Faculty Research Abroad Program, I.R.E.X. and the National Academy of Sciences. Funding from the A.C.L.S. and the Russian and East European Center of the University of Illinois permitted me to present an earlier version of this paper at the 12th International Congress of Anthropological and Ethnological Sciences held in July 1988 in Zagreb, Yugoslavia.

NOTES

1 The basic data set on Late Pleistocene hominid remains from northern Eurasia used in this chapter were obtained from Oakley et al. (1971). All entries in Oakley et al. (1971) which on chronometric, morphological and archaeological criteria unambiguously dated between the last interglacial and last glacial cycle before some 20 000 BP (defined as the boundary for EAMHs) were utilized. All ambiguous cases or those where relative dating spanned all of the Upper Paleolithic were not considered. For example, I did not include hominid remains from Magrite where lithic inventories were assigned to the Aurignacian but were accompanied by pottery sherds. Similarily I did not include remains from Oetrange where archaeological dating indicated too broad an age estimate from the Aurignacian to the Magdalenian. Individuals identified by Oakley et al. (1971) as adolescents were treated as adults while those termed juveniles were grouped together with children. Hominid remains whose exact classification is uncertain at present were likewise eliminated (e.g. Dzerava Skala or Ohaba, see Smith 1984). The data in Oakley et al. (1971) were updated as follows: (A) for the Neanderthals: 1 adult (MNI and MaxNI) at St. Cesaire as per Day (1976); 1 adult and 2 juveniles (MNI) or 10 adults and 3 juveniles (MaxNI) as per Giacobini & de Lumley 1988 at Fate; and a total of 5 juveniles and 2 adults (MNI and MaxNI) at Zaskal'naya V and VI as per Kolosov (1983, 1986). Estimates for numbers of Neanderthals found at Krapina were calculated as per Wolpoff (pers. comm.), White (1988) and Trinkaus (1988). (B) for the EAMHs: For Italy 3 adults (both MNI and MaxNI) as per Mussi (1986a, b); for Czechoslovakia 5 adults (MNI and MaxNI) from Dolní Věstonice II as per Klima (1987) and Svoboda (1987) and 1 adult (MNI and MaxNI) for Podbaba as per Smith (1984); for Germany 1 adult each (MNI and MaxNI) from Kelsterbach, Hahnöfersand and Sande/Paderborn as per Smith (1984). Data for the USSR (MNI and MaxNI) was augmented by 1 infant for Kostenki 12 as per Anikovich (1987) and ammended to 2 juveniles and 3 adults for Sungir as per Soffer (1985), 1 juvenile each for Kostenki XV and XVIII as per Praslov & Rogachev (1982), Soffer (1985) and 1 adult and 1 juvenile for Starosel'ye as per Alekseev (1976).

2 Data on Middle Pleistocene hominids from northern Eurasia (excluding Java) came from Cook et al. (1982), Pope (1982) and Stringer et al. (1984).

REFERENCES

Alekseev, V.P. 1976. Position of the Starosel'ye find in the hominid system. *J. Hum. Evol.* 5: 413-421.

Anikovich, M.V. 1987. Novaia antropologicheskaia nakhodka v sele Kostenki Arkheologicheskiie aspekty. *Kratkie Soobscheniia Instituta Arkheologii* 189: 3-8.

Ben-Itzhak, S., P. Smith & R.A. Bloom 1988. Radiographic study of humerus in Neanderthals and *Homo sapiens sapiens*. *Am. J. Phys. Anthropol.* 77: 231-242.

Binford, L.R. 1985. Human ancestors: Changing views of their behavior. *J. Anthropol. Archaeol.* 4: 292-327.

Binford, S.R. 1968. A structural comparison of disposal of the dead in the Mousterian and Upper Paleolithic. *Southwestern J. Anthropol.* 24: 139-151.

Boriskovskij, P.I. 1984. *Paleolit SSSR*. Nauka Publishing Co., Moscow.

Brennan, M.U. 1987. Nonspecific indicators of stress in the Upper Palaeolithic. Paper presented at the 86th Annual Meeting of the American Anthropological Association, November, Chicago.

Brennan, M.U. 1988. Harris lines and dental hypoplasia in the French Palaeolithic. *Am. J. Phys. Anthropol.* 75: 190.

Brothwell, D.R. 1963. The macroscopic dental pathology of some earlier human populations. In D.R. Brothwell (ed.), *Dental Anthropology,* New York: McMillan, pp. 271-287.

Chase, P. & H. Dibble 1987. Middle Paleolithic Symbolism: A review of current evidence and interpretations. *J. Anthropol. Archaeol.* 6: 263-296.

Clark, G.A. & J.M. Lindly 1989. Observations on the biocultural transition in Europe and Western Asia. In P. Mellars & C. Stringer (eds), *The Human Revolution: Behavioural and Biological Perspectives on the Origins of Modern Humans.* Edinburgh: Edinburgh University Press, pp. 626-676.

Conkey, M. 1983. On the origins of Paleolithic art: A review and some critical thoughts. In E. Trinkaus (ed.), The Mousterian Legacy, *Br. Archaeol. Rep. Int. Ser.* 164: 201-227.

Cook, J., C.B. Stringer, A.P. Currant, H.P. Schwarcz & A.G. Wintle 1982. A review of chronology of European Middle Pleistocene hominid record. *Yrbk. Phys. Anthropol.*, pp. 19-67.

Day, M.H. 1976. *Guide to Fossil Man.* Chicago: Univ. Chicago Press.

Formozov, A.A. 1958. Perschernaia Stoianka Starosel'e i ee mesto v Paleolite. *Matieriali i Issledovaniia po Arckheologii No. 71.* Moscow: Izdatel'stvo Akademii Nauk SSSR.

Frayer, D.F. 1981. Body size, weapon use, and the natural selection in the European Upper Paleolithic and Mesolithic. *Am. Anthropol.* 83: 57-73.

Frayer, D.F. 1984. Biological and cultural change in European Late Pleistocene and Early Holocene. In F. Smith & F. Spencer (eds), *The Origins of Modern Humans.* New York: Alan R. Liss, pp. 221-250.

Frayer, D.F. 1986. Cranial variation of Mladec and the relationship between Mousterian and Upper Palaeolithic hominids. In V.V. Novotný & A. Mizerova (eds), *Fossil Man: New Facts – New Ideas.* Brno: Anthropos Institute, Moravian Museum.

Frayer, D.F. & M.H. Wolpoff 1985. Sexual dimorphism. *Ann. Rev. Anthropol.* 14: 429-473.

Geneste, J.-M. 1988. Systemes d'approvisionnement en matières premieres au Paléolithique moyen et au Paléolithique supérieur en Aquitaine. In M. Otte (ed.), L'Homme de Neanderthal. Vol. 8 La Mutation. Liege: *Et. Rech. Arch. l'Univ. Liege,* no. 35: 61-70.

Gerasimov, I.P. & A.A. Velichko 1982. *Paleogeografiia Evropy za poslednije sto tysiach let.* Moscow: Nauka.

Harrold, F.B. 1980. A comparative analysis of Eurasian Paleolithic burials. *World Archaeol.* 12: 195-211.

Isaac, G.L. 1983. Aspects of human evolution. In D.S. Bendall (ed.), *Evolution from Molecules to Man.* Cambridge: Cambridge Univ. Press, pp. 509-543.

Jacobs, K. 1985a. Climate and the hominid postcranial skeleton in Würm and Early Holocene Europe. *Curr. Anthropol.* 26: 512-514.

Jacobs, K. 1985b. Evolution in the postcranial skeleton of Late Glacial and Postglacial European hominids. *Zeitschrift für Morphologie und Anthropologie* 75: 307-326.

Jochim, M. 1981. *Strategies for Survival.* New York: Academic Press.

Jochim, M. 1988. Optimal foraging and the division of labor. *Am. Anthrop.* 90: 130-136.

Klima, B. 1987. A triple burial from the Upper Paleolithic of Dolní Věstonice, Czechoslovakia. *J. Hum. Evol.* 16: 831-835.

Kolosov, Yu.G. 1983. *Must'erskiie Stoianki raiona Belogorska.* Kiev: Naukova Dumka.

Kolosov, Yu.G. 1986. *Akkajskaia Must'erskaia Kul'tura.* Kiev: Naukova Dumka.

Kolosov, Yu.G. 1988. K metodike poiska pozdnepaleoliticheskikh stoianok v Krymu (v sviazi s prirondnoj obstanovkoj v pozdnem Pleistotsene). Problemy Vzaimosviazi Prirody i Obschestva v Kamennom Veke v Srednej Azii. *Tezisy Dokladov Koferentisii Posviaschennoj 50-Letiiu Otkrytiya Teshik-Tasha.* Tashkent: Fan, pp. 45-47.

Kozlowski, J.K. 1990. A multiaspectual approach to the origins of the Upper Paleolithic in Europe. In P. Mellars (ed.), *The Emergence of Modern Humans.* Edinburgh: Edinburgh Univ. Press, pp. 419-437.

Lovejoy, O. 1981. The origin of man. *Sci.* 211: 341-351.

Marshack, A. 1988. The Neanderthals and the human capacity for symbolic thought: Cognitive and problem-solving aspects of Mousterian symbol. In M. Otte (ed.), L'Homme de Neanderthal, Vol. 5 La Pensée. Liege: *Et. Rech. Arch. l'Univ. Liege,* no. 30.

Mellars, P. & C. Stringer (eds) 1989. *The Human Revolution: Behavioural and Biological Perspectives on the Origins of Modern Humans*. Edinburgh: Edinburgh Univ. Press.

Molnar, S. & I.M. Molnar 1985. The incidence of enamel hypoplasia among the Krapina Neanderthals. *Am. Anthropol.* 87: 536-549.

Musil, R. 1985. Palaeobiography of terrestrial communities in Europe during the Last Glacial. *Acta Musei Nationalis Pragae* XLIB, no. 1-2.

Mussi, M. 1986a. Italian Paleolithic and Mesolithic burials. *Hum. Evol.* 1: 545-556.

Mussi, M. 1986b. On the chronology of the burials found in the Grimaldi caves. *Anthropologia Contemporanea* 9: 95-104.

Oakley, K.P., B.G. Campbell, T.I. Molleson 1971. *Catalogue of Fossil Hominids*. Pt. II: Europe. London: The British Museum (Natural History).

Pope, G.G. 1982. Hominid Evolution in East and Southeast Asia. Ph.D. dissertation, University of California Berkeley.

Praslov, N.D. & A.N. Rogachev 1982. *Paleolit Kostenkovsko-Borshchevskogo raiona na Donu 1879-1979*. Leningrad: Nauka.

Robroeks, W., J. Kolen & E. Rensink 1988. Planning depth, anticipation and the organization of Middle Paleolithic technology: The 'archaic natives' meet 'Eve's descendants'. *Helinium* 28: 17-34.

Simek, J.F. 1987. Spatial order and behavioural change in the French Paleolithic. *Antiquity* 61: 25-40.

Skinner, M. 1981. Dental attrition in immature hominids of the Late Pleistocene: Implications for adult longevity. *Am. J. Phys. Anthropol.* 54: 278-279.

Smith, F.H. 1983. Behavioral interpretaton of changes in craniofacial morphology across the Archaic/Modern *Homo sapiens* transition. In E. Trinkaus (ed.), The Mousterian Legacy. *Br. Archaeol. Rep. Int. Ser.* 164: 141-163.

Smith, F.H. 1984. Fossil hominids from the Upper Pleistocene of central Europe and the origin of modern Europeans. In F.H. Smith & F. Spencer (eds), *The Origins of Modern Humans*. New York: Alan R. Liss, pp. 137-210.

Smith, F.H. & F. Spencer (eds) 1984. *The Origins of Modern Humans*. New York: Alan R. Liss.

Soffer, O. 1985. *The Upper Paleolithic of the Central Russian Plain*. New York: Academic Press.

Soffer, O. 1989. The Middle to Upper Paleolithic transition on the Russian plain. In P. Mellars & C. Stringer (eds), *The Human Revolution: Behavioural and Biological Perspectives on the Origins of Modern Humans*. Edinburgh: Edinburgh Univ. Press, pp. 714-742.

Straus, L. 1985. Stone age prehistory of northern Spain. *Sci.* 230: 501-507.

Stringer, C.B., J.-J. Hublin & B. Vandermeersch 1984. The origin of anatomically modern humans in western Europe. In F.H. Smith & F. Spencer (eds), *The Origins of Modern Humans*. New York: Alan R. Liss, pp. 51-136.

Svoboda, J. 1988. A new male burial from Dolní Věstonice. *J. Hum. Evol.* 16: 827-830.

Trinkaus, E. (ed.) 1983a. The Mousterian Legacy. *Br. Archaeol. Rep. Int. Ser.* 164.

Trinkaus, E. 1983b. Neanderthal postcrania and the adaptive shift to modern humans. In E. Trinkaus (ed.), The Mousterian Legacy. *Br. Archaeol. Rep. Int. Ser.* 164: 165-200.

Trinkaus, E. 1984. Western Asia. In F. Smith & F. Spencer (eds), *The Origins of Modern Humans*. New York: Alan R. Liss, pp. 251-293.

Trinkaus, E. 1986. The Neandertals and modern human origins. *Ann. Rev. Anthropol.* 15: 193-218.

Trinkaus, E. 1987. Bodies, brawn, brains and noses: Human ancestors and human predation. In M.H. Nitecki & D.V. Nitecki (eds), *The Evolution of Human Hunting*. New York: Plenum Publishing Co., pp. 147-176.

Trinkaus, E. 1988. Hominid postcrania from Krapina. Paper presented at the 12th International Congress of Anthropological and Ethnological Sciences, July, Zagreb.

Trinkaus, E. & F.H. Smith 1985. The fate of the Neandertals. In E. Delson (ed.), *Ancestors: The Hard Evidence*. New York: Alan R. Liss, pp. 325-333.

Velichko, A.A. & M. Pecsi (in press). *Palaeogeographic Atlas of the Northern Hemisphere*. Budapest: Kiado.

Velichko, A.A. & N.D. Praslov 1978. *Arkehologiya i Paleogeografiia Rannego Paleolita Kryma i Kavkaza-Putevoditel'*. Moscow: Nauka.

Whallon, R. 1989. Elements of cultural change in the Later Paleolithic. In P. Mellars & C. Stringer (eds), *The Human Revolution: Behavioural and Biological Perspectives on the Origins of Modern Humans*. Edinburgh: Edinburgh Univ. Press, pp. 433-454.

White, R. 1982. Rethinking the Middle-Upper Paleolithic transition. *Curr. Anthropol.* 23: 169-192.
White, R. 1985. Upper Palaeolithic land use in the Perigord. *Br. Archaeol. Rep. Int. Ser.* 253.
White, R. 1989. Production complexity and standardisation in early Aurignacian bead and pendant manufacture: Evolutionary implications. In P. Mellars & C. Stringer (eds), *The Human Revolution: Behavioural and Biological Perspectives on the Origins of Modern Humans*. Edinburgh: Edinburgh Univ. Press, pp. 366-390.
White, T. 1988. The minimum number of Neanderthals in the Krapina assemblage. Paper presented at the 12th International Congress of Anthropological and Ethnological Sciences, July, Zagreb.
Wolpoff, M.H. 1979. The Krapina dental remains. *Am. J. Phys. Anthropol.* 50: 67-114.
Wolpoff, M.H. 1980. *Paleoanthropology.* New York: Alfred A. Knopf, Inc.
Zihlman, A.L. 1981. Women as shapers of the human adaptation. In F. Dahlberg (ed.), *Woman the Gatherer.* New Haven: Yale University Press, pp. 75-120.

Middle Palaeolithic chronology and the transition to the Upper Palaeolithic in southwest Asia

Ofer Bar-Yosef
Department of Anthropology, Peabody Museum, Harvard University, Cambridge, USA

1 INTRODUCTION

Since their discovery in the late 1920's and early 1930's, human fossils from southwest Asia have occupied an important place in human evolution and been a subject of constant controversies. At present, the main disagreements are centered around the dating of the fossils, the definition of the various morpho-types in terms of presence/absence of phylogenetic relationships, and the place of the fossils within the evolutionary and cultural sequence of the entire Old World. An additional problem, no less important but poorly understood, concerns the transition from the Middle Palaeolithic or the Mousterian Industrial Complex to the Upper Palaeolithic, not only in terms of what population was responsible for this major change but also 'why' and 'how' the shift occurred.

What follows is an updated summary of stratigraphical and archaeological observations from western Asia together with the implications for the geochronological and cultural interpretations. The last part of the paper raised questions regarding the 'transition' to the Upper Palaeolithic.

This summary takes into account the available ^{14}C, Th/U and recent thermoluminescence and ESR dates. Thus the sites which contained the fossils are described in a chronological order that reflects the favoured interpretation of the author (Bar-Yosef & Vandermeersch 1981, Vandermeersch 1982, Bar-Yosef 1988, 1989a, b). Different chronological interpretations of the same sites have prevailed in the literature since the late 1950's (e.g. Howell 1959, Jelinek et al. 1973, Jelinek 1981, 1982, Farrand 1979, 1982, Trinkaus 1984).

2 SITES, STRATIGRAPHIES AND DATES

2.1 *Zuttiyeh Cave*

The fragmentary Zuttiyeh (Galilee) skull was uncovered, according to F. Turville-Petre, at the base of the archaeological sequence in a shallow pit dug into the sterile sandy deposit which filled the lower part of the cave. It was Garrod (1962) who noted that the human specimen was actually found under a layer which contained an Acheulo-Yabrudian assemblage similar to the industry of Tabun E. This was later confirmed by excavations in the remaining brecciated deposits (Gisis & Bar-Yosef 1974). Various stalagmitic layers, above and under the Acheulo-

Yabrudian, were dated by Th/U. The layers above the Acheulo-Yabrudian provided dates of 95 ± 10 and 97 ± 13 kyr, and below it are determinations of 148 + 6 kyr and 164 + 21 kyr (Schwarcz et al. 1980).

2.2 *Qafzeh Cave*

The Mousterian sequence at the cave's entrance (layers V-XXIV in the Vandermeersch excavations or G-L in Neuville's stratigraphy) can be subdivided into two parts. The upper part, which is rich in lithic industry and animal bones, is poor in microvertebrates and lacks human remains. The lower part is poor in lithic industry, preserves the remains of a few hearths, and is rich in microvertebrates. It is in this part of the sequence that all the human burials were uncovered. The lithic industry of the entire sequence is Mousterian, heavily dominated by Levallois technique with clear preference for radial preparation (Boutie 1989).

The late G. Haas already noted in his first reports that the Qafzeh Mousterian is probably earlier than Tabun C (e.g. Haas in Jelinek et al. 1973). His preliminary observations were supported by Tchernov's (1981) large scale seriation of Middle to Upper Pleistocene microvertebrates, which defined the lower levels of Qafzeh in 'Early Mousterian' age. A reinterpretation of the depositional sequence of Qafzeh Cave and its geochronological position has led to the estimation that the age of the fossils is 100 kyr to 80 kyr (Bar-Yosef & Vandermeersch 1981). This unexpected antiquity was rejected by Jelinek (1982) on the basis of the thickness/width ratio of complete flakes (which demonstrates a linear developmental sequence in Tabun Cave and was thus used as a yardstick to indicate that the Qafzeh assemblages are very late in the Mousterian). Additionally Jelinek argued that the two archaic rodent species in Qafzeh only represent a different environmental setting from Mt. Carmel. The recently obtained thermoluminiscence dates (Valladas et al. 1988) indicate an average age of 92 kyr and clearly support the first claim for the relative antiquity of the Qafzeh hominids. Additional ESR dates suggest a range of ages from 73.7 through 118.0 kyr (not in a stratigraphic order) which, calculated on the basis of linear uptake, gives and average of 92 ± 5 kyr; based on early Uranium uptake the average is 115 ± 15 kyr (Schwarcz et al. 1988).

2.3 *Tabun Cave*

At the time of writing this paper, the dating of the Tabun assemblages is mainly based on the geo-chronological interpretation of the cave's stratigraphic sequence. On the whole the sequence is characterized by three depositional units (Farrand 1979). The lower sandy unit comprises layers G, F and the lower part of E. The upper part of E is siltier. Two sliding events were recorded, one followed the deposition of Layer G and the second near the top of Layer F. The middle unit is the loam (or loess) of layer D. Its upper part was channelled and slumped. Layer C marks a change in the depositional regime and contains red earth (washed-in terra rossa) intermingled with large ashy lenses and an increasing number of angular rock fragments. The upper unit, Layer B and the fill of the chimney, is mainly washed-in terra rossa with cobbles, indicating the full opening of the chimney. The original chronological interpretation of Farrand (1979), which placed the entire sequence of Tabun cave in the Upper Pleistocene, has recently been modified (Bar-Yosef & Goldberg 1988, Bar-Yosef 1989). The new version places the lower sandy accumulation (G, F and part of E) in Isotope Stages 7 and 6 and proposes an earlier date for the Early Mousterian of Tabun D in the range of 115-90 kyr. The erosional gap between layer D and C, of unknown temporal length, enables us to suggest a date in the range of 90-60 kyr for layer C which, according to Garrod, is the provenience of the Tabun 1 female Neanderthal burial.

2.4 *Kebara Cave*

The current excavations in Kebara (Bar-Yosef et al. 1986) have reached bedrock and exposed a depositional sequence (Units XV-I) which begins with a sandy-silt sterile layer. Following a subsidence in the swallow-hole and the erosion of the upper part of the lower Unit (XIV), the cave was occupied intensively by humans and the rate of sedimention increased (Goldberg & Laville 1988, Laville & Goldberg 1989). While Unit XIII contains the remains of many well preserved round and oval hearths (but few artifacts), the overlying Units XII through VI are rich in hearth remains, artifacts and bones. The lithic industry is Mousterian, mostly produced by Levallois technique (with a dominant uni-directional, convergent reduction strategy). This industry resembles that of Tabun B-type (Meignen & Bar-Yosef 1988a, b, 1989, Copeland 1975).

A series of thermoluminiscence dates indicates that Units XII through VI span the range of 60 kyr to 48 kyr (Valladas et al. 1987). ESR dates (Schwarcz et al., in press) suggest a range of 60-64 kyr for units X-XI. If the estimated age of the first human occupation is around 65 kyr then the subsidence in the swallow-hole (possibly due to the lowering of the water table) occurred during Isotope Stage 4.

The adult skeleton exposed in its original grave in Unit XII (Bar-Yosef et al. 1986) is classified as a Neanderthal as are the two infants uncovered by M. Stekelis in Unit X (Rak & Arensburg 1987, Smith & Arensburg 1977, Bar-Yosef et al. 1986, Tillier et al. 1988, Smith & Tillier 1989).

2.5 *Amud Cave*

Amud Cave, a small site about 1 km upstream from Zuttiyeh cave, provided a Neanderthal burial embedded in typical Mousterian industry mostly produced by Levallois technique (Suzuki & Takai 1970, Ohnuma, pers. comm.). Earlier claims for the presence of early Upper Palaeolithic assemblage commonly known as the 'Transitional Industry' at the site were recently refuted. The age of the deposits cannot presently be ascertained.

2.6 *Skhūl Cave*

This small rockshelter, which was entirely emptied by the 1930's excavations, contained a large number of burials, all incorporated in layer B (Garrod & Bate 1937). Remains of an earlier layer (C) were found in small pockets in the original bedrock, indicating that a period of erosion separated the two sedimentary units. One may speculate that this erosional phase can be correlated with the erosional phase which truncated the top of Tabun D. This would place the Skhūl fossils in the time span of 60-80 kyr. Recently obtained ESR dates on bovine teeth from collections stored in the British Museum suggested average ages which, based on linear uptake of Uranium, are calculated as 81 ± 15 kyr and based on an early uptake gave an average date of 101 ± 12 (Stringer et al. 1989). The temporal range of the dates is larger than in Qafzeh or Kebara, where a similar thickness of deposits (approximately two meters) were dated. The range of the ESR dates in Skhūl (from 54.6 ± 10.3 through 101.0 ± 19.0 kyr for an early uptake or 77.2 ± 15.7 through 119.0 ± 25.1 kyr for linear uptake) possibly indicates that the human burials represent multiple and sparsely spaced events. Moreover, a brief examination of the lithic industry from layer B, to which the hominids are related, indicates a larger techno-typological heterogeneity than in the two former sites.

2.7 *Shanidar Cave*

The deep stratigraphy of Shanidar cave, as yet unpublished in detail, provided two series of hominids and a few radiocarbon dates (Solecki 1963, Trinkaus 1983). The Mousterian industry is of Zagrosian character with low frequencies of Levallois products. It is quite similar to the Mousterian of Karain cave, near Antalya, in the western Taurus mountains in Turkey (Yalçinkaya 1988). The medium and small size mammalian fauna is dominated, through the various periods (including the Upper Palaeolithic) by sheep and goats (Evins 1982). A rough estimate would place the Shanidar hominids in the 50-70 kyr time span but this estimate needs further confirmation.

2.8 *The Middle Palaeolithic/Upper Palaeolithic boundary*

In the past three decades several ^{14}C dates were obtained from charcoal samples associated with Mousterian and early Upper Palaeolithic assemblages. These dates, around 42-30 kyr, are at the maximum range of measurement by most conventional radiocarbon laboratories. Radiocarbon dates from Boker Tachtit (in the Negev), an indisputable sequence of archaeological assemblages of the early Upper Palaeolithic period, classified as 'Transitional Industry', provided the following readings: 47 280 ± 9050 (SMU-580), 44 930 ± 2420 (SMU-259 and > 45 055 ± 4100 (SMU-184) BP (Marks 1983a). It seems that a cautious estimate would place the Middle/Upper Palaeolithic chronological boundary around 40/43 kyr although it is not inconceivable that it should be even slightly earlier, i.e. ca. 43/45 kyr. Thus the major technological and typological shift as reflected in the lithic industry, does not correspond to any known important climatic event, nor to positive evidence for a change in hominid morpho-type. While comparing the Levantine transition to the European one, which took place at a somewhat later time, two points should be made clear: (a) The transitional character of the Levantine assemblages is recorded in the core reduction strategies (Marks 1983a, b). The shift is from a low frequency of Levallois cores and blade cores (from which Levallois points were removed by bipolar reduction instead of the traditional uni-polar convergent reduction) to a fully blade industry. (b) Since the earliest level of this techno-typological transition the dominant tool types are those which serve as a typological criteria for defining the Upper Palaeolithic, namely, the end scrapers and burins. Thus the so-called 'Transitional Industry', sometime referred to as Emiran, marks the onset of the Upper Palaeolithic period.

3 THE ARCHAEOLOGICAL SEQUENCE

The archaeological sequence in the Levant is described as a series of superimposed entities or industrial complexes. Wherever a stratified site has been excavated, or assemblages embedded within a series of fluviatile-lacustrine layers or marine shorelines were studied, the order of industries remains the same. The sequence begins with a Late (or Upper) Acheulian, either with or without Levallois technique, continues with the Acheulo-Yabrudian (also known as the Mugharan Tradition as defined by A. Jelinek) and ends with the Mousterian (sometimes still referred to as 'Levalloiso-Mousterian', as suggested by Garrod, due to the relative high frequencies of Levallois products when compared to the average European Middle Palaeolithic assemblage).

The Late (or Upper) Acheulian is poorly dated. A K/Ar date of 233 kyr was obtained for a lava flow which covers a Late Acheulian occurrence in Birket Ram (Goren-Inbar 1985). As in East Africa, this industrial complex seems to have had a longer time span than was suggested by early research. In this context the Th/U date of earlier than 300 kyr for the rich Acheulian scatters, near Ma'ayan Baruch, in the Hulah Valley seems reasonable (Schwarcz et al. 1980). The geographic distribution of Upper Acheulian sites covers the entire area of southwest Asia including the desertic parts of the Levant. This stands in contrast with the distribution of the Acheulo-Yabrudian which is located in only the central and northern Levant. Environmental constraints can be excluded as an explanation for this dispersal pattern as the Acheulo-Yabrudian sites are situated both in the coastal forested region as well as in semi-arid localities such as Yabrud (in the Anti-Lebanon mountains) and the El-Kowm oasis (northeast Syria). No sites or isolated finds were recorded in similar environments in the southern Levant (south of the Mt. Carmel area) despite intense surveys in recent years.

The Acheulo-Yabrudian is a distinct lithic industry with mostly small, oval or pointed bifaces and scrapers shaped on thick flakes. Blank production on the whole did not utilize the Levallois technique. This was not caused by limitations imposed by size, quality or accessibility of raw material as demonstrated by the presence of Acheulo-Yabrudian in sites rich in flint of good quality, such as Tabun Cave and El-Kowm. It resembles the Charentian (Mousterian of Quina-type), hence, when the Bordian model of the four contemporary Mousterian entities prevailed, there was a need to try to date it independently in order to find out whether it was or was not contemporary with other Mousterian assemblages (Farrand 1979). As mentioned above, the stratigraphic evidence indicates that in every stratified locality the Acheulo-Yabrudian precedes the Mousterian.

The Mousterian Industrial Complex is often subdivided, on the basis of Tabun Cave as a key site, into three or two entities: 'Tabun D-type', 'C-type' and 'B-type' (Copeland 1975) or 'Mousterian 1' and '2-3' (Ronen 1979, Jelinek 1981, 1982).

The 'Tabun D-type' Mousterian is described as an industry dominated by high frequencies of points (often elongated), blades and flakes. Levallois technique served as the method for obtaining a large amount of blanks and mainly uni-directional core reduction strategy resulted in a proliferation of convergent scar patterns.

The 'Tabun C-type' industry is largely flake-dominated with low frequencies of points. Levallois technique was often employed in radial and bi-directional core reduction methods.

The 'Tabun B-type' industry was commonly produced by the uni-directional or convergent methods, which resulted in proliferation of flakes but with fair amount of blades and short and broad Levallois points.

The industry of Tabun D is as yet unknown in El-Kowm (northern Syria) but possibly appears in Jerf Ajla in the Palmyra basin (Schroeder 1969). In El-Kowm another lithic industry which occupies the same stratigraphic position as Tabun D, was found and named by the excavators – 'Hummalian' (Copeland & Hours 1983, Hours 1982). The Hummalian shares with the 'Tabun D' industry as proliferation of blades and points produced without employing the Levallois technique.

The Mousterian of the Zagros and the Taurus differs from the Levantine Mousterian. It is predominantly a non-Levallois industry (Smith 1986, Dibble 1984, Yalçinkaya 1988) with rare occurrences of Levallois products. It, therefore, resembles many of the European Mousterian industries. The paucity of Levallois products in the Zagros region is attributable to the small size of available raw material. The picture changes northward in the Caucasus ranges where Upper Acheulian is overlain by Mousterian rich in Levallois products.

The early Upper Palaeolithic assemblages in the Levant share a dominance of end scrapers, including the chamfered pieces (*chanfreins*) and burins over the typical Mousterian side scrapers and points. However, the core reduction techniques in the earliest Upper Palaeolithic occurrences indicate a clear shift from the Levallois technique to blade cores, best represented in the stratigraphic sequence of Boker Tachtit from the Negev (Marks 1983a, b). The frequencies of Levallois blanks decrease rapidly from Level 1 through Levels 2 and 3 to Level 4 (all separated by sterile deposits) and are mostly produced by blade core reductions. A number of Levallois points were shaped into Emireh points (with the buts and platforms removed by bifacial retouch). Thus the evidence from Boker Tachtit restored the place of this point as a chronological marker as suggested originally by D. Garrod and R. Neuville, although the possibility of its occurrence in late Mousterian sites such as Shovakh cave (in Wadi Amud), may indicate that its temporal range is longer than is accepted at present.

The technological transition is somewhat blurred in the Levantine cave sites because (1) raw material is smaller than in the Negev, (2) repeated occupations has caused some mixing of levels, (3) natural post-depositional effects led to similar results, and (4) no conjoining of cores and blanks were ever tried by the excavators.

However, the published reports from Ksar Akil (Azoury 1986, Bergman 1987) indicate that the transitional features, namely the dominance of Upper Palaeolithic tool forms and the decreasing presence of Levallois blanks, prevails in Layers 25-21.

It is agreed by all scholars that blade industries developed from the Transitional Industries and are commonly found in sites dated to 38-30 kyr. Two fragmentary skulls from Qafzeh cave, as yet unpublished, and a fragmentary skeleton ('Egbert') uncovered in Ksar Akil Layer 17, are at present the only remains from this time span.

Finally, the issues involved in a comparison between the cultural attributes of the Middle and the Upper Palaeolithic can now be discussed. The Levantine Mousterian sequence as a whole provides ample evidence for mortuary practices such as dug-out graves, semi-flexed burials and occasional grave offerings. Rare use of red ochre (e.g. Qafzeh) and a few occurrences of marine shells (Qafzeh and Skhūl) may add to the limited signs of symbolic behaviour. Rounded and oval hearths (e.g. Kebara cave) and spatial distribution of artifacts and bones reflect some form of organization within habitation sites. Procurement of raw material for tool production seems to indicate a rather limited radius of about 5-15 km in the Levant, while in Europe and Africa cases of rare artifacts brought from distances of 60 up to 200 km are already recorded (Roebrooks et al. 1988). However, the entire subject of raw material economy is currently under investigation in various regions including the Levant and future results may change this first impression.

The early Upper Palaeolithic sites do not seem to differ in size from the Mousterian ones although when the distribution of occupational debris is compared in caves such as El-Wad, Kebara and Hayonim, the Upper Palaeolithic remains cover more limited surfaces and indicate, perhaps, a higher degree of mobility. Mousterian open air sites are often large and demonstrate dense accumulations of artifacts which are interpreted as resulting from repeated occupations (e.g. Rosh Ein Mor, Ain Aqev, Bikat Quneitra). Upper Palaeolithic open-air sites are of smaller sizes and lend the impression of having a more ephemeral nature. Spatial distribution of artifacts and the presence of hearths indicate a clearer intra-site organization in the Upper Palaeolithic (e.g. Marks 1983a, b, Bar-Yosef & Belfer 1977). On the whole, open-air early Upper Palaeolithic sites seem to reflect a highly mobile society, composed of small groups of people (Marks 1983b). Less is known about the Levantine cave sites which were mostly excavated in the 1930's and 1950's.

With the exception of Ksar Akil, no early Upper Palaeolithic burials are known, and only one (Nahal Ein Gev I, see Arensburg 1977) is recorded for the period of 30-20 kyr BP. The use of red ochre is well represented in early Ahmarian (blade industries) in Qafzeh where a grinding stone and hand stone smeared with ochre were found as well as in the Lagaman sites (northern Sinai). Marine shells occur sporadically in the early Upper Palaeolithic and increase considerably around 20 kyr BP. Similarly bone and antler tools are rarely noted in early Upper Palaeolithic assemblages but clearly increase in the Levantine Aurignacian, only to decrease in later industries.

To sum up, although the shift from the Middle Palaeolithic to the Upper Palaeolithic is clearly recognizable in the Levantine archaeological record, it took at least an additional 10 kyr before the entire set of cultural attributes which characterize modern humans can be fully seen.

4 CHRONOLOGICAL AND CULTURAL IMPLICATIONS

Based on the above survey, the following conclusions can be made:

A. There is no direct relationship between the hominid types and the Middle Palaeolithic industries in western Asia. The fossils from Shanidar-Amud-Tabun-Kebara caves occur with various assemblages, including the Zagros non-Levallois Mousterian, 'Tabun C-type', Tabun 'B-type' and another as yet unclassified variant of the 'Levalloiso-Mousterian' industry (in Amud cave).

B. The 'Proto-Cro-Magnons' were embedded in 'Tabun C-type' industry in Qafzeh and in a somewhat similar assemblage in Skhūl (although the techno-typological range deserves a thorough study).

C. No fossils of the 'Tabun D-type' were found. For the Acheulo-Yabrudian we only have the fragmentary skull from Zuttiyeh and the femur from Tabun E. No human fossils have been discovered in an Upper Acheulian site, with the possible exception of the femur in Gesher Benot Ya'acov (Geraads & Tchernov 1983). However, in most of the known Levantine Acheulian occurrences bones were not preserved.

D. Placing the hominids in a chronological sequence, without taking into account the chronology still employed by those who do not accept the TL and ESR dates and the palaeontological sequence of the rodents, would be as mentioned in Table 1.

If the differences between the two Mousterian human types were important, as suggested by several scholars, then the length of the time span when both were contemporary is of relevance to our discussion. Table 1 suggests that this period was perhaps no longer than the Holocene and therefore it is not inconceivable that the two populations remained isolated from each other albeit producing similar lithic industries. This conclusion may sound strange to those who would like to equate a certain type of industry with a distinct human type, but as discussed above, no such correlation existed during the Middle Palaeolithic time in western Asia.

E. If the group of Tabun-Amud-Shanidar and Kebara is defined as western Asian Neanderthals; then, it seems that their later arrival in the Mousterian sequence could be related to an eastward expansion of south-eastern European populations. This was possibly caused by the rapid accumulation of continental and mountain glaciers and the considerable enlargement of periglacial areas for which Mousterian groups did not have the necessary technology to ensure

Table 1. Chronology of sites and fossils in southwest Asia.

Dates	Industry	Fossils
36-28 kyr	Levantine Aurignacian	
43/40-36 kyr	Early U.P.	Qafzeh, Ksar Akil (unpubl.)
55-45/43 kyr	Mousterian	Skhūl Cave (Later burials?)
80/65-55 kyr	Mousterian	Kebara, Tabun C, Amud(?) and Shanidar
120-90/80 kyr	Mousterian	Qafzeh, Skhūl (Early burials?)
130/125-120 kyr	Mousterian/Hummalian	Tabun D, El Kowm
160-130 kyr	Acheulo-Yabrudian	Zuttiyeh, Tabun E
600/500-160 kyr	Upper (or Late) Acheulian	Gesher Benot Ya'acov, Tabun F

a long-term survival (Bar-Yosef 1988). This could have begun in the time of Isotope Stage 5d (around 115 kyr) or during each of the successive colder periods, especially Isotope Stage 4.

F. Cultural attributes which later characterize modern humans of Upper Palaeolithic times were already present during the Mousterian sequence. The Qafzeh dates indicate that the burials exposed there are the earliest recorded mortuary practices (Vandermeersch 1988). Similar observations can be made for the rare use of red ochre or collection of a few marine shells.

Well-delineated hearths and intensive accumulations of ashes are known not only from Tabun, Kebara, Qafzeh and Hayonim caves but also from Klasies River Mouth in South Africa. Similar phenomena were rarely recorded in the European Mousterian.

In generalizing about the Mousterian world we tend to ignore archaeological attributes which signify what could have been Middle Palaeolithic 'cultural zones'. These perhaps reflect common communication systems and ranges of mobility within large mating systems (Wobst 1976). The 'Levantine Mousterian of Levallois facies' is one example and the Aterian (which extends from Atlantic Morocco to the eastern Sahara) is another case. Similarly, the Mousterian with bifacial tools in eastern Europe or the Mousterian of Acheulian Tradition in western Europe may be other examples of 'cultural zones'.

G. Mousterian environmental adaptations were limited in their spatial distribution. Cold and warm deserts remained outside their regularly visited habitats and ephemeral occupations in these ecozones took place only when the local conditions were drastically improved. There is no marked difference in the spatial adaptations between the Neanderthals and the so-called 'Proto-Cro-Magnons' thus indicating that the shift to the Upper Palaeolithic period emerged from a major behavioural or neurological change around 45 kyr (Klein 1989).

H. It is quite possible that the overall contemporaneity between Neanderthals and modern looking hominids is a false impression caused by the paucity of skeletal remains, dating inaccuracies and our tendency to group fossils, sites, assemblages into large time spans. We have no evidence for the presence of Middle Palaeolithic modern humans in the Zagros and therefore the problem of the allegedly 'comtemporaneity' is limited to the Levant and not the entire Near East. One can easily suggest various scenarios in order to explain the supposed contradiction of having two human types surviving in the same general region. One scenario will see the Neanderthals as pushing the modern looking humans into northeast Africa or alternately into the marginal steppic belt of the Near East. The second scenario would suggest a short period of coexistence, say, 5000 years and the disappearance of Neanderthals from the Levant. The third alternative would view the Levant as the boundary zone between the two different hominids with adjacent shifting territories, and so on. We should keep in mind that accepting the current

TL and ESR dates means that we have no human fossils from the Late Mousterian deposits, namely from 58/57 kyr through 45 kyr.

I. The emergence of the Upper Palaeolithic cultural system took place earlier in the Levant than in Europe (and resembles in its overall pattern the origins and dispersals of agricultural subsistence strategies). It cannot be directly related to the appearance of modern looking hominids in Qafzeh or to any known climatic event. The techno-typological changes observed in the lithic assemblages are major shifts in the concepts of core reduction techniques and in tool forms. Such an important departure from older knapping techniques probably reflects additional important cultural changes which are not necessarily preserved in the archaeological record. Whether these behavioural changes were caused by the appearance of modern languages which enabled better communication, storage of information and planning depth, is as yet unknown (Whallon 1989). The emergence of more definable cultural areas in the early Upper Palaeolithic can perhaps be interpreted as indicating both population growth and smaller spatial distribution of mating systems. The recognizable variability among the tool kits still does not indicate a better environmental adaptation.

5 CONCLUDING REMARKS

The above revised chronology for the Mousterian fossils in southwest Asia necessitates a re-examination of suggested hominid phylogenetic relationships. Movements of humans across large distances and gene flow were offered as mechanisms for explaining the available evidence. New dating techniques and more readings with existing techniques will improve the chronological seriation of the hominids, but additional fossils are needed in order to fill the 'blank areas'. Cross cultural comparisons between Middle Palaeolithic sites will clarify the geographic distribution of archaeological attributes and crucial regions or sites for further exploration will stand out.

The emergence of anatomically modern humans occurred biologically during the Mousterian Industrial Complex but cannot explain the major cultural change which marks the beginning of the Upper Palaeolithic. While we can temporally and spatially locate the first appearance of Upper Palaeolithic lithic industries in various regions we have still no idea why it appeared during this particular time span. If the dispersals of the new populations resemble the dispersals of agricultural systems at the end of the Pleistocene and early Holocene, then we need to know the source area. The limiting factors in improving our knowledge are the limits of the radiocarbon dating technique (including the AMS) and the degree of inaccuracy in the TL and ESR techniques. Available models (e.g. Binford 1968, Gilman 1984, Mellars 1989) do not satisfactorily explain the rapid shift from the Mousterian world of populations struggling for survival to the conquering, successful populations of the Upper Palaeolithic, which, in the author's view, remain an important challenge for future research.

ACKNOWLEDGEMENTS

I would like to thank my colleagues B. Arensburg, A. Belfer-Cohen, P. Goldberg, H. Laville, L. Meignen, Y. Rak, E. Tchernov, A.M. Tillier and B. Vandermeersch in the French-Israeli team who currently excavate Kebara Cave for the numerous discussions we had on the dating of early modern humans. I also thank Lynne Schepartz and Fred Smith, David Pilbeam and John Shea who made useful comments on an early draft of this paper. I, however, am solely responsible for the views expressed here.

REFERENCES

Arensburg, B. 1977. New Upper Palaeolithic remains from Israel. *Eretz Israel* 13: 208-215.

Azoury, I. 1986. In C.A. Bergman & L. Copeland (eds), *Ksar Akil, Lebanon,* Vol. I: Levels XXV-XII. Oxford: BAR International Series 289.

Bar-Yosef, O. 1988. The date of the Southwest Asian Neanderthals. In M. Otte (ed.), *L'Homme de Neanderthal,* Vol. 3. Liège: Université de Liège, pp. 31-38.

Bar-Yosef, O. 1989a. Upper Pleistocene human adaptations in Southwest Asia. In E. Trinkaus (ed.), *Patterns and Processes in Later Pleistocene Human Emergence,* Cambridge: Cambridge University Press, pp. 154-179.

Bar-Yosef, O. 1989b. Geochronology of the Levantine Middle Palaeolithic. In P.A. Mellars & C.B. Stringer (eds), *The Human Revolution: Behavioural and Biological Perspectives on the Origins of Modern Humans.* Edinburgh: University of Edinburgh Press, pp. 589-610.

Bar-Yosef, O. & A. Belfer 1977. The Lagaman industry. In O. Bar-Yosef & J.L. Phillips (eds), *Prehistoric Investigations in the Gebel Maghara, Northern Sinai.* Qedem 7. Jerusalem: Hebrew University Institute of Archaeology, pp. 42-88.

Bar-Yosef, O. & P. Goldberg 1988. An outline of the chronology of the Middle Palaeolithic in the Levant. In M. Otte (ed.), *L'Homme de Neanderthal,* Vol. 2. Liège: Université de Liège.

Bar-Yosef, O. & B. Vandermeersch 1981. Notes concerning the possible age of the Mousterian layers in Qafzeh Cave. In J. Cauvin & P. Sanlaville (eds), *Prehistoire du Levant,* Paris: CNRS, pp. 281-286.

Bar-Yosef, O., B. Vandermeersch, B. Arensburg, P. Goldberg, H. Laville, L. Meignen, Y. Rak, E. Tchernov & A.M. Tillier 1986. New data concerning the origins of modern man in the Levant. *Curr. Anthropol.* 27: 63-64.

Bergman, C.A. 1987. *Ksar Akil, Lebanon,* Vol. II.: Levels XIII-VI. Oxford: BAR International Series 329.

Binford, S. 1968. Early Upper Palaeolithic adaptations in the Levant. *Am. Anthropol.* 70: 707-717.

Boutie, P. 1989. Etude technologique de l'industrie mousterienne de la grotte de Qafzeh (pres de Nazareth, Israel). In O. Bar-Yosef & B. Vandermeersch (eds), *Prehistoric investigations in the southern Levant.* Oxford: BAR International Series.

Copeland, L. 1975. The Middle and Upper Palaeolithic of Lebanon and Syria in the Light of Recent Research. In. F. Wendorf & A.E. Marks (eds), *Problems in Prehistory: North Africa and the Levant. Dallas: SMU Press, pp. 317-350.*

Copeland, L. & F. Hours 1983. Le Yabroudien d'El Kowm (Syrie) et sa place dans le Paleolithique du Levant. *Paleorient* 9: 21-38.

Dibble, H. 1984. The Mousterian industry from Bisitun Cave (Iran). *Paleorient* 10: 23-34.

Evins, M.A. 1982. The fauna from Shanidar Cave: Mousterian wild goat exploitation in northeastern Iraq. *Paleorient* 8: 37-56.

Farrand, W.R. 1979. Chronology and palaeoenvironment of Levantine prehistoric sites as seen from sediment studies. *J. Archaeol. Sci.* 6: 369-392.

Farrand, W.R. 1982. Environmental Conditions During the Lower/Middle Palaeolithic Transition in the Near East and the Balkans. In A. Ronen (ed.), *The Transition from the Lower to the Middle Palaeolithic and the Origin of Modern Man.* BAR International Series 151: 105-108.

Garrod, D.A.E. 1962. The Middle Palaeolithic of the Near East and the problem of Mount Carmel man. *J. Roy. Anthropol. Inst.* 92: 232-259.

Garrod, D.A.E. & D. Bate 1937. *The Stone Age of Mount Carmel,* Vol. 1. Oxford: Clarendon.

Geraads, D. & E. Tchernov 1983. Fémurs humains du Pleistocene moyen de Gescher Benot Ya'acov (Israël). *L'Anthropol.* 87: 138-141.

Gilman, A. 1984. Explaining the Upper Palaeolithic Revolution. In M. Spriggs (ed.), *Marxist Perspective in Archaeology.* Cambridge: Cambridge University Press, pp. 115-126.

Gisis, I. & O. Bar-Yosef 1974. New excavations in Zuttiyeh Cave, Wadi Amud, Israel. *Paleorient* 2: 175-180.

Goldberg, P. & H. Laville 1988. Le contexte stratigraphique des occupations paleolithiques de la Grotte de Kebara (Israel). *Paleorient* 14(2): 117-123.

Goren-Inbar, N. 1985. The lithic assemblage of the Berekhat Ram Acheulian site, Golan Heights. *Paleorient* 11: 7-28.

Hours, F. 1982. Un novelle industrie en Syrie entre l'Acheuleén superieur et le Levalloiso-Mousterien. *Archaeologie du Levant*. Lyon: Maison de l'Orient, pp. 33-46.

Howell, F.C. 1959. Upper Pleistocene stratigraphy and early man in the Levant. *Proc. Am. Phil. Soc.* 103: 1-65.

Jelinek, A.J. 1981. The Middle Palaeolithic of the Southern Levant from the perspective of the Tabun Cave. In J. Cauvin & P. Sanlaville (eds), *Prehistoire du Levant*. Paris: CNRS, pp. 265-280.

Jelinek, A.J. 1982. The Middle Palaeolithic in the southern Levant with comments of the appearance of modern *Homo sapiens*. In A. Ronen (ed.), *The Transition from the Lower to the Middle Palaeolithic and the Origin of Modern Man*. Oxford: BAR International Series 151: 57-104.

Jelinek, A.J., W.R. Farrand, G. Haas, A. Horowitz & P. Goldberg 1973. New excavations at the Tabun Cave, Mount Carmel, Israel 1967-1972. *Paleorient* 1: 151-183.

Klein, R.G. 1989. *The Human Career*. Chicago: University of Chicago Press.

Laville, H. & P. Goldberg 1989. The collapse of the Mousterian sedimentary regime and the beginning of the Upper Palaeolithic at Kebara Cave, Mount Carmel. In O. Bar-Yosef & B. Vandermeersch (eds), *Investigations in South Levantine Prehistory*. Oxford: BAR International Series 497, pp. 75-96.

Marks, A.E. (ed.) 1983a. *Prehistory and Palaeoenvironments in the Central Negev, Israel*, Vol. III: The Avdat/Aqev Area Part 3. Dallas: SMU Press.

Marks, A.E. 1983b. The Middle to Upper Palaeolithic Transition in the Levant. In F. Wendorf & A.E. Close (eds), *Advances in World Archaeology* 2: 51-98.

Meignen, L. & O. Bar-Yosef 1988a. Variabilité technologique au Proche Orient: l'example de Kebara. In M. Otte (ed.), *L'Homme de Neanderthal*, Vol. 4. Liège: Université de Liège, pp. 81-95.

Meignen, L. & O. Bar-Yosef 1988b. Kebara et le Paleolithique Moyen du Mont Carmel (Israel). *Paleorient* 14(2): 123-136.

Meignen, L. & O. Bar-Yosef 1989. Nouvelles recherches sur le Paléolithique Moyen d'Israel: La Grotte de Kebara, unites VII-XII. In O. Bar-Yosef & B. Vandermeersch (eds), *Investigations in South Levantine Prehistory*. Oxford: BAR International Series 497, pp. 169-184.

Mellars, P. 1989. Major issues in the emergence of modern humans. *Curr. Anthropol.* 30: 349-385.

Ohnuma, K. & T. Akazawa (in press). A Techno-typological Study of the Lithic Artifacts from Layer B of the Amud Cave, Israel. *Paleorient*.

Rak, Y. & B. Arensburg 1987. Kebara 2 Neanderthal pelvis: First look at a complete inlet. *Am. J. Phys. Anthropol.* 73: 227-231.

Roebrooks, W., J. Kolen & E. Resnik 1988. Planning depth anticipation and the organization of Middle Palaeolithic technology: The 'archaic natives' meet Eve's descendants. *Helinium* 28(1): 17-34.

Ronen, A. 1979. Palaeolithic industries. In A. Horowitz (ed.), *The Quaternary of Israel*. New York: Academic Press, pp. 296-307.

Schroeder, B. 1969. The Lithic Industries of Jeft Ajla and Their Bearing on the Problem of a Middle to Upper Palaeolithic Transition. Unpublished Ph.D. Thesis, Columbia University: University Microfilms International.

Schwarcz, H., W.M. Buhay, R. Grun, H. Valladas, E. Tchernov, O. Bar-Yosef & B. Vandermeersch (in press). ESR dating of the Neanderthal site, Kebara Cave, Israel. *J. Archaeol. Sci.*

Schwarcz, H., P. Goldberg & B. Blackwell 1980. Uranium series dating of archaeological sites in Israel. *Israel J. Earth Sc.* 29: 157-165.

Schwarcz, H., R. Grun, B. Vandermeersch, O. Bar-Yosef, H. Valladas & E. Tchernov 1988. ESR Dates for the Hominid Burial Site of Qafzeh in Israel. *J. Hum. Evol.* 17: 733-737.

Smith, P.E.L. 1986. *Palaeolithic Archaeology in Iran*. Philadelphia: University of Pennsylvania Museum.

Smith, P. & B. Arensburg 1977. A Mousterian skeleton from Kebara Cave. *Eretz Israel* 13: 164-176.

Smith, P. & A.-M. Tillier 1989. Additional infant remains from the Mousterian strata, Kebara Cave (Israel). In O. Bar-Yosef & B. Vandermeersch (eds), *Prehistoric investigations in South Levantine Prehistory*. Oxford: BAR International Series 497, pp. 323-336.

Solecki, R.S. 1963. Prehistory in the Shanidar Valley, Northern Iraq. *Science* 139: 179-193.

Stringer, C.B., R. Grün, H.P. Schwarcz & P. Goldberg 1989. ESR dates for the hominid burial site of Es Skhūl in Israel. *Nature* 338: 758.

Suzuki, H. & F. Takai (eds) 1970. *The Amud Man and his Cave Site*. Tokyo: University of Tokyo Press.

Tchernov, E. 1981. The biostratigraphy of the Middle East. In J. Cauvin & P. Sanlaville (eds), *Prehistoire du Levant*. Paris: CNRS, pp. 67-98.

Tillier, A.-M., B. Arensburg, Y. Rak & B. Vandermeersch 1988. Les sépultures néanderthaliennes du Proche – Orient: état de la question. *Paleorient* 14: 130-136.

Trinkaus, E. 1983. *The Shanidar Neandertals*. New York: Academic.

Trinkaus, E. 1984. Western Asia. In F. Smith & F. Spencer (eds), *The Origins of Modern Humans: A World Survey of the Fossil Evidence*. New York: Liss, pp. 251-293.

Valladas, H., J.L. Joron, G. Valladas, B. Arensburg, O. Bar-Yosef, A. Belfer-Cohen, P. Goldberg, H. Laville, L. Meignen, Y. Rak, E. Tchernov, A.-M. Tillier & B. Vandermeersch 1987. Thermoluminescence dates for the Neanderthal burial site at Kebara (Mount Carmel), Israel. *Nature* 330: 159-160.

Valladas, H., J.L. Reyss, J.L. Joron, G. Valladas, O. Bar-Yosef & B. Vandermeersch 1988. Thermoluminescence dating of the Mousterian proto-Cro-Magnon remains of Qafzeh Cave (Israel). *Nature* 331: 614-616.

Vandermeersch, B. 1982. The first *Homo sapiens sapiens* in the Near East. In A. Ronen (ed.), *The Transition from the Lower to the Middle Palaeolithic and the Origin of Modern Man*. Oxford: BAR International Series 151: 297-300.

Vandermeersch, B. 1988. A propos des plus anciennes sépultures. *Bull. Soc. Prehist. Fran.* 85(5): 98-99.

Whallon, R. 1989. Elements of cultural change in the Later Palaeolithic. In P. Mellars & C. Stringer (eds), *The Human Revolution*. Edinburgh: Edinburgh University Press, pp. 433-454.

Wobst, H.M. 1976. Locational relationships in Palaeolithic Society. *J. Hum. Evol.* 5: 49-58.

Yalçinkaya, I. 1988. Resultats récents des fouilles á Karain en Anatolie. In M. Otte (ed.), *L'Homme de Neanderthal*. Liège: Université de Liège. Vol. 8: 257-272.

The origin of anatomically modern humans in east Asia

Phillip J. Habgood
Department of Anthropology, University of Sydney, Sydney, Australia

There are two extreme explanations as to the origin of anatomically modern (AM) humans (Stringer & Andrews 1988). The Rapid Replacement Hypothesis argues for a single origin for AM humans, usually taking place in Africa and/or western Asia, and an outward migration and rapid replacement of all existing archaic populations (Stringer & Andrews 1988, Stringer et al. 1984). The Regional Continuity Hypothesis proposes that AM humans evolved in a number of geographical regions from already differentiated ancestral populations and argues for both morphological continuity within the differentiated geographical regions and the presence of distinct regional morphological features (Stringer & Andrews 1988, Wolpoff et al. 1984).

Most discussions of the origin of AM humans concentrate on the western half of the Old World, but in recent years more attention has been paid to the eastern sector (Groves 1989, Habgood 1989a, b, Wolpoff 1985, 1989, Wolpoff et al. 1984). It has also been argued on cladistic grounds that Asian *Homo erectus* material did not contribute to the origin of AM humans (Andrews 1984, Stringer 1984, Wood 1984). There are, however, significant problems with both the method used and the conclusion reached by these authors (Habgood 1989c).

Elsewhere I have argued that there is a combination of at least four morphological features (a long and sagittally flat frontal bone, a posterior position of minimum frontal breadth, eversion of the lower border of the malar, and presence of a malar-zygomaxillary tuberosity) that, although they are not found on the earliest AM crania from the region (Niah Cave and Lake Mungo 1 and 3), indicate some degree of morphological continuity within Australasia (Habgood 1989a, b). This paper will examine skeletal material from another geographic region, east Asia, so as to ascertain if there is evidence for morphological continuity or discontinuity in this sample. Genetic data will not be dealt with, but discussions of it may be found in Stringer & Andrews (1988), Wolpoff (1989) and Wolpoff et al. (1984).

Weidenreich felt that

Not less than twelve peculiarities, characteristics of the cranial and limb bones of 'Sinanthropus', and of the teeth also – some having a high percentage of occurrence in the available specimens – occur with a relatively high degree of frequency in Mongolian groups.

(1943:276-277)

These features are the presence of: A mid-sagittal crest and parasagittal depression, metopic suture, Inca bone, infraorbital margin rounded and even with the floor of the orbit, more perpendicularly orientated nasal bones, wide nasal bones with little difference between upper and middle breadths, high profile angle of the nasal floor, forwarding-jutting malars, maxillary exostosis, exostosis of external auditory meatus, mandibular torus, a high degree of platymerism of the femur combined with a strong deltoid tuberosity of the humerus, and shovel-shaped upper lateral incisors (Weidenreich 1943:277).

Weidenreich's list of 'regional features' has been adopted and added to (Aigner 1976, 1978a, Coon 1962, Thoma 1964, Wolpoff 1980, 1985, Wolpoff et al. 1984, Wu Rukang 1986a, Wu Rukang & Wu Xinzhi 1982, Wu Xinzhi & Wu Maolin 1985). Coon (1962) added the presence on 'Sinanthropus' 3, the adolescent skull from Locus E (Black 1930), of a fissure in the tympanic plate, known as the infantile gap of foramen of Huschke, which remained open. He argued that in European crania this fissure closed before the age of 5 years, whereas in Mongoloids it tended to remain open into adulthood. Wu Xinzhi & Wu Maolin (1985) added: The contact of the frontonasal and frontomaxillary sutures are on almost the same level and form a slight arc, flattened nasal bones, and the angular rather than rounded junction of the zygomatic bone and the zygomatic process of the maxilla. Wolpoff et al. (1984) added: An early reduction in facial size, especially height, and posterior tooth size, relatively small frontal sinuses restricted to the interorbital area, and third molar agenesis. Wolpoff (1985) added: Lack of anterior facial projection and low degree of prognathism, and the shape of the orbits. Other features, such as the presence of a marginal process below the zygomaticofrontal suture, mentioned by Weidenreich (1943:87, Fig. 167) as being common on 'Sinanthropus' remains and on Mongoloid crania could also be added to the list of possible 'regional features'.

These 'regional features' are said to have a higher frequency of occurrence in prehistoric Chinese fossils and modern Mongoloid crania than on populations from other geographic regions. Weidenreich expressed this view when he wrote

All features, with very few exceptions, used for discriminating races can be found in many races; the question is to what percentage or degree of development do they occur in given groups. (1943:80)

Wolpoff et al. concur, suggesting that

Although all of these characteristics are not absolutely absent in other populations of the Pleistocene world, they have much lower frequencies and are distributed discontinuously in regions other than China. (1984:435)

Wolpoff et al. (1984) also suggest that some of the proposed 'regional features' change through time. Some, such as the degree of sagittal keeling, and the frequencies of Inca bones and mandibular exostoses, become less common, but they are still said to be found in higher frequencies on Mongoloid crania than on other crania.

If these proposed 'regional features' are to be of use in documenting morphological continuity in east Asia as has been suggested (Aigner 1976, 1978a, Coon 1962, Jelinek 1982, Kwang-chih Chang 1962, Smith 1985, Weidenreich 1943, 1947, Wolpoff 1980, 1985, Wolpoff et al. 1984, Wu Rukang 1986a, Wu Rukang & Dong Xingren 1985, Wu Rukang & Wu Xinzhi 1982, Wu Xinzhi & Wu Maolin 1985, Wu Xinzhi & Zhang Zhenbiao 1985), they cannot be primitive retentions that are commonly found on *Homo erectus* and/or archaic *Homo sapiens* crania from other regions. Their frequencies must also always be higher in Mongoloid populations than in other populations. A recent detailed evaluation of these traits concluded that they were not of use as 'regional features' in that many were primitive retentions, and because most modern Mongoloid populations did not consistently have a closer relationship, as reflected by the frequencies of the various features, to Chinese *Homo erectus* material than populations from other geographical regions (Groves 1989).

Two traits will be discussed in detail so as to demonstrate the variation in frequencies that occur with most of the proposed 'regional features'.

1. Agenesis of the third molar. Wu Rukang (1986a) described the congenital absence of the third molar on the *Homo erectus* mandible from Lantian, and on the cranium from Liujiang. Groves (1989) cites an early jaw from Omo as lacking third molars. The Tabon mandible from Palawan Island, the Philippines, also lacks third molars (Barker 1978). The Cave of Hearths mandibular fragment appears to have lacked a right third molar (Tobias 1971). The Niah cranium from Sarawak possibly displays a retarded third molar eruption (Bulbeck 1981, compare the aging of the specimen by Birdsell [1979] and Brothwell [1960]). Agenesis of the third molar is therefore, found on early hominid crania and mandibles from throughout the Old World and not just those from east Asia.

A cline of decreasing frequency of third molar agenesis from northwest Asia to Melanesia and Australia has been suggested (Brothwell et al. 1963). Some Mongoloid samples, such as Mokapu Hawaiians (Bulbeck 1981) and Eskimo samples (Brothwell et al. 1963, Cadien 1972), do have high frequencies for this trait, but others, such as some American Indian, Hawaiian, Burmese, Japanese and Nicobarese samples (Brothwell et al. 1963, Bulbeck 1981, Cadien 1972, Ossenberg 1976), have relatively low frequencies that are exceeded by European, Melanesian and African samples (Brothwell et al. 1963, Bulbeck 1981, Cadien 1972, Ossenberg 1976). Again, a high frequency of this feature is not always typical of all Mongoloid samples and populations.

2. Persistence at maturity of a foramen of Huschke. The frequency of persistence is quite variable. De Villiers (1968) found that in South African Negro crania it was rare, being found in only 1.2% of her sample, whereas in eastern Cape Bushman crania it occurred in 30.7% of her samples. De Villiers (1968) also gives the frequency of its occurrence in some other populations which ranges from 3% of an Hawaiian sample to 32% of a sample from Guam.

Berry & Berry (1967) also provide data on the geographical distribution of this trait. They found that its highest frequencies were 30.4% of a sample from Nigeria, 24.5% of a sample from Burma, and 46.3% of a sample from Peru. In samples from ancient Egypt the frequency varied from 3.9% up to 38% (Berry et al. 1967). Berry's (1975) study provided the frequency for males and females which he found to vary a good deal. A sample from Burma demonstrated a range from 11.5% for males to 38% for females.

Kellock & Parsons (1970a, b) provide data for Australia and the Pacific. The highest frequencies they found were 17.1% in their Melanesian sample, 14.7% for a combined Pacific Ocean sample, 10.6% for a sample from northwestern Australia (consisting predominantly of the Northern Territory and northern Western Australia), and 10.1% for a Northern Territory sample, while only 7.7% of their Polynesian sample retained at maturity a foramen of Huschke.

There is, therefore, a great deal of variation in the frequency of this trait. There is no consistent pattern of Mongoloid samples having higher frequencies than other samples. There appears to be a great deal of overlap between samples from different geographical regions, and so it would appear that this feature is more population specific and not geographically and/or racially specific.

Most of the other proposed 'regional features' display a similarly variable frequency distribution with only some, but not all, Mongoloid samples displaying higher frequencies than samples from other geographical regions.

The usefulness of auditory exostosis as a 'regional feature' must also be doubted because it may not be totally genetic in origin, but possibly a behavioral attribute (Kennedy 1986, Pardoe 1986). Kennedy states that

While the use of auditory exostosis frequencies cannot be justified in estimates of genetic distance, their moderate to high frequency in a population does point strongly to the exploitation of cold water resources. (1986:412)

A high degree of platymerism of the femur is now known to be a general feature of *Homo erectus* and not unique to Chinese remains (Day 1971, 1984, Howell 1978, Kennedy 1983, 1984a, b). This may also be the case with a strong deltoid tuberosity on the humerus, which Howell (1978) lists as being typical of *Homo erectus* in general.

Some of the features proposed as being 'regional' are also not of use because they do not appear to reflect the pattern that is actually present on the cranial material.

Wolpoff et al. (1984) suggest that in China there is an earlier reduction in facial size, especially height, and posterior tooth size than in other regions, such as Australasia. Weidenreich (1943:Table 32) gives the reconstructed upper facial height of 'Sinanthropus' as 77 mm, which is greater than that which Thorne & Wolpoff (1981) provide for Sangiran 17 (74.5 mm). On 'face' value this would seem to negate the suggestion made by Wolpoff et al. (1984) for an earlier reduction in east Asia. The splanchnocranium of Sangiran 17, however, has been squashed up under the neurocranium and so the upper facial height has probably been underestimated. The clival part of the nasoalveolar region of KNM-ER 3733 has been eroded (Leakey & Walker 1985), but it does not appear as though the upper facial height would be much greater than that which Weidenreich (1943) provides for 'Sinanthropus'. The bimaxillary breadth of 'Sinanthropus' (Weidenreich 1943) is smaller than that of Sangiran 17 (Thorne & Wolpoff 1981) and KNM-ER 3733. The range for the upper posterior dentition of 'Sinanthropus' (Weidenreich 1937) overlaps with the respective tooth sizes of KNM-ER 3733 (Leakey & Walker 1985), KNM-WT 15 000 (Brown et al. 1985), *Pithecanthropus* 4 (Day 1977) and Sangiran 17 (Day 1977), and so there would not appear to be any major justification for suggesting an earlier reduction in the size of posterior dentition in east Asia.

Wolpoff et al. (1984) also contend that relatively small frontal sinuses that are restricted to the interorbital area are a 'regional feature', and that the absolute frontal sinus size reduces through time. In general, 'Sinanthropus' had small frontal sinuses confined to the interorbital or glabellar region, although Skull 3 had larger sinuses that extended laterally to the medial orbital walls where they continued into the orbital roof (Weidenreich 1943). The Maba calotte, however, does not fit the suggested typical pattern in that it has frontal sinuses that are 'very large, extending laterally well into the supraorbitals' (Wu Xinzhi & Wu Maolin 1985:97-8). The Ziyang (Tzeyang) calvaria is also described as having large frontal sinuses (Woo Jukang 1958). Both of these hominids have frontal sinuses that are larger than those of the earlier *Homo erectus* material. If one takes this as evidence, frontal sinuses increase in size through time until at least the Early Upper Pleistocene period in China. There is also great variation in the size of frontal sinuses in modern crania (Weidenreich 1943).

Wolpoff (1985) states that the shape of the orbits is a distinct east Asian facial feature, but he does not describe the form of the orbit to which he is referring. Both Weidenreich (1943) and Coon (1962) suggest that the orbits of Chinese *Homo erectus* are not as low as they initially appear. Weidenreich (1943:Table 33) gave the orbital index for 'Sinanthropus' as 81.9 when the maxillofrontale is used as the median landmark for orbital breadth, and 90 when the lacrimale is used. Stringer (pers. comm.), using Howells' (1973) definitions, records the values of 35 mm and 41 mm for the orbital height and breadth, respectively, on Skull 3 from Zhoukoudian Locality 1, which provides an orbital index of 85.4 for this individual. This index is higher than that obtained for KNM-ER 3733 (80.9) and KNM-ER 3883 (81.4) when Howells' (1973) definitions are also used. Sangiran 17 has an even higher orbital index of 93.3(?). Wu Xinzhi (1986) argues that Chinese Upper Palaeolithic crania have low orbits. It is, therefore, difficult to ascertain to what shape Wolpoff is referring. In any case, low orbits are commonly found on archaic *Homo sapiens* crania, such as Petralona, Arago 21, Bodo 1, Jebel Irhoud 1 and Skhūl 5, and early modern crania from Europe including Cro-Magnon 1, Mladeč 1, Předmost 3, and Oberkassel 1. Also, Maba has a round orbit, like the Neanderthals, with an orbital index of 88 (Wu Xinzhi & Wu Maolin 1985) which cannot be classified as low. In China the height of the orbit also increases from the Late Pleistocene into the Holocene (Wu Xinzhi & Zhang Zhenbiao 1985).

Wolpoff (1985) also argues that east Asia is characterised by a lack of anterior facial projection and a low degree of prognathism. This feature is difficult to evaluate as few *Homo erectus* crania with complete facial skeletons have been recovered. The splanchnocranium of Sangiran 17 has been squashed up under the neurocranium and so it is difficult to accurately ascertain its degree of facial prognathism. Using the measurements provided by Weidenreich (1943:Tables 26, 32), a gnathic index of 108.1 is obtained for 'Sinanthropus'. This index is only a little less than that for KNM-ER 3733 (110.3) and is greater than that for Kabwe 1 (105). The Yingkou and Dali crania appear to have only limited facial prognathism, but Yingkou is heavily reconstructed (using Dali as the model), and the splanchnocranium of Dali has been pushed and twisted up under the neurocranium. However, the Maba calotte may have been more prognathic than Dali and Yingkou (Smith 1985). There would, therefore, seem to be little justification in arguing that east Asia is typified by a lack of anterior facial projection and a low degree of prognathism.

Some of the proposed 'regional features' which are not totally dependent on population/sample frequencies may thus be examined on isolated hominid crania. This was attempted for nine of the features (Table 1) so as to ascertain if these features are unique to east Asia or have a wider distribution and may be regarded as primitive retentions.

1. Sagittal crest (keel) with parasagittal depressions. Weidenreich (1943:252) admits that this feature 'is somewhat changed in modern skulls', and that it is also commonly found on Melanesian and Australian crania. Although most Australian crania lack parasagittal depressions, most, as Weidenreich stated, do have sagittal keeling. Fenner (1939) found this feature in 86% of his female and 93% of his male sample. Larnach & Macintosh (1966, 1970) found sagittal keeling on 86% of their female and 98.5% of their male sample from coastal New South Wales (93% of total sample), and on 89.4% of their female and 98.6% of their male sample from Queensland. Groves (1989) argues that most Mongoloid crania lack this feature and those, like Eskimos, that have a sagittal crest lack the parasagittal depressions. Many archaic *Homo sapiens* crania, including Kabwe 1, Bodo 1, Petralona and Arago 21, display a sagittal crest or sagittal keeling (Table 1; Weidenreich 1943). Weidenreich (1943) also mentions that this feature is seen on the *Pithecanthropus* 4 *Homo erectus* from Indonesia. Howell (1978:194) states that *Homo erectus* usually has a 'distinct sagittal thickening...often associated with marked parasagittal depression'.

2. A non-depressed nasal root. This feature can be seen on the Keilor cranium from Australia, the frontofacial fragment from Zuttiyeh, Israel and Klasies River Mouth No. 16425 and Border Cave 1 and possibly Florisbad from sub-Saharan Africa (Table 1).

3. Non-projecting and more perpendicularly oriented nasal bones. The nasal bones are often not well preserved on fossil crania, but those from Europe and Africa that do preserve them appear to have more prominent nasal bones (Table 1). KNM-ER 3883 does, however, appear to have had non-projecting nasal bones (Leakey & Walker 1985). The Keilor and Lake Tandou

Table 1. Some proposed 'regional features' for east Asia examined on crania from around the world. Numbers of features follow text. Y = present, - = absent, ? = questionable identification or area not preserved. * Thorne (1976) states that the Kow Swamp crania do not display sagittal keeling.

Crania	Feature 1	2	3	4	5	6	7	8	9
Homo erectus									
KNM-ER 3733	-	-	-	-	?	?	Y	?	-
KNM-ER 3883	-	-	Y?	-?	?	?	?	?	-
Olduvai H9	-	?	?	?	?	?	?	?	?
Sangiran 17	-	-	-?	Y	?	?	-	-	Y/-
Ngandong 11	-	-	?	?	?	?	?	?	?
Zhoukoudian (reconstruction)	Y	Y	Y	Y	?	Y	Y	Y	Y
East Asia									
Yingkou	Y?	-?	Y?	Y	Y	Y	?	?	?
Dali	Y	Y?	Y?	Y	?	Y	Y?	-?	Y?
Maba	Y	Y	Y	Y	?	?	?	?	?
Liujiang	-	Y	Y	Y	Y	Y?	Y	Y	Y
UC 101	Y	-	-	Y	Y	-?	Y	Y	Y
UC 102	Y	Y	-	Y	?	Y?	Y	Y	Y
UC 103	Y	Y	-	Y	?	Y?	Y	Y	Y
Australasia									
Wadjak 1	Y	Y	Y	Y	?	Y?	Y	-	Y
Wadjak 2	?	Y	Y	Y	?	?	?	?	?
Keilor	Y	Y	Y	Y	?	Y?	Y	-	Y
Kow Swamp 1*	Y/-	-	-?	-	?	-?	Y	-	Y
Kow Swamp 5*	Y/-	-	-	-	?	-	Y	-	Y
Kow Swamp 15*	Y/-	-	?	-	?	-	Y	-	Y
Africa									
Jebel Irhoud 1	Y	-	?	Y	?	-?	?	Y	Y
Bodo 1	Y	-	-	Y	?	Y?	?	-	Y
Ndutu	-?	-	-	?	?	-?	?	?	?
Kabwe 1	Y	-	-	-	?	Y/-	-	-	Y
Omo 1	-?	-?	?	?	?	?	?	?	Y
Border Cave 1	Y	Y	?	?	?	?	?	?	Y
Florisbad	-	Y?	?	?	?	?	?	-?	Y
Western Asia									
Zuttiyeh	-	Y	?	Y	?	-?	-	?	-
Skhūl 4	-	?	?	Y	?	?	?	-	Y
Skhūl 5	-	-	?	Y	?	?	-	-	Y?
Jebel Qafzeh 9	-	?	?	Y	?	?	-	-	Y
Amud 1	-	-	-	-	?	?	?	?	-
Europe									
Petralona	Y	-	-	-	?	Y/-?	-?	-	Y
Arago 21	Y	-	-	-	?	Y/-	-	-	Y
Steinheim	-	-	-	-	?	-?	-	-	Y
La Ferrassie 1	-	-	?	-	?	?	-	-	Y
Cro-Magnon 1	-	-	-	-	?	?	-	-	Y
Oberkassel 1	Y	-	-	-	?	?	-	-	Y
Oberkassel 2	Y	-	-	-	?	?	-	-	-
Dolní Věstonice 3	-	-	-	-	?	?	-	-	-
Předmost 3	Y	-	-	-	?	?	-	-	-
Předmost 4	-	-	-	-	?	?	-	-?	-
Mladeč 1	-	-	-	-	?	?	-	-	-

crania from Australia also appear to have more perpendicularly oriented nasal bones.

4. Flat upper faces with forward-jutting malar bones. The extent to which this feature is expressed is variable within Mongoloid samples, with some overlapping with sub-Saharan African and European samples (Bulbeck 1981, Groves 1989). A flat face is not common on archaic *Homo sapiens* crania from Europe, but is found on Bodo 1 from Ethiopia, Jebel Irhoud 1 from North Africa, and Zuttiyeh, Skhūl 4, Skhūl 5, Jebel Qafzeh 9 and Jebel Qafzeh 6 from western Asia (Table 1). Sangiran 17 also has a relatively flat face. Howell (1978) suggests that 'forward-facing malar facies' are commonly found on *Homo erectus* crania. Plio-Pleistocene hominids from Africa also display flat faces (Groves 1989, Rak 1983, 1985). It has also been argued that flat faces are a primitive feature (Smith & Paquette 1989, Stringer 1989).

5. Shovel-shaped incisors, especially upper laterals. Many authors provide data on the percentage of this feature for various populations (cf. Birdsell 1967, Bulbeck 1981, Cadien 1972, Campbell 1925, Carbonell 1963, Hrdlička 1920, Riesenfeld 1956, Turner 1987). In general, Mongoloid populations have the highest frequencies, but there is a large range, with some Mongoloid populations, such as Hawaiian samples (cited in Bulbeck 1981, Carbonell 1963, contra Hrdlička 1920) and various southeast Asian populations (Turner 1987) having quite low frequencies. There also appears to be geographical variation in occurrence of this feature in east Asia itself (cited in Carbonell 1963), and a decrease in frequency from west to east across the Pacific Ocean (Riesenfeld 1956). The frequency of shovel-shaped incisors in some non-Mongoloid groups (Cadien 1972, Riesenfeld 1956). As Carbonell (1963) states, 'it is evident that shovel-shaped incisors vary considerably in different recent populations'.

Wu Rukang (1986a:152) contends that 'all fossil human upper incisors so far found in China, without exception, are shovel-shaped'. This feature, however, is also commonly found on fossil crania from outside of east Asia (Carbonell 1963). It is commonly found on: Australopithecine crania (Robinson 1956); *Homo erectus* material (Howell 1978) including that from Rabat (Coon 1962) and also Sangiran 4 (von Koenigswald 1978); and Neanderthal crania (Coon 1962) including La Quina H18 (Martin 1926), Le Moustier (Carbonell 1963, Weidenreich 1937), Teshik-Tash (Weidenreich 1945), Tabun 1 (McCown & Keith 1939), Shanidar 2 (Trinkaus 1983) and the material from Krapina (Smith 1976), and, early modern crania such as Skhūl 5 (Carbonell 1963), Grotte des Enfants 6 (Coon 1962), and north African Mesolithic material from Jebel Sahaba (Anderson 1968) and Wadi Halfa (Greene & Armelagos 1972). Wu Rukang & Wu Xinzhi (1982) suggest that shovel-shaped incisors are common features of early hominids, while Cadien (1972:208) states that shovel-shaped incisors are 'an ancient characteristic that today shows a large amount of geographical variation'.

This feature is, therefore, a primitive retention found on many early crania throughout the Old World and not unique to east Asia (Table 1).

6. A relatively horizontal course of the frontonasal and frontomaxillary sutures. Weidenreich (1943) stated that this form was not only common among Mongoloid populations, but also Negro populations. He also contended that Kabwe 1 had a similar configuration, but examination of the original specimen reveals that only the medial segment of the suture is relatively horizontal while the remainder angles downwards. Murrill (1975) contends that on Petralona the frontonasal and frontomaxillary sutures form a continuous, almost horizontal line. The nasal region is often damaged on hominid crania and so it can be difficult to clearly identify the exact configuration of the frontonasal and frontomaxillary sutures, but it would appear that a relatively horizontal course for them is not commonly found on archaic *H. sapiens* crania (Table 1).

7. The presence of a marginal process below the zygomaticofrontal suture. Weidenreich stated that a marginal process is

...found as an occasional structure in all races of mankind. But there cannot be any doubt that it is more frequent and more pronounced in Mongolian races than in others. (1943:87)

Weidenreich's 'Sinanthropus'-Mongoloid pattern, however, is also found on the *Homo erectus* KNM-ER 3733 crania, and on prehistoric Australian crania including Keilor, and Kow Swamp 1, 5, 7 and 15 (Table 1). Murrill (1975) contends that Petralona has a *processus marginalis*. A marginal process does not appear to be commonly found on archaic or early modern crania from Europe (Table 1). Fenner (1939) found a moderate to great development of a marginal process on 70% of his male and 49% of his female sample of Australian Aboriginal crania. Larnach & Macintosh (1966, 1970) found a medium to large marginal process on 55.2% of their female and 56.1% of their male sample from coastal New South Wales (55.7% of total sample), and 88.9% of their female and 81.5% of their male sample from Queensland (84.5% of total sample). However, they stated that

No skull in this series (coastal New South Wales) showed such great development of the marginal process as can be seen in some skulls from the Chatham Island, the Jordan Valley and in some Chinese. (1966:36)

These data suggest that a developed marginal process is not unique to modern east Asian crania.

8. An angular rather than rounded junction of the zygomatic bone and zygomatic process of the maxilla forming a deep notch. Weidenreich (1943:83) referred to this notch as the *incisura malaris*. Krogman seems to have been referring to a similar pattern when he stated that

In a considerable number of (Australian) skulls, particularly males, the inferior margin of the maxilla is concave superiorly, so that the lower portion of the maxillo-malar suture (facial aspect) is situated on a downward projection lateral to the concavity. (1932:407)

Fenner (1939) found this feature on 47% of males and 22% of females in his series of Australian Aboriginal crania, whereas Larnach & Macintosh (1966, 1970) found it on 57.8% of males and 49% of females from their coastal New South Wales sample (54% of total sample), and 77.6% of males and 53.3% of females from their Queensland sample (67% of the total sample). Larnach & Macintosh (1966:36) also stated that 'Although no figures are available, this concavity does not seem uncommon in other races'. A similar notch is found on Jebel Irhoud 1.

9. Rounding of the infraorbital margin of the orbit. This feature is also suggested to be a 'regional feature' for Australasia (Thorne & Wolpoff 1981). This feature is present on most *Homo erectus* and archaic *Homo sapiens* crania as well as on Australian Aboriginal crania (Habgood 1989a).

From this discussion it is evident that none of the proposed 'regional features' can be said to be documenting 'regional continuity' in east Asia as they are commonly found on modern crania from outside of this region (especially evident on Australian Aboriginal crania), and are consistently found on archaic *Homo sapiens* and/or *Homo erectus* crania throughout the Old World. These features cannot, therefore, be used to document a 'morphological clade' in east Asia because they are not unique to this region.

However, a number of the features, such as a non-depressed nasion, more perpendicularly oriented nasal bones, frontonasal and frontomaxillary sutures on almost the same level, and an angular rather than a rounded junction of the zygomatic bone and the zygomatic process of the maxilla, appear to have only a limited distribution outside of east Asia. They are also not all commonly found on single individuals from outside of east Asia (Table 1). The Keilor cranium does, however, have three of these features (Table 1). All of these features can be related to facial flatness, and the greater retention of this 'primitive' character may typify the east Asian region, as all Middle Pleistocene and early Upper Pleistocene hominid crania recovered from there have all of the features expressed to a marked degree (Table 1). As mentioned before, it has been suggested that a flat face was the primitive condition for archaic *Homo sapiens* (Smith & Paquette 1989, Stringer 1989). For example, Bodo 1 from Ethiopia has a flat face, but does not display the remaining features (except possibly a relatively horizontal course for the frontonasal and frontomaxillary sutures), whereas the Dali cranium has all of the features. A greater retention of particular features that are related to facial flatness may typify morphological development in east Asia, and so document a certain degree of 'regional continuity'.

One problem with this suggestion is that some late Pleistocene-early Holocene crania from China do not display all of these features (Table 1).

The three Zhoukoudian Upper Cave crania, which are dated to between 18 865 ± 420 BP (ZK-136-0[1]) and 10 470 ± 360 BP (ZK-136-0[3]) with a placement closer to the latter date favoured on stratigraphical grounds (Wu Xinzhi & Zhang Zhenbiao 1985, Zhou Mingzhen et al. 1982, contra Aigner [1972], who on faunal grounds dates them earlier), are often seen as a discontinuity in any evolutionary sequence in China. Weidenreich stated

The most surprising fact is, however, that each one of the skulls offer quite a different physiognomy which cannot be considered as a mere expression of sexual differences because they concern not only the general form and shape of the brain case and face but also some characteristic details. (1939:163)

He suggested that the three crania typified different 'racial elements', namely primitive Mongoloid or possibly Europoid (UC 101), Melanesoid (UC 102), and Eskimoid (UC 103) types. Weidenreich felt that

...we look in vain for special features which might indicate a closer relationship between 'Sinanthropus' and the population of the 'Upper Cave' of Choukoutien. Therefore, it seems at first glance that no such line leads from Sinanthropus to any special racial group of the Far East...But such a conclusion would be too hasty. (1943:251)

Weidenreich (1939:171) clarified this by suggesting that his study of the Upper Cave crania demonstrated that a 'direct relationship between 'Sinanthropus' and the Upper Cave people cannot be excluded'.

Neumann (1956:380) studied casts of the Upper Cave crania and concluded that they were proto-Mongoloids which displayed features that 'appear repeatedly in various early American Indian populations'. Wu Xinzhi also re-examined the Upper Cave material. He concluded that

...the Upper Cave Man represents the primitive Mongoloid type and is closely related to the ancestors of modern Chinese, Eskimo and Amerindians though there are still some typical Mongolian characters not yet well developed in that time. (1961:203)

Coon (1962) also felt that Upper Cave 101 resembled Amerindian crania. He argued that

...the Upper Cave skulls from Choukoutien approach the end of the 'Sinanthropus'-Mongoloid line, bearing the same kind of relationship to the modern Chinese that the Upper Palaeolithic skulls of Europe do to modern Europeans. (1962:475)

Howells (1983) found that in a discriminant function analysis Upper Cave 101 was placed near Plains Indians and, at a slightly further distance, northern Europeans. He felt that

Such an assessment, essentially making the Upper Cave individuals pre-Chinese, 'unspecialized' Mongoloids resembling American Indians, is logical and satisfying. (1983:301)

Other multivariate analyses comparing Upper Cave 101 with modern populations also found a closer affinity with European samples than with Mongoloid populations (R.V.S. Wright, pers. comm., Kamminga & Wright, in prep., Wang Linghong & Bräuer 1984), whereas, when it was compared with late Pleistocene and early Holocene crania it grouped with other eastern Asian crania such as Upper Cave 103, Liujiang and Wadjak 1, which suggests some degree of regionality (Habgood 1985, 1986). The Upper Cave 102 crania appears to have suffered cranial deformation which caused the flat frontal and increased cranial height (Coon 1962, Weidenreich 1939, questioned by Wolpoff 1980). The three Zhoukoudian Upper Cave crania, although they display individual variation which may be due to chronological differences (Howells 1983), share many morphological features (Coon 1962, Weidenreich 1939, Wu Xinzhi 1961, Wu Xinzhi & Zhang Zhenbiao 1985). Wolpoff et al. suggest that

All of these common features are in concordance with Mongoloid affinities except the low orbit and broad nose, and these can be placed in the context of the regional evolutionary sequence. (1984:429)

The late Pleistocene or early Holocene cranium from Liujiang fills a similar position to that of the Upper Cave material in that it displays many so called Mongoloid features, although it does not have the total morphological pattern (Table 1, Habgood 1985, 1986, Wolpoff et al. 1984, Wu Xinzhi & Zhang Zhenbiao 1985). Wolpoff feels that

It is likely that this sample represents the populations from which living Asians, Amerinds, and Eskimos directly evolved... (1980:324)

which seems a reasonable conclusion when all of the material is considered and when fully modern Mongoloid crania are not expected during the late Pleistocene or early Holocene periods in east Asia. This material also appears to demonstrate morphological continuity with later Holocene samples from China (Wu Xinzhi & Zhang Zhenbiao 1985). The links that Thorne (1977, 1980) and Wolpoff et al. (1984) have seen between Chinese late Pleistocene and/or early Holocene crania and the Australian late Pleistocene crania from Lake Mungo and Keilor are very broad and general, and so do not demonstrate a close phylogenetic relationship.

The question of whether the origin of anatomically modern *Homo sapiens* in east Asia can be explained in terms of regional continuity or rapid replacement cannot, at present, be fully answered. The stone tool material does not suggest the type of developmental discontinuity one would expect if there was a major influx of people from the west (Aigner 1978b, Qiu Zhonglang 1985). The hominid material, due to a lack of complete crania and poor dating, is more difficult to successfully explain. The features that are suggested to be documenting 'regional continuity' in east Asia (Aigner 1976, Coon 1962, Thoma 1964, Weidenreich 1943, Wolpoff 1980, 1985, 1989, Wolpoff et al. 1984, Wu Rukang 1986a, Wu Rukang & Wu Xinzhi 1982, Wu Xinzhi & Wu Maolin 1985), either do not have a consistently high frequency of occurrence in all Mongoloid samples, or can be found on *Homo erectus* and/or archaic *Homo sapiens* crania from other regions. However, a number of the features related to facial flatness appear to have only a limited distribution outside of east Asia, and are not commonly found on single individuals outside of the region. These features, which may be documenting a certain degree of 'regional continuity' are: A non-depressed nasion, more perpendicularly oriented nasal bones, frontonasal and frontomaxillary sutures on almost the same level, and an angular rather than a rounded junction of the zygomatic bone and the zygomatic process of the maxilla.

There are similarities between east Asian crania and hominids from other parts of the world, but these are generally 'grade' similarities (cf. Wolpoff 1986 and Tattersall 1986 for a discussion of 'grade'), or general features that do not necessarily document phylogenetic proximity (but see Stringer & Andrews 1988). An example of the latter would be the pinched nasal bones with a central ridge seen on Zhoukoudian Upper Cave 101 and Předmost 1 from Europe. This feature is also found on the Dali and Maba hominids, which means that an external source is not needed to explain its presence on latter Chinese crania.

As with Australasia (Habgood 1989a, b), therefore, there is a case for a certain degree of 'regional continuity' in east Asia (morphologically and technologically), although the case is not exceptionally robust. There would not seem to be any direct evidence for multiple migrations (the initial *Homo erectus* migration[s] being followed by migrations of archaic *Homo sapiens* and finally anatomically modern *Homo sapiens)* into the region that some authors postulate (Groves 1989, Stringer, pers. comm.).

In this paper only two possible explanations for the origin of AM humans have been discussed and evaluated, with neither fully explaining the morphological pattern identified in east Asia. Bräuer (1984b, 1989) has proposed what he calls the 'hybridization and replacement model' that combines gene flow (migrants) from outside of east Asia and local morphological continuity to explain the evidence for both morphological continuity and discontinuity in the east Asian hominid sample (see also Stringer et al., [1984]). This model is consistent with the morphological data that indicates both continuity and discontinuity in east Asia; its agreement with genetic data, however, remains controversial (see Stringer & Andrews 1988, Bräuer, this volume). It is worth noting, however, that Wolpoff et al. (1984) argue for gene flow into east Asia in their 'multiregional evolution' hypothesis, but feel that there is no major disruption to morphological continuity in the region. Some combination of the two extreme hypotheses (Stringer & Andrews 1988), such as regional continuity with some gene flow or gene flow and migration with some regional continuity, would be the most parsimonious explanation of the east Asian hominid sample at present. However, it is currently impossible to ascertain whether regional morphological continuity or migration/gene flow is the driving force in the origin of AM humans in east Asia.

It should be noted that conclusions about the evolutionary sequence in east Asia could be greatly influenced by refinement of the chronological positions of the *Homo erectus* fossils from China and many of the poorly dated archaic *Homo sapiens* remains from Africa and Europe (Aigner 1986, 1987, Bräuer 1984a, b, Cook et al. 1982, Wu Xinzhi & Wang Linghong 1985). If it can be conclusively demonstrated that some of the archaic *Homo sapiens* hominids from Africa (Bodo, Kabwe, Saldanha, Ndutu) and Europe (Petralona, Arago, Mauer, Vértesszöllös) overlapped in time with *Homo erectus* material from east Asia, the case for a movement of archaic *Homo sapiens* into east Asia would be more feasible in that they must be earlier outside of east Asia if they are to have migrated into the area (the same must also be true for AM humans).

ACKNOWLEDGEMENTS

I would like to thank Günter Bräuer and Fred Smith for inviting me to submit this paper for the Zagreb Conference publication and for their helpful comments on an earlier draft of the paper. My research has benefited from discussions with many colleagues. Those I would especially like to thank are G. Bräuer, C. Groves, G.P. Rightmire, F.H. Smith, C.B. Stringer, A.G. Thorne and M.H. Wolpoff. I also greatly appreciated the assistance provided by C.B. Stringer & R.G. Kruszynski during my many visits to the British Museum (Natural History) where much of the data that this paper is based on was collected. Financial support for this research was provided by the Carlyle Greenwell Research Fund, the University of Sydney and the Australian Institute of Aboriginal Studies.

REFERENCES

Aigner, J.S. 1972. Relative dating of North Chinese faunal and cultural complexes. *Arctic Anthropol.* 9: 36-79.

Aigner, J.S. 1976. Chinese Pleistocene cultural and hominid remains: A consideration of their significance in reconstructing the pattern of human bio-cultural development. In *Le Paléolithique Inferieur et Moyen en Inde, en Asie Centrale, en Chine et dans le sud-est Asiatique*. U.I.S.P.P. IXe Congrès, Nice, Colloque VII, pp. 65-90.

Aigner, J.S. 1978a. Pleistocene faunal and cultural stations in south China. In F. Ikawa-Smith (ed.), *The Early Palaeolithic of South and East Asia*. The Hague: Mouton, pp. 129-160.

Aigner, J.S. 1978b. The Paleolithic of China. In A.L. Bryan (ed.), *Early Man in America from a Circum-Pacific Perspective*. Occasional Papers No. 1 of the Department of Anthropology, University of Alberta, Canada, pp. 25-41.

Aigner, J.S. 1986. The age of Zhoukoudian Locality 1: The newly proposed O18 correspondence. *Anthropos (Brno)* 23: 157-173.

Aigner, J.S. 1987. Correlations 18O et Localite 1 de Chou-Kou-Tien. *L'Anthropol.* 91: 733-748.

Anderson, J.E. 1968. Late Paleolithic skeletal remains from Nubia. In F. Wendorf (ed.), *The Prehistory of Nubia*. Dallas: Southern Methodist University Press, pp. 996-1040.

Andrews, P. 1984. An alternative interpretation of the characters used to define *Homo erectus*. *Cour. Forsch-Inst. Senckenberg* 69: 167-175.

Barker, B.C.W. 1978. Dental features of the Tabon mandible. *Arch. Phys. Anthropol Oceania* 13: 160-166.

Berry, A.C. 1975. Factors affecting the incidence of non-metrical skeletal variants. *J. Anat.* 101: 361-379.

Berry, A.C. & R.J. Berry 1967. Epigenetic variation in the human crania. *J. Anat.* 101: 361-379.

Berry, A.C., R.J. Berry & P.J. Ucko 1967. Genetical change in ancient Egypt. *Man* 2: 551-568.

Birdsell, J.B. 1967. Preliminary data on the trihybrid origin of Australian aborigines. *Arch. Phys. Anthropol. Oceania* 2: 100-155.

Birdsell, J.B. 1979. A reassessment of the age, sex and population affinities of the Niah cranium (abstract). *Am. J. Phys. Anthropol.* 50: 419.

Black, D. 1930. On an adolescent skull of *Sinanthropus pekinensis*. *Palaeont. Sin. Ser.* D, 7.

Bräuer, G. 1984a. A craniological approach to the origin of anatomically modern *Homo sapiens* in Africa and implications for the appearance of modern Europeans. In F.H. Smith & F. Spencer (eds), *The Origin of Modern Humans: A World Survey of the Fossil Evidence*. New York: Alan R. Liss, pp. 327-410.

Bräuer, G. 1984b. The Afro-European *sapiens*-hypothesis, and hominid evolution in East Asia during the late Middle and Upper Pleistocene, *Cour. Forsch. Inst. Senckenberg* 69: 145-166.

Bräuer, G. 1989. The evolution of modern humans: A comparison of the African and non-African evidence. In P. Mellars & C.B. Stringer (eds), *The Origin and Dispersal of Modern Humans: Behavioural and Biological Perspectives*. Edinburgh: University of Edinburgh Press, pp. 123-154.

Brothwell, D.R. 1960. Upper Pleistocene human skull from Niah Caves. *Sarawak Mus. J.* 9: 323-349.

Brothwell, D.R., V.M. Carbonell & D.H. Goose 1963. Congenital absence of teeth in human populations. In D.R. Brothwell (ed.), *Dental Anthropology*. Oxford: Pergamon Press, pp. 179-190.

Brown, F., J. Harris, R. Leakey & A. Walker 1985. Early *Homo erectus* skeleton from west Lake Turkana, Kenya. *Nature* 316: 788-792.

Bulbeck, F.D. 1981. Continuities in Southeast Asian Evolution since the Late Pleistocene. Unpublished M.A. thesis, Australian National University, Canberra.

Cadien, J.D. 1972. Dental variation in man. In S.L. Washburn and P. Dolhinow (eds), *Perspectives on Human Evolution*. New York: Rinehart and Winston, pp. 199-222.

Campbell, T.D. 1925. *The dentition and palate of the Australian Aboriginal.* Adelaide: Sheridan Foundation, University of Adelaide.

Carbonell, V.M. 1963. Variation in the frequency of shovel-shaped incisors in different populations. In D.R. Brothwell (ed.), *Dental Anthropology.* Oxford: Pergamon Press, pp. 211-234.

Cook, J., C.B. Stringer, A.P. Currant, H.P. Schwarcz & A.G. Wintle 1982. A review of the chronology of the European Middle Pleistocene hominid record. *Yrbk. Phys. Anthropol.* 25: 19-65.

Coon, C.S. 1962. *The Origin of Races.* New York: Alfred A. Knopf.

Day, M.H. 1971. Postcranial remains of *Homo erectus* from Bed IV, Olduvai Gorge, Tanzania. *Nature* 232: 383-387.

Day, M.H. 1977. *Guide to Fossil Man: A Handbook of Human Palaeontology* (3rd revised edition). London : Cassell.

Day, M.H. 1984. The postcranial remains of *Homo erectus* from Africa, Asia and possibly Europe. *Cour. Forsch-Inst. Senckenberg* 69: 113-121.

Fenner, F.J. 1939. The Australian Aboriginal skull: Its non-metric morphological characters. *Trans. Roy. Soc. Australia* 63: 248-306.

Greene, D.L. & G. Armelagos 1972. *The Wadi Halfa Mesolithic Population.* Department of Anthropology, University of Massachusetts, Amherst, Research Report No. 11.

Groves, C.P. 1989. A regional approach to the problem of the origin of moderrn humans in Australasia. In P. Mellars & C.B. Stringer (eds), *The Origin and Dispersal of Modern Humans: Behavioural and Biological Perspectives.* Edinburgh: University of Edinburgh Press, pp. 274-285.

Habgood, P.J. 1985. The origin of the Australian Aborigines: An alternative approach and view. In P.V. Tobias (ed.), *Hominid Evolution: Past, Present and Future.* New York: Alan R. Liss, pp. 367-380.

Habgood, P.J. 1986. The origin of the Australian Aborigines: A multivariate approach. *Arch. Oceania* 21: 130-137.

Habgood, P.J. 1989a. The evolution of modern humans: Evidence from Australasia seen in a global context. In P. Mellars & C.B. Stringer (eds), *The Origin and Dispersal of Modern Humans: Behavioural and Biological Perspectives.* Edinburgh: University of Edinburgh Press, pp. 245-273.

Habgood, P.J. 1989b. The origin of modern humans in east and southeast Asia. In G. Giacobini (ed.), *Proceedings of the 2nd International Congress on Human Palaeontology, Turin, Italy, 1987.* Jaca: Milan. pp. 431-435.

Habgood, P.J. 1989c. An investigation into the usefulness of a cladistic approach to the study of the origin of anatomically modern humans. *Human Evolution* 4: 241-252.

Howell, F.C. 1978. Hominidae. In V.J. Maglio & H.B.S. Cooke (eds), *Evolution of African Mammals.* Cambridge (Mass.): Harvard University Press, pp. 154-248.

Howells, W.W. 1973. Cranial Variation in Man: A Study by Multivariate Analysis of Patterns of Differences Among Recent Human Populations. *Peabody Museum Papers* Vol. 67.

Howells, W.W. 1983. Origins of the Chinese people: Interpretations of the recent evidence. In D.N. Keightley (ed.), *The Origin of Chinese Civilization.* Berkeley: University of California Press, pp. 297-319.

Hrdlička, A. 1920. Shovel-shaped teeth. *Am. J. Phys. Anthropol.* 3(os): 429-465.

Jelinek, J. 1982. The East and Southeast Asian way of regional evolution. *Anthropos (Brno)* 21: 195-212.

Kellock, W.R. & P.A. Parsons 1970a. Variation of minor non-metrical cranial variants in Australian Aborigines. *Am. J. Phys. Anthropol.* 32: 409-421.

Kellock, W.R. & P.A. Parsons 1970b. A comparison of the incidence of minor non-metrical cranial variants in Australian Aborigines with those of Melanesia and Polynesia. *Am. J. Phys. Anthropol.* 33: 235-240.

Kennedy, G.E. 1983. Some aspects of femoral morphology in *Homo erectus*. *J. Hum. Evol.* 12: 587-616.

Kennedy, G.E. 1984a. The emergence of *Homo sapiens:* The postcranial evidence. *Man* 19: 94-110.

Kennedy, G.E. 1984b. Are the Kow Swamp hominids 'archaic'? *Am. J. Phys. Anthropol.* 65: 163-168.

Kennedy, G.E. 1986. The relationship between auditory exostoses and cold water: A latitudinal analysis. *Am. J. Phys. Anthropol.* 71: 401-415.

Koenigswald, G.H.R. von 1978. The palate of *Pithecanthropus modjokertensis.* In P.M. Batter & K.A. Joysey (ed.), *Development, Function and Evolution of Teeth.* London: Academic Press, pp. 353-357.

Krogman, W.M. 1932. The morphological character of the Australian skull. *J. Anat.* 66: 399-413.

Kwang-chih Chang 1962. New evidence on fossil man in China. *Science* 136: 749-760.

Larnach, S.L. & N.W.G. Macintosh 1966. The Craniology of the Aborigines of Coastal New South Wales. *Oceania Monographs* No. 13.

Larnach, S.L. & N.W.G. Macintosh 1970. The Craniology of the Aborigines of Queensland. *Oceania Monographs* No. 15.

Leakey, R.E.F. & A.C. Walker 1985. Further hominids from the Plio-Pleistocene of Koobi Fora, Kenya. *Am. J. Phys. Anthropol.* 67: 135-164.

McCown, T.D. & A. Keith 1939. *The Stone Age of Mount Carmel* 2. The Fossil Human Remains from the Levalloiso-Mousterian. Oxford: Clarendon Press.

Martin, H. 1926. *Recherches sur l'Evolution du Mousterien dans le gisement de La Quina, Charente,* 4. L'Enfant de La Quina. Angouleme: Impr. Ouvrieve.

Murrill, R.I. 1975. A comparison of the Rhodesian and Petralona upper jaws in relation to other Pleistocene hominids. *Z. Morph. Anthropol.* 66: 176-187.

Neumann, G.K. 1956. The Upper Cave skulls from Choukoutien in the light of Paleo-Amerind material (abstract). *Am. J. Phys. Anthropol.* 14: 380.

Ossenberg, N.S. 1976. Within and between race distances in population studies based on discrete traits of the human skull. *Am. J. Phys. Anthropol.* 45: 701-716.

Pardoe, C. 1986. Prehistoric economy, aquatic resources and the sexual division of labour (abstract). In G.K. Ward (ed.), *Archaeology at A.N.Z.A.A.S., Canberra.* Canberra: Canberra Archaeological Society, pp. 254-255.

Qiu Zhonglang 1985. The Middle Palaeolithic of China. In Wu Rukang & J.W. Olsen (eds), *Palaeoanthropology and Palaeolithic Archaeology in the People's Republic of China.* Orlando: Academic Press, pp. 187-210.

Rak, Y. 1983. *The Australopithecine Face.* New York: Academic Press.

Rak, Y. 1985. Systematic and functional implications of the facial morphology of Australopithecus and early Homo. In E. Delson (ed.), *Ancestors: The Hard Evidence.* New York: Alan R. Liss, pp. 168-170.

Riesenfeld, A. 1956. Shovel-shaped incisors and a few other dental features among native peoples of the Pacific. *Am. J. Phys. Anthropol.* 14: 505-521.

Robinson, J.T. 1956. The Dentition of Australopithecinae. *Transvaal Museum Memoir* No. 9.

Smith, F.H. 1976. *The Neanderthal remains from Krapina: A descriptive and comparative study.* University of Tennessee, Department of Anthropology, Reports of Investigations No. 15.

Smith, F.H. 1985. Continuity and change in the origin of modern *Homo sapiens*. *Z. Morph. Anthropol.* 75: 197-222.

Smith, F.H. & S. Paquette 1989. The adaptive basis of Neanderthal facial form with some thoughts on the origin of modern humans. In E. Trinkaus (ed.), *The Emergence of Modern Humans: Biocultural Adaptations in the Later Pleistocene.* Cambridge: University of Cambridge Press, pp. 181-210.

Stringer, C.B. 1984. The definition of *Homo erectus* and the existence of the species in Africa and Europe. *Cour. Forsch.-Inst. Senckenberg* 69: 131-144.

Stringer, C.B. 1989. The origin of early modern humans: A comparison of European and non-European evidence. In P. Mellars and C.B. Stringer (eds), *The Origin and Dispersal of Modern Humans: Behavioural and Biological Perspectives.* Edinburgh: University of Edinburgh Press, pp. 232-244.

Stringer, C.B. & P. Andrews 1988. Genetic and fossil evidence for the origin of modern humans. *Science* 239: 1263-1268.

Stringer, C.B., J.J. Hublin & B. Vandermeersch 1984. The origin of anatomically modern humans in western Europe. In F.H. Smith & F. Spencer (eds), *The Origin of Modern Humans: A World Survey of the Fossil Evidence.* New York: Alan R. Liss, pp. 51-135.

Tattersall, I. 1986. Species recognition in human palaeontology. *J. Hum. Evol.* 15: 165-175.

Thoma, A. 1964. Die Entstehung der Mongoliden. *Homo* 15: 1-22.

Thorne, A.G. 1976. Morphological contrasts in Pleistocene Australians. In R.L. Kirk & A.G. Thorne (eds), *The Origin of the Australians.* Canberra: A.I.A.S., pp. 95-112.

Thorne, A.G. 1977. Separation or reconciliation? Biological clues to the development of Australian society. In J. Allen, J. Golson & R. Jones (eds), *Sunda and Sahul: Prehistoric Studies in Southeast Asia, Melanesia and Australia.* London: Academic Press, pp. 187-204.

Thorne, A.G. 1980. The arrival of man in Australia. In A. Sherratt (ed.), *The Cambridge Encyclopedia of Archaeology.* Cambridge: Cambridge University Press, pp. 96-100.

Thorne, A.G. & M.H. Wolpoff 1981. Regional continuity in Australasian Pleistocene hominid evolution. *Am. J. Phys. Anthropol.* 55: 337-349.

Tobias, P.V. 1971. Human skeletal remains from the Cave of Hearths, Makapansgat, Northern Transvaal. *Am. J. Phys. Anthropol.* 34: 335-368.

Trinkaus, E. 1983. *The Shanidar Neanderthals*. New York: Academic Press.

Turner, C.G. II 1987. Late Pleistocene and Holocene population history of East Asia based on dental variation. *Am. J. Phys. Anthropol.* 73: 305-321.

Villiers, H. de 1968. *The Skull of the South African Negro*. Johannesburg: Witwatersrand University Press.

Wang, Linghong & G. Bräuer 1984. A multivariate comparison of the human calva from Huanglong County, Shaanxi Province. *Acta Anthropol. Sin.* 3: 313-321.

Weidenreich, F. 1937. The Dentition of *Sinanthropus pekinensis*. *Palaeontol. Sin. n.s.* D, No. 1 (whole series No. 101).

Weidenreich, F. 1939. On the earliest representatives of modern mankind recovered on the soil in East Asia. *Peking Nat. Hist. Bull.* 13: 161-174.

Weidenreich, F. 1943. The skull of *Sinanthropus pekinensis:* A comparative study on a primitive hominid skull. *Palaeontol. Sin. n.s.* D, No. 10 (whole series No. 127).

Weidenreich, F. 1945. The Palaeolithic child from the Teshik-Tash Cave in southern Uzbekistan (Central Asia). *Am. J. Phys. Anthropol.* 3: 151-162.

Weidenreich, F. 1947. Facts and speculations concerning the origin of *Homo sapiens*. *Am. Anthropol.* 49: 187-203.

Wolpoff, M.H. 1980. *Paleoanthropology*. New York: Alfred A. Knopf.

Wolpoff, M.H. 1985. Human evolution at the peripheries: The pattern at the eastern edge. In P.V. Tobias (ed.), *Hominid Evolution: Past, Present and Future*. New York: Alan R. Liss, pp. 355-366.

Wolpoff, M.H. 1986. Describing anatomically modern *Homo sapiens* a distribution without a definable difference. *Anthropos (Brno)* 23: 41-53.

Wolpoff, M.H. 1989. Multiregional evolution: The fossil alternative to Eden. In P. Mellars & C.B. Stringer (eds), *The Origin and Dispersal of Modern Humans: Behavioural and Biological Perspectives*. Edinburgh: University of Edinburgh Press, pp. 62-108.

Wolpoff, M.H., Wu Xinzhi & A.G. Thorne 1984. A general theory of hominid evolution involving the fossil evidence from East Asia. In F.H. Smith & F. Spencer (eds), *The Origin of Modern Humans: A World Survey of the Fossil Evidence*. New York: Alan R. Liss, pp. 411-484.

Woo, Jukang 1958. Tzeyang Palaeolithic man: Earliest representative of modern man in China. *Am. J. Phys. Anthropol.* 16: 459-471.

Wood, B. 1984. The origin of *Homo erectus*. *Cour. Forsch.-Inst. Senckenberg* 69: 99-112.

Wu, Rukang 1985. New Chinese *Homo erectus* and recent work at Zhoukoudian. In E. Delson (ed.), *Ancestors: The Hard Evidence*. New York: Alan R. Liss, pp. 245-248.

Wu, Rukang 1986a. Chinese human fossils and the origin of Mongoloid racial group. *Anthropos (Brno)* 23: 151-155.

Wu, Rukang 1986b. A fossil skeleton from Yingkou, Liaoning Province, People's Republic of China, archaic *Homo sapiens* or *Homo erectus? The World Archaeological Congress Precirculated Papers, Vol. 1 The Pleistocene Perspective*. London: Allen and Unwin.

Wu, Rukang & Dong Xingren 1985. *Homo erectus* in China. In Wu Rukang & J.W. Olsen (eds), Palaeoanthropology and Palaeolithic Archaeology in the People's Republic of China. Orlando: Academic Press, pp. 79-89.

Wu, Rukang & Wu, Xinzhi 1982. Comparison of Tautavel man with *Homo erectus* and early *Homo sapiens* in China. *Prét. 1er Cong. Int. Paléont. Hum., Nice*. Nice: CNRS, pp. 605-616.

Wu, Xinzhi 1961. Study on the Upper Cave man of Choukoutien. *Vert. PalAsiat.* 3: 202-211.

Wu, Xinzhi 1986. Upper Palaeolithic man in China and their relation with populations of neighbouring areas. *The World Archaeological Congress Precirculated Papers, Vol. 1, The Pleistocene Perspective*.

Wu, Xinzhi & Wang Linghong 1985. Chronology in Chinese palaeoanthropology. In Wu Rukang & J.W. Olsen (eds), *Palaeoanthropology and Palaeolithic Archaeology in the People's Republic of China*. Orlando: Academic Press, pp. 29-51.

Wu, Xinzhi & Wu Maolin 1985. Early *Homo sapiens* in China. In Wu Rukang & J.W. Olsen (eds), *Palaeoan-*

thropology and Palaeolithic Archaeology in the People's Republic of China. Orlando: Academic Press, pp. 91-96.

Wu, Xinzhi & Zhang Zhenbiao 1985. *Homo sapiens* remains from late Palaeolithic and Neolithic China. In Wu Rukang & J.W. Olsen (eds), *Palaeoanthropology and Palaeolithic Archaeology in the People's Republic of China*. Orlando: Academic Press, pp. 107-133.

Zhou, Mingzhen, Li Yanxian & Wang Linghong 1982. Chronology of the Chinese fossil hominids. *Prét 1er Cong. Int. Paléont. Hum.* Nice: CNRS, pp. 593-604.

The human colonisation of the Australian continent

Rhys Jones
Department of Prehistory, The Australian National University, Canberra, Australia

1 INTRODUCTION

Australia has never been connected by land to Asia, but is separated from it by a series of island arcs and deep oceanic troughs. Its mammalian fauna evolved separately from those of other continents with a radiation of monotremes and marsupials. During the Ice Ages, Indo-Malay and Australian archipelagos were periodically joined to their mother continents. From a biogeographic point of view, it is the configuration of the continental shelves which is decisive. During periods of glacial low sea levels, the present island of Java, Sumatra and Borneo were all connected as Sundaland to southeast Asia (Hooijer 1975) and conversely New Guinea, Aru and Tasmania were all part of a Sahul or 'Greater Australia'. In between were the true oceanic islands of Wallacea to be colonised by flightless mammals across water (Diamond 1987). Before the arrival of humans, of the land mammals of the teeming Asiatic fauna, only the rodents – rats and mice – managed to cross all of the water barriers to Australia.

Two species of the elephant-like Stegodontidae, one species of elephant, an extinct rhinoceros, a cervid and a bovid managed to cross the single water barrier to the island of Luzon in the Philippines; two species of stegodon got to Timor and Flores along the Lesser Sunda chain (Hooijer 1975). On the mighty Sepik river in northern New Guinea, islands of vegetation, several scores of metres wide, and containing trees of sago and palms still standing on them, have been carried down-stream. In wet season floods, some are swept out to sea many kilometres off shore. Yet such phenomena on similar Asiatic rivers did not serve as mechanisms for the successful carriage of colonising populations across the Wallacea sea barriers. The Pliocene and Pleistocene fossil record of southeast Asia has included amongst the primates the orangutang, macaques and colobine monkeys (Weitzel 1987). These biogeographic facts indicate that the arrival of humans on the Sahul-Australian continent necessitated the use of water craft of a sea-crossing capacity of a different order to any such naturally occurring floats. Clearly the simplest craft would be rafts. A suitable timer would be provided by bamboo or bundles of thick reeds, but there are many other soft woods in the tropical forests.

2 ASIAN HOMINIDS

Eugene Dubois' discovery in 1891 of the skull cap of what he called *Pithecanthropus erectus*, now referred to as *Homo erectus*, at Trinil on the banks of the Solo River in central Java, established southeast Asia as one of the cradles of humankind. Subsequent research has shown that the oldest Javanese hominids from the upper part of the Pucangan beds at Sangiran can be dated by the fission track method of just over 1 million years ago (Watanabe & Kadar 1985, Aimi & Aziz 1985). Other fossils, including the important Sangiran 17 skull, which has the greatest extent of the upper face for any *erectus* skull (Thorne & Wolpoff 1981), was found in the lower lower Kabuh beds, dated to about three quarters of a million years old. The Ngandong or Solo fossils are now believed to be much younger, perhaps extending into the Upper Pleistocene. A re-excavation at the original find-site revealed animal bones, from the same basal layers where the hominid samples were originally obtained. A series of uranium/thorium dates showed these bones to be dated to between 75 and 100 kyr old (Bartstra et al. 1988). Javanese fossil hominid remains document the radiation of a pioneering archaic lineage of humankind into the tropical lowlands of southeast Asia, where they persisted relatively little changed in morphology over the best part of a million years.

In contrast to this long fossil record, stone tool assemblages, or indeed any stratified archaeological sites showing evidence of human domestic occupation structures of any high antiquity, have proved notoriously difficult to obtain, not only in Java, but also anywhere else in island or mainland southeast Asia (Bartstra 1982). Previous claims for a Middle Pleistocene antiquity for the Patjitanian assemblages, characterised by large flat flakes with lateral retouch and by high domed, keel-like cores have not been confirmed by recent systematic field work. Indeed, Bartstra, who has worked extensively on these assemblages in Java, is convinced that they are quite recent, within the later stages of the Upper Pleistocene. He has argued that the typological 'largeness' or 'crudity' of the assemblages has led observers to deduce a great antiquity for them, an ascription not based on any convincing stratigraphic evidence. Equally, the Hoabinian assemblages of mainland southeast Asia, with their large, retouched pebble-choppers and steeply retouched flake tooks, are no longer believed to be Pleistocene in age, but rather date from the early Post-Glacial or Holocene period (Anderson 1987).

An important site in this regard is the limestone cave of Lang Rongrien in peninsular Thailand. Within its sequences are heavy stone tools of Hoabinian type; but these are securely dated to Early Holocene times, i.e. between nine and a half and seven and a half thousand years ago (Anderson 1987). The earliest evidence for human occupation at this site has been dated to about 37 000 years ago, the earliest securely dated archaeological finds in the whole of southeast Asia. The stone tools from the lowest levels consist of unifacially retouched flake tools. Niah Cave in Sarawak and therefore on the Sunda continental shelf, which was joined to Asia during low sea level times, has an archaeological sequence going back probably to some 40 000 years ago, with basal levels containing flaked tools similar to those from the Thai site (Majid 1982). Equally, in Tabon Cave on present day Palawan Island, again once part of Sundaland, the lowest archaeological material dates back to some 26 000 years ago at least (Fox 1970).

On the truly oceanic islands of Wallacea itself, the oldest reliable dated evidence for human presence is at the Leang Burung 2 cave site on Sulawesi (Celebes), with stone tools stratified to levels securely dated to some 31 000 years ago (Glover & Presland 1985). On Timor, the oldest direct dates so far obtained are those of about 14 000 years old at the Uai Bobo 2 cave site (Glover 1987), although there may also be other sites of potential interest within some of the gravel deposits on the island (Glover & Glover 1970).

It can be seen from this brief review that Wallace's Line was crossed by humans reasonably late in the Pleistocene, say 40 000 plus years ago, but that once this had happened, movement was rapid throughout the entire archipelago and onto the Sahul shelf itself.

3 AUSTRALIA: THE PROCESS OF DISCOVERY

Only some 25 years ago in 1962, the oldest direct carbon date for human occupation of Australia/New Guinea was just over 10 000 years ago from Kenniff Cave in southern Queensland (Mulvaney 1964). This discovery was a land-mark since it passed a kind of a psychological barrier in prehistoric archaeology; 10 000 years ago being conventionally taken to denote the end of the Pleistocene geological period. Some scholars had long advocated a Pleistocene antiquity for human colonisation of the continent, but unequivocal direct evidence for this proved difficult to obtain. In this survey of world prehistory, published in 1962, the great Cambridge prehistorian, Grahame Clark wrote that 'the scientific pursuit of prehistory is still in its infancy over large areas of the vast Australian territories... (and) ... It may first be said that there is no convincing evidence for the immigration of man into Australia before Neothermal times' (1962:243). A technique that was to transform this picture was radiocarbon dating, which came into general use in archaeology in the early 1960's. At Kenniff Cave, Mulvaney obtained further dates of 14 000 and then 19 000 years from his basal levels (Mulvaney & Joyce 1965). By 1966, dates of just over 20 000 years were obtained from the base of excavated sequences in sandstone or quartzite rock shelters on the coast of New South Wales and on the East Alligator River in the Northern Territory, and from a deep limestone cave called Koonalda in the Nullarbor Plain (see Jones 1979 and 1989 for further references). An equivalent antiquity came from Papua New Guinea with a date of 26 000 years for stone tools from a swamp buried by volcanic ash at Kosipe in the eastern Highlands (White et al. 1970).

3.1 *Lake Mungo*

A major turning point revolved around the discoveries at Lake Mungo and associated fossil lake systems of the Willandra creek in western New South Wales between 1968 and 1972. Geomorphological work by J.M. Bowler (1976) on the concave-curved sand dunes on the eastern lee shores of these now dry lake beds had enabled him to establish that the lakes had been full some 30 000-50 000 years ago. Conversely during the height of the last Ice Age at 18 000 years ago Bowler determined that they had dried out completely, and strong cold, dry winds had remobilised sand dunes into a great anti-clockwise continental swirl which can ben seen on satellite photographs of Australia today (Jones & Bowler 1980:10).

Archaeological research at Lake Mungo in 1969 showed the presence of stone tools, hearths containing animal bones and shell middens of freshwater mussels, all stratified in deposits dated to between 24 000 and 30 000 years ago (Bowler et al. 1970). Of prime importance was the discovery of human remains – those of a young woman who had been cremated, her bones then smashed and buried in a shallow pit. These were dated to 26 000 years BP (Bowler et al. 1970). Recently obtained A.M.S. dates have confirmed this value (R. Gillespie, pers. comm.). From the Mungo finds one could reconstruct aspects of a society: women collecting mussels from the lakeshore muds, other people fishing for golden perch (*Plectroplites ambiguous*), perhaps using woven barrier nets; with other food such as emu eggs and small marsupials being brought in from the dry scrubs behind the lakes. Stone tools, manufactured from silicified quartzites, consisted of steep edged scrapers, some showing concavities perhaps used for smoothing the shafts of wooden spears or digging sticks. These were also distinctive dome-shaped 'horse hoof' cores, some probably used as core tools for heavy duty scraping or chopping. Small pieces of hydrated iron oxide were used as red ochre pigment. The evidence for the rites of cremation were the oldest anywhere in the world and implied a profound sense of religious beliefs.

Elsewhere in the Willandra fossil lakes system, a fresh water mussel midden from the Lake Arumpo lunette gave a date of 37 000 years (Bowler 1976:59). West of the lower Darling River, and some 100 km west of the Willandra, are other fossil stream systems which lead into Lake Tandou. Some freshwater mussel middens have been dated here back to about 36 000 years ago (Hope et al. 1983 and J. Hope, pers. comm.). Altogether in the Willandra/lower Darling region about 300 samples have been radiocarbon assayed, about half of which are from archaeological sites. Of these, some 30% date from the period 15 000 to 27 000 years ago, with a few extending backwards into the oldest time bracket discussed above. Further deeper excavations in the Mungo lunette itself, showed the presence of stone tools in basal gravels for which a tiny charcoal sample gave a value indistinguishable from background (Shawcross 1975, Shawcross & Kaye 1980).

When re-constituted, the Mungo I cranium proved to be that of an extremely gracile person, with a thin cranial vault and a rounded profile to the skull (Thorne 1971). In contrast to this, some other fossils of about the same antiquity in Australia are extremely rugged, and show other 'primitive' features such as large jaws, heavy brow-ridges, supra-orbital constrictions, sloping foreheads and rugged occipital regions. An extreme example of the latter combination of traits is the Willandra Lakes 50 specimen (W.L.H. 50), eroded from the surface of a small sand lunette within the floor of the Garnpang fossil lake, just north of Lake Mungo. Primary attempts at dating the W.L.H. 50 specimen using electon spin resonance techniques have suggested the same order of antiquity as the Mungo I (W.L.H. 1) hominid, i.e. some 20-30 000 years ago (Caddie et al. 1987:175). The cranial thickness of the W.L.H. 50 specimen is some 15 mm thick (S.G. Webb, pers comm.) an astonishing value, which can be compared with the same measurement on the Mungo specimen of only 2 mm. In terms of a series of simple key parameters, there is greater variation amongst the series of hominids recovered from the Willandra Lakes region and dated to roughly the same time period of the late Pleistocene, than now exists within the whole of present-day humanity right across the world. One school of thought sees this variability as being accommodated within a single population having a single genetic ancestry; the range of variability becoming much less during Holocene times (e.g. Brown 1987). The others, and I count myself amongst these (Thorne 1971, Jones 1979) are not convinced by such a cosy argument and feel that some more radical explanation is called for. Do we have here representatives from two quite distinct episodes of human colonisation of the continent with some genetic memory of an initial colonisation by peoples related in some ways to Solo Man? (cf. Thorne & Wolpoff 1981).

3.2 *Swan River and Devil's Lair, southwestern Australia*

A carefully controlled excavation by Pearce and Barbetti (1981) within the Guilford terrace of the Upper Swan River, northeast of Perth resulted in the recovery of some 200 artefacts *in situ*, some of which could be con-joined. A ^{14}C date of 39 500 ± ca 2000 (SUA 1500) was obtained and this was confirmed by another sample taken during a later season's excavation which gave a value of 37 100 ± ca 1500 (SUA 1665) and one of about 35 000 BP from a sample 5 cm higher. The excavators concluded that there was clear evidence of the site being occupied at about 38 000 years ago, and they noted that the actual count rates in the ^{14}C machine were close to the 'lower limit of our best counter' (Pearce & Barbetti 1981:177). The artefacts were flakes, some retouched, mostly from quartz and quartzite, with a few made from a bryozoan chert, a soure of raw material only exposed during Pleistocene low sea phases and now drowned by the sea several dozen kilometres west of the present coast.

South of Perth is the limestone cave of Devil's Lair. A careful and extensive excavation by C. Dortch revealed almost 5 m of stratified deposit dated by a series of 27 carbon dates (Dortch 1984). Reasonably regular use of the cave with hearths, stone and bone artefacts and charred and smashed bones of hunted macropods was indicated down to layer 30, dated to between 30 590 ± ca 2000 (GX 7255) and 32 480 ± 1250 (SUA 585) BP. However, some artefacts such as a small struck flake of opal came from layer 33/34. Layer 31 has a date of 35 160 ± 1800 BP (SUA 586) and the lowest charcoal was dated to 37 750 ± 2500 BP (698) from layer 39. It is hard to imagine how charcoal could have gotten into this enclosed cave site without some form of human agency. From layer 39 down to 51, there is no evidence for human presence. To sum up the Devil's Lair data, there is proof of human presence down to about 32 000 years ago, with strong presumptive evidence of sparse or intermittent presence back to ca 38 000 years ago.

3.3 *Huon Peninsula terraces, Papua New Guinea*

Possibly the oldest data come from the coast of the Huon Peninsula in north eastern New Guinea. Rapid uplift has resulted in the formation of a spectacular flight of uplifted terraces of old coral rock, the highest being some 400 m above sea level and dated to Middle Pleistocene times (Chappell 1974). Behind each reef was a back lagoon, with high productivity of shell fish, fish and estuarine plant species, and thus potentially easily exploitable by prehistoric hunter/gatherers. In the walls of a small gully cutting into the old lagoon deposits associated with reef terrace IIIa, several stone tools were found *in situ* (Groube et al. 1986). The reef itself had been dated by uranium series methods to be between 45 and 53 thousand years old. The tools were within a deposit of volcanic ash which although mantling the terrace IIIa, was not present over the later terraces IIIb and II, which were then still under water or which had not yet been formed. The later terraces were dated to 40 000 years and 28-30 000 years old respectively. An independent assessment of the age of quartz particles within the tephra at the find-site yielded a thermoluminescent age of greater than 40 000 years (Groube et al. 1986). Thus on a conservative estimate, these artefacts are 40 000 years old and quite probably considerably older.

4 LIMITATIONS OF OUR DATING METHODS

While being able confidently to document an antiquity of 38-40 000 years at widely scattered locations on the continent, an additional question arises whether this apparent basement represents something close to the actual time of the colonising episode or whether it is largely an artefact of our dating methods (cf. Jones 1982). Conventional radiocarbon dating reaches its effective limits at about this time range. If the true age of a sample was say 80 000 years, this would still give a finite reading. Translated into radiocarbon years, this might be presented as being within the 35 to 40 000 years time bracket and with a large standard error.

One research development is the accelerator mass spectrometer (A.M.S.) technology, which directly weighs the isotopic fractions of the carbon molecules. Such an antiquity is paradoxically too young for other absolute physical methods such as the potassium-argon or most of the uranium series methods. A new technique which shows great promise is that of thermoluminescence (T.L.) dating of naturally accumulated sands (Wintle et al. 1984). There are several sites in Australia where stone artefacts have been found in sequences below the lowest presence of charcoal and which would be prime targets for such renewed dating research.

One site is the Mungo lunette itself where tools were found, water-rolled in gravels forming the base of the Mungo stratigraphic unit and several metres below the lowest datable charcoal. Some of the terraces of the Nepean River west of Sydney dated by the T.L. method to about 40 000 years ago may contain stone artefacts (Nanson et al. 1987). On the Victorian coast at Warnambool in a sequence of fossil, partially carbonated sand dunes, there are broken gastropod shells (*Subninella undulata*) which look very much like an eroded shell midden and which are considerably older than 35 000 years (J. Sherwood, pers. comm., personal observation December 1986, Bowler 1987).

Finally there are sites in the Kakadu region of western Arnhem Land. The Lindner Site, Nauwalabila I in Deaf Adder Gorge, was formed by a large rock which had fallen off the cliff face and had landed obliquely on the valley floor below (Jones 1985). This valley became mantled in sands washed down from the escarpment and which accumulated up to four metres deep in the centre of the valley (Hope et al. 1985). In the shelter we carefully excavated a 2.5 m depth of sand in a series of almost 80 superimposed excavation units, and every one contained some stone artefacts (Jones & Johnson 1985). Several carbon dates were obtained from this sequence, with a date of about 20 000 years corresponding to a depth of 1.8-2.0 m (Kamminga & Allen 1973). Extrapolation of the depth/age curve to the base of the sand suggests a date for the latter of possibly 30 000 years old (Jones & Johnson 1985:182). Under the sands is a further half metre of heavily weathered and partially cemented rubble within a sand matrix. Weathered stone flakes exist to the very base of this rubble. T.L. dates have successfully been carried out on similar sand sheets in the Magela Creek on the East Alligator River, just 60 km north of our site. Results indicate dates of about 35 000 years for the base of the sand there (East et al. 1987). The Lindner Site would be an excellent location in which to pursue the dating of the basal deposits with the T.L. technique. Access to Deaf Adder Gorge is at present restricted due to the recent death of a prominent traditional Aboriginal owner.

5 FIRE AND THE EMPTY CONTINENT

Australia and the Americas share one major factor in common in terms of their human histories. Both were continents from which people were totally absent during the long period of the Early and Middle Pleistocene, when hominids and especially *Homo sapiens* evolved physically, and slowly developed cultural and technological capacities. Viewed from this long term timescale, they were both colonised suddenly and recently within the Upper Pleistocene and by biologically modern humans with his and her full technological and social capacities. The impact of such a new element in the environment has been the subject of recent inquiry, particularly because through hunting – humans were predators at the top of the food chain and also significantly affected the landscape, especially with the wide-spread use of fire.

Ethnographic accounts indicate that at the time of European contact over most of the Australian continent, Aborigines used fire in a systematic way to alter the vegetation in their landscape. They did this for a variety of reasons, such as to clear dried grass for ease of access in travelling, to 'clean up' the country, for hunting, signalling and also to promote the growth of new green grass for attracting game (e.g. see Jones 1969). The frequency of fire in any one locus is a major ecological determinant in the local vegetation succession, favouring some more fire resistant species at the expense of other more vulnerable ones, maintaining certain vegetation structures, especially tending to remove young seedlings and trees in the shrub stage, etc. In the tropical north, a high fire frequency helped to maintain the typical savanna of the region, with mature, well spaced eucalypt trees forming the cover over seasonal grasses; fire sensitive species are often restricted to 'fire shadow' areas such as on escarpment cliff-edges (Russel-Smith 1985). Natural fire due to lightning has of course always been a feature of the Australian ecosystem, but the advent of people with their fire sticks would have increased the probability of a fire occurring in any one area of their foraging ranges by several orders of magnitude. I have maintained that the advent of human impact should be perceptible in the palaeo-environmental record, in terms of changes both in vegetation and in local geomorphological regimes due to water borne erosion following fire events (Jones 1968, 1985:297).

The deep pollen profile obtained by Singh and his co-workers from sediments in the bed of Lake George in the south eastern tablelands of New South Wales may illustrate such a phenomenon (Singh & Geissler 1985). The full sequence goes back beyond 700 000 years ago and documents a series of glacial and interglacial events. In summary, during glacial times there were few trees in the area, but conversely, during the warmer interglacials the area was wooded. In the upper part of the sequence, zone F, there was a marked change unmatched in any of the previous sequence. Firstly, there was a sudden appearance of large amounts of charcoal fragments, which persisted in the sequence from then on, even during the relatively tree-less Last Glacial maximum. Secondly, the species composition of the trees themselves exhibited a marked change. Previously, the trees were dominated by *Casuarina* sp., podocarps and tree ferns; subsequently these species declined and were replaced by eucalyptus, other fire adapted trees and grasses. Many of the earlier dominant species such as the casuarina and eventually the podocarps, became extinct, even in climatic conditions that in previous interglacial cycles had favoured them, Singh could find no plausible natural explanation for this, but postulated that it could be accounted for as the result of the arrival of humans with their fire regime impact (Singh & Geissler 1985:437-439).

The problem lay with the date of zone F, which Singh and Geissler believed on correlation with deep sea palaeotemperature cores to correspond to the Last Interglacial, i.e. dating to about 120 000 years ago. This level of antiquity is beyong any hard archaeological evidence for human presence in Australia. However the argument that these changes are humanly related was so compelling, that it caused Richard Wright (1986) to re-examine the chronology of the pollen sequence and see if zone F had to be that old. There was a series of ^{14}C dates carried out on the upper part of the sequence, the oldest reliable date being ca 26 000 BP. Undertaking an age-depth comparison, Wright found a correlation which when extended back to the depth of zone F, giving a value of the order 60 000 years ago.

It would be a profound irony if the landscape of Australia, long seen as the quintessence of the 'natural landscape' should turn out to have been fundamentally affected by human impact; initially by the distant ancestors of the Aborigines, the first human colonists to the continent.

A key question concerns the extinction of the so-called 'giant marsupials'. Such animals include Diprotodontids, which were the size and shape of a rhinoceros; giant kangaroos up to 3 m tall; and several genera of kangaroo-like animals larger than any living macropod, such as the Sthenurids and Protemnodons (Gillieson & Mountain 1983, Hope 1978). All became extinct in Upper Pleistocene times, some possibly surviving into the last 25 000 years. The question arises, why suddenly did about a third of all genera of marsupials, and overwhelmingly the larger animals, become extinct in geologically such a short time. Together with several other scholars, I have invoked the influence of the arrival of humans, either directly as a predator, or indirectly by changing the existing ecological relationships.

6 ECOLOGICAL ADAPTATIONS AND TECHNOLOGY

The first immigrants to Australia must originally have lived on the coastal forested lowlands of southeast Asia. Many of the animal and plant resources on the northern coast of Sahul would have been familiar to them. Whereas the marsupial fauna would have been unfamiliar to the colonists, presumably they carried hunting technologies which has been developed and used on the placental fauna of Sunda Land. Modern ethnographic accounts from northern Australia show Aborigines exploiting many wild plant foods which are common food sources in southeast Asia. These include *Oryza* rice, *Dioscorea* yams, *Colocasia* taro, *Eleocharis* water chestnut, *Impomoea*, tubers, cycads, the pith within palm-tree trunks, and a variety of tree fruits such as *Terminalia, Syzygium, Canarium, Morinda, Ficus* and palms. In the wetter rain forest edges of the New Guinea part of the Sahul continent, there would have been wild bananas, sugar cane, coconuts, pandanuts, breadfruit and perhaps sago. Many of these plants require leaching and baking processes to remove toxins or to render the food palatable.

From the original beach-heads, we may consider two of the major ecological zones into which colonising groups would have had to move. The first were the rainforests of New Guinea, which with the mountainous topography had strong altitudinal zonation, providing a great deal of floral diversity. In the high intermontane valleys at 2000 m above sea level, there may have been grasslands at the boundaries between the forest and the high Alpine conditions of the mountain tops (Hope & Hope 1976). There is archaeological evidence that people had penetrated to these high valleys by 30 000 years ago and were hunting the now extinct giant marsupial fauna there by at least 25 000 years ago, prior to the peak of the Last Glacial phase (Hope 1986, Gillieson & Mountain 1983).

The other major area away from the Sahul coastal front was the tropical savanna, with its strongly marked wet and dry seasons. The Kakadu sites show that successful occupation of this monsoonal savanna region had occurred at least 30 000 years ago. A major problem was whether or not the occupation of the desert core of Australia had also been effected in Pleistocene times (e.g. Bowdler 1977). At Puritjarra, a large rock shelter in a dunefield of the MacDonnell ranges, northeast of Alice Springs, evidence has recently been obtained demonstrating human occupation, with stratified hearths and stone tools extending back to beyond 22 000 years ago (Smith 1987, Jones 1987a). Other sites with a similar antiquity from arid areas have been found within the Pilbara (Brown 1987), as well as from the Nullarbor, demonstrating that people had managed to penetrate into the fully dry areas of Australia and develop economic techniques to subsist there. Whether or not initial occupation had occurred during a generally more humid period than now, i.e. prior to about 25-30 000 years ago (Jones & Bowler 1980) awaits further investigation.

6.1 *The Bismarck and Solomon Archipelagos*

The truly oceanic islands of the Bismarck Archipelago east of New Guinea, namely New Britain and New Ireland are large and have rugged topographies, with mountains in the latter rising up to 2000 metres. They are separated by deep (greater than 1000 fathoms) straits from each other and from New Guinea. Although covered in tropical rain forest, they have a depauperate land fauna typical of such oceanic islands. A terminal Pleistocene date of 12 000 years was obtained from New Britain at the cave site of Misisil in 1981 (Specht et al. 1981). In 1985, a series of excavations in caves developed within uplifted coral reefs on the east coast of New Ireland revealed three sites with basal occupation extending back to Pleistocene times; ranging from 15 000 years at the inland cave at Panakiwuk, to 33 000 years from the base of a shell midden with fish bones within the coast-edge cave of Matenkupkum (Allen et al. 1988).

About 175 km to the southeast of New Ireland across an open ocean is Buka, the northernmost island of the Solomon chain. Here at Kilu cave, a shell midden containing stone artefacts and fish bones had a basal unit dated to between 20 and 28 kyr BP (Wickler & Spriggs 1988). The crossings to the other main islands of the Solomon chain would have been easy as far as San Christobal, 1000 km to the southeast especially with a glacial low sea level (Jones 1989).

The colonists into these islands clearly had an excellent water-crossing technology and considerable skills in exploiting marine resources such as fish and molluscs. On the islands, there may have been plant foods such as yams, sugar cane and tree fruits. Conversely, there would have been a paucity of land animals, the latter being confined to a few reptiles and rodents. The human occupation of the large oceanic islands of the western Pacific may have been effected almost as soon as our oldest presently-obtained dates from the Australian continent itself.

6.2 *Tasmania*

A sequence extending back to about 23 000 years was obtained in 1974 at Cave Bay Cave on the present-day Hunter Island off the north west coast of Tasmania (Bowdler 1974). At that time, it was joined as part of a continuous land bridge connecting Tasmania to Wilson's Promontary on the Australian continent. This had been formed when the sea level began to drop at the beginning of the full phase of the last Ice Age some 24 000 years ago.

A number of searches into the Ordovician limestone belts on the Franklin-Gordon River system in southwest Tasmania from 1981 onwards has revealed a series of caves, occupied by hunters during the height of the last Ice Age. Sites such as Kutikina (formerly Fraser) Cave and Deena Reena were first occupied about 20 000 years ago and were abandoned at about 12 500-13 500 years ago, when terminal glacial conditions allowed a re-spread of the rain-forests of southwest Tasmania and the drowning of the low Bassian Plain (Kiernan et al. 1983, Jones 1987b). Recently, research into limestone caves on the Florentine and Weld rivers, somewhat to the East, have shown occupation extending back to 30 kyr ago (Cosgrove 1989).

At the height of occupation, the inhabitants of these caves were living within a few kilometres of valley glaciers under tundra-like conditions. They hunted red-necked wallabies and for stone used both local quartzites and also highly localised sources such as the impactite glass from the Darwin Crater, formed by a meteorite impact. Discoveries during 1986 and 1987 showed that they had sometimes penetrated deep into some caves, such as Ballywinne on the Maxwell River and Judd's Cavern on the Cracroft River and decorated the white limestone walls with hand stencils and daubs of pigment. Tests have shown that some of the pigment at Judd's Cavern consists of human blood dated by the accelerator mass spectrometer method to about 10 700 ± 800 BP (RIDDL 1268) (Jones et al. 1988).

7 FLAKED STONE TOOLS AND GROUND AXES

Many of the Australian Pleistocene sites had flaked steep edged scrapers and core tools as described for the Mungo Site. However some of the tools found in the Huon and elsewhere in New Guinea and tropical Australia are of special interest. In the oldest Huon terrace site were large core tools called 'waisted blades' (Golson 1971). Made from volcanic rocks, they had been flaked on both faces, resulting in a ovoid axe-like shape with a pronounced hammer-dressed waist or groove in their middle. They are large, some weighing up to 2.9 kg. It seems obvious that these were hafted in some ways and were probably used as heavy-duty cutting tools. They were also present at the Kosipe open site mentioned (White et al. 1970) and in caves in the New Guinea Highlands, such as Nombe dated back to at least 25 000 years ago (Gillieson & Mountain 1983). In northern Australia, small hatchet heads, also made from volcanic rocks, were found in the Kakadu sites of Malangangerr and Nawamoyn back to about 24 000 years ago (Schrire 1982).

Les Groube has suggested that these kinds of tools were used to ring-bark and otherwise modify the vegetation in the tropical primary forest, especially at their edges. This would enhance natural processes such as the trees falling down and forming temporary glades. In such conditions, light can reach down to the floor of the forest and allow the growth of colonising plants. In New Guinea, many of these plants are also human foods, such as sugar cane, wild bananas, yams, probably taro and many of the tree fruits such as *Canarium, Terminalia*, some of the palms, *Pandanus*, etc. Thus manipulating the forest in this way could enhance the growth of favoured food plants. Groube obviously sees such a scenario as a possible precursor to the horticultural practices that had allowed people to dig large water-control drains in the Waghi Valley near Mount Hagen at least 9000 years ago, probably for the production of taro (*Colocasia esculenta*) (Golson 1977). Accepting the reality of the Huon data, it is quite astonishing to consider that people in New Guinea were using such tools and modifying their environment in such a way at a time when there were still Neanderthal people occupying western Europe.

We are at present poised on the edge of a 38-40 000 year antiquity for humans in Australia. Imagine if some of the present hints of significantly older occupation were to be confirmed – say to 60 000 years or beyond? These considerations make the Australian data one of the crucial theatres within which we might investigate key developments in the cultural and biological emergence of modern people. Sir John Barrow wrote at the beginning of the 19th century, that after the world had been divided into its four proper continents, a fifth quarter, namely Australia was discovered. Almost two hundred years later, the same can be said about its prehistory.

REFERENCES

Aimi, M. & F. Aziz 1985. Vertebrate fossils from the Sangiran Dome, Mojokerto, Trinil and Sambungmacau areas. In N. Watanabe & D. Kadar (eds), *Quaternary Geology of the Hominid Fossil Bearing Formations in Java*, pp. 155-197. Geological Research and Development Centre, Special Publication No. 4: Ministry of Mines and Energy, Bandung, Republic of Indonesia.

Allen, J., C. Gosden, R. Jones & J.P. White 1988. Pleistocene dates for the human colonisation of New Ireland, Northern Melanesia. *Nature* 331: 707-9.

Anderson, D.P. 1987. A Pleistocene-Early Holocene rock shelter in Peninsula Thailand. *National Geographic Research* 3: 184-98.

Barbetti, M. & H. Allen 1972. Prehistoric man at Lake Mungo, Australia by 32 000 years BP. *Nature* 240: 46-48.

Bartstra, G.-J. 1982. *Homo erectus*: The search for his artefacts. *Curr. Anthropol.* 23: 318-20.

Bartstra, G.-J., S. Soegondho & A. van der Wijk 1988. Ngandong man: Age and artifacts. *J. Hum. Evol.* 17: 325-337.

Bowdler, S. 1974. Pleistocene date for man in Tasmania. *Nature* 252: 697-8.

Bowdler, S. 1977. The coastal colonisation of Australia. In J. Allen, J. Golson & R. Jones (eds), *Sunda and Sahul: Prehistoric Studies in Southeast Asia, Melanesia and Australia*, pp. 205-46. London: Academic Press.

Bowler, J.M. 1976. Recent developments in reconstructing late Quaternary environments in Australia. In R.L. Kirk & A.G. Thorne (eds), *The Origins of the Australians*, pp. 55-77. Canberra: Australian Institute of Aboriginal Studies.

Bowler, J.M. 1987. Edmund Dwen Gill 1908-1986. *Aus. Arch.* 24: 48-52.

Bowler, J.M., R. Jones, H.R. Allen & A.G. Thorne 1970. Pleistocene human remains from Australia: A living site and human cremation from Lake Mungo. *World Archaeology* 2: 39-60.

Brown, P. 1987. Pleistocene homogeneity and Holocene size reduction: The Australian human skeletal evidence. *Arch. Oceania* 22(2): 41-67.

Caddie, D.A., P.J. Pomeny & H.J. Hall 1987. The ageing chemist – can electron spin resonance (E.S.R.) help? In W.R. Ambrose & J.M.J. Mummery (eds), *Archaeometry: Further Australasian Studies*, pp. 167-76. Canberra: Department of Prehistory, Research School of Pacific Studies, Australian National University.

Campbell, K.S.W. (ed.) 1975. *Gondwana Geology*. Australian National University Press.

Chappell, J. 1974. Geology of coral terraces on Huon Peninsula, New Guinea: A study of Quaternary tectonic movements and sea level changes. *Bulletin of the Geological Society of America* 85: 553-70.

Clark, G. 1962. *World prehistory, an outline*. Cambridge: Cambridge University Press.

Cosgrove, R. 1989. Thirty thousand years of human colonization in Tasmania: New Pleistocene dates. *Science* 243: 1706.

Diamond, J.M. 1987. How do flightless mammals colonise oceanic islands? *Nature* 327: 374.

Dortch, C. 1984. *Devil's Lair, a study in prehistory*. Perth: Western Australian Museum.

East, T.J., A.S. Murray, G.C. Nanson & R.L. Clark 1987. Late Quaternary evolution of Magela Creek sand-bed channels. In Supervising Scientist for the Alligator Rivers Region (ed.), *Alligator Rivers Region Research Institute, Annual Research Summary for 1985-86*. Canberra: Australian Government Publishing Service.

Fox, R.B. 1970. *The Tabon caves*. National Museum Monographs No. 1, Manila.

Gillieson, D.S. & M.J. Mountain 1983. Environmental history of Nombe rockshelter, Papua New Guinea Highlands. *Arch. Oceania* 18(1): 53-62.

Glover, I. 1987. *Archaeology in eastern Timor 1966-67*. Terra Australis No. 11. Canberra: Department of Prehistory, Research School of Pacific Studies, Australian National University.

Glover, I. & E.A. Glover 1970. Pleistocene flaked tools from Timor and Flores. *Mankind* 7: 188-90.

Glover, I. & G. Presland 1985. Microliths in Indonesian stone industries. In P. Bellwood & V.N. Misra (eds), *Recent advances in Indo-Pacific prehistory*. New Delhi.

Golson, J. 1971. Both sides of the Wallace Line: Australia, New Guinea and Asian prehistory *Arch. Phys. Anthropol. Oceania* 6: 124-44.

Golson, J. 1977. No room at the top: Agricultural intensification in the New Guinea Highlands. In J. Allen, J. Golson & R. Jones (eds), *Sunda and Sahul: Prehistoric studies in southeast Asia, Melanesia and Australia,* pp. 601-38. London, San Francisco: Academic Press.

Groube, L., J. Chappell, J. Muke & D. Price 1986. A 40 000 year old occupation site at Huon Peninsula, Papua New Guinea. *Nature*. 324: 453-455.

Hooijer, D.A. 1975. Quaternary mammals east and west of Wallace's Line. *Modern Quaternary Research in Southeast Asia* 1: 37-46.

Hope, G. 1982. Pollen from archaeological sites: A comparison of swamp and open archaeological site pollen spectra at Kosife Mission, Papua New Guinea. In W. Ambrose & J.M.J. Mummery (eds), *Archaeometry: An Australasian Perspective*, pp. 211-19. Canberra: Department of Prehistory, Australian National University.

Hope, G., P.J. Hughes & J. Russel-Smith 1985. Geomorphological fieldwork and the evolution of the landscape of Kakadu National Park. In R. Jones (ed.), *Archaeological research in Kakadu National Park,* pp. 229-40. Special Publication No. 13. Canberra: Australian National Parks and Wildlife Service.

Hope, J.H. 1978. Pleistocene mammal extinctions: The problem of Mungo and Menindee, New South Wales. *Alcheringa* 2: 65-82.

Hope, J.H., A. Dare-Edwards & M.L. McIntyre 1983. Middens and megafauna: Stratigraphy and dating of Lake Tandau lunette, western New South Wales. *Arch. Oceania* 18: 38-45.

Hope, J.H. & G.S. Hope 1976. Palaeoenvironments for man in New Guinea. In R.L. Kirk & A.G. Thorne (eds), *The origin of the Australians*, pp. 29-54. Canberra: Australian Institute of Aboriginal Studies.

Jones, R. 1968. The geographical background to the arrival of man in Australia and Tasmania. *Arch. Phys. Anthropol. Oceania* 3: 186-215.

Jones, R. 1969. Fire-stick farming. *Aust. Nat. Hist.* 16: 224-28.

Jones, R. 1977. Man as an element of a continental fauna: The case of the sundering of the Bassian Bridge. In J. Allen, J. Golson & R. Jones (eds), *Sunda and Sahul: Prehistoric studies in southeast Asia, Melanesia and Australia*, pp. 317-86. London: Academic Press.

Jones, R. 1979. The fifth continent: Problems concerning the human colonization of Australia. *Ann. Rev. Anthropol.* 8: 445-66.

Jones, R. 1982. Ions and eons: Some thoughts on archaeological sicence and scientific archaeology. In W. Ambrose & J.M.J. Mummery (eds), *Archaeometry: An Australasian perspective*, pp. 22-35. Canberra: Department of Prehistory, Research School of Pacific Studies, Australian National University.

Jones, R. (ed.) 1985. *Archaeological Research in Kakadu National Park*. Special Publication No. 13. Canberra: Australian National Parks and Wildlife Service.

Jones, R. 1987a. Pleistocene life in the dead heart of Australia. *Nature* 328: 666.

Jones, R. 1987b. Ice-age hunters of the Australian wilderness. *Aust. Geo.* 8: 26-45.

Jones, R. 1989. East of Wallace's Line: Issues and problems in the colonization of the Australian continent. In P. Mellars & C.B. Stringer (eds), *The Human Revolution: Behavioural and Biological Perspectives in the Origins of Modern Humans*. Edinburgh: University of Edinburgh Press, pp. 743-782.

Jones, R. & J. Bowler 1980. Struggle for the savanna: Northern Australia in ecological and prehistoric perspective. In R. Jones (ed.), *Northern Australia: Options and implications*. School Seminar Series Vol. 1. Canberra: Research School of Pacific Studies, Australian National University.

Jones, R., R. Cosgrove, J. Allen, S. Crane, K. Kiernan, S. Webb, T. Loy, D. West & E. Sadler 1988. An archaeological reconnaissance of Karst caves within the Southern Forests region of Tasmania. *Aus. Arch.* 26: 1-23.

Jones, R. & I. Johnson 1985. Deaf Adder Gorge: Lindner Site, Nauwalabila 1. In R. Jones (ed.), *Archaeological Research in Kakadu National Park*, pp. 165-227. Special Publications No. 13. Canberra: Australian National Parks and Wildlife Service.

Kamminga, J. & H.R. Allen 1973. *Report of the archaeological survey*. Alligator Rivers Environmental Fact-finding Study. Canberra: Australian Government Publishing Service.

Kershaw, A.P. 1985. An extended late Quaternary vegetation record from northeastern Queensland and its implications for seasonal tropics of Australia. *Proceedings of the Ecological Society of Australia* 13: 179-89.

Kiernan, K., R. Jones & D. Ranson 1983. New evidence from Fraser Cave for glacial age man in southwest Tasmania. *Nature* 301: 28-32.

Majid, Z. 1982. The west mouth, Niah, in the prehistory of southeast Asia. *Sarawak Museum Journal* 31.

Merrilees, D. 1968. Man the destroyer: Late Quaternary changes in the Australian marsupial fauna. *J. Roy. Soc. West. Australia* 51: 1-24.

Mulvaney, D.J. 1964. The Pleistocene colonization of Australia. *Antiquity* 38: 263-7.

Mulvaney, D.J. & E.B. Joyce 1965. Archaeological and geomorphological investigations on Mt. Moffet Station, Queensland. *Proceedings of the Prehistoric Society* 31: 147-212.

Nanson, G.C., R.W. Young & E.D. Stockton 1987. Chronology and palaeoenvironment of the Cranebrook Terrace (near Sydney) containing artefacts more than 40 000 years old. *Arch. Oceania* 22: 72-78.

Pearce, R.H. & M. Barbetti 1981. A 38 000 year old archaeological site at Upper Swan, Western Australia. *Arch. Oceania* 16(3): 173-178.

Russell-Smith, J. 1985. Studies in the jungle: People, fire and monsoon forest. In R. Jones (ed.), *Archaeological research in Kakadu National Park*, pp. 229-267. Special Publication No. 13. Canberra: Australian National parks and Wildlife Service.

Schrire, C. 1982. *The Alligator Rivers: Prehistory and ecology in western Arnhem Land*. Terra Australia No. 7. Canberra: Department of Prehistory, Research School of Pacific Studies, Australian National University.

Shawcross, F.W. 1975. Thirty thousand years and more. *Hemisphere* 19: 26-31.

Shawcross, F.W. & M. Kaye 1980. Australian archaeology: Implications of current interdisciplinary research. *Interdis. Sci. Rev.* 5: 112-128.

Singh, G. & E.A. Geissler 1985. Late Cainozoic history of vegetation, fire, lake levels and climate, at Lake George, New South Wales, Australia. *Transactions of the Royal Society of London* 311: 379-447.

Smith, M.A. 1987. Pleistocene occupation in arid central Australia. *Nature* 328: 710-11.

Specht, J., I. Lilley & J. Norman 1981. Radiocarbon dates from west New Britain, Papua New Guinea. *Aus. Arch.* 12: 13-15.

Thorne, A.G. 1971. Mungo and Kow Swamp: Morphological variation in Pleistocene Australians. *Mankind* 8: 85-89.

Thorne, A.G. & M.H. Wolpoff 1981. Regional continuity in Australasian Pleistocene hominid evolution. *Am. J. Phys. Anthropol.* 55: 337-49.

Watanabe, N. & D. Kadar (eds) 1985. *Quaternary Geology of the Hominid Fossil Bearing Formations in Java.* Special Publication No. 4. Bandung: Geological Research and Development Centre.

Weitzel, V. 1987. The mangrove men. *Australian Primatology* 2(2):1.

White, J.P., K.A.W. Crook & B.P. Ruxton 1970. Kosipe: A late Pleistocene site in the Papuan Highlands. *Proceedings of the Prehistoric Society* 36: 152-70.

Wickler, S. & M. Spriggs 1988. Pleistocene human occupation of the Solomon Islands, Melanesia. *Antiq.* 62: 703-706.

Wintle, A.G., N.J. Shackleton & J. Lautridou 1984. Thermoluminescence dating of periods of loess deposition and soil formation in Normandy. *Nature* 310: 491.

Wright, R. 1986. How old is Zone F at Lake George? *Arch. Oceania* 21(2): 138-39.

Wyrwoll, K-H. & C.E. Dortch 1978. Stone artefacts and an associated Diprotodontid mandible from the Greenough River, Western Australia. *Search* 9: 411-13.

Author index

Subject index